Emergent Pollutant Treatment in Wastewater

Emerging Materials and Technologies
Series Editor Boris I. Kharissov

For more information about this series, please visit:
https://www.routledge.com/Emerging-Materials-and-Technologies/book-series/CRCEMT

Emergent Pollutant Treatment in Wastewater

S.K. Nataraj

CRC Press
Taylor & Francis Group
Boca Raton London New York

CRC Press is an imprint of the
Taylor & Francis Group, an **informa** business

First edition published 2022
by CRC Press
6000 Broken Sound Parkway NW, Suite 300, Boca Raton, FL 33487-2742

and by CRC Press
4 Park Square, Milton Park, Abingdon, Oxon, OX14 4RN

CRC Press is an imprint of Taylor & Francis Group, LLC

ISBN: 978-1-032-10324-2 (hbk)
ISBN: 978-1-032-10325-9 (pbk)
ISBN: 978-1-003-21478-6 (ebk)

DOI: 10.1201/9781003214786

Typeset in Times
by SPi Technologies India Pvt Ltd (Straive)

Contents

Preface

Emerging pollutants are entering the environment every day. To understand where these emerging pollutants come from, we need to think about our modern lifestyles, pattern of water supply or consumption routes and release of used water and their constituents in the environment. This book is an attempt to understand the origin, fate and environmental risk associated with these pollutants due to their presence in aquatic systems and the environment. The book also attempts to evaluate the best methods and techniques currently in practice for their detection, identification and treatment in various sources. Recent decades have seen the extensive use and discarding of synthetic chemical-based products such as surfactants, plastics, medicines, personal care or household cleaning products, pesticides and agricultural additive products into water bodies. On the other hand, the increased industrial usage of water and the unregulated release of partially treated or untreated wastewater is causing serious contamination to fresh water bodies and the ecosystem at large.

Even though several domestic and industrial treatment practices have been in use for decades now, a majority of these techniques or methods fail to address the emerging pollutant problem, holistically. Even though industrial scale treatment processes and some of the advanced treatment techniques in practice are designed to treat specific categories of emerging pollutants, they fail to remove them all at once. Eventually, they end up in the aquatic systems and environment. Recent studies have estimated that emerging pollutants have reached a sufficient saturation point to enter the food chain, putting even non-aquatic species at risk. Nevertheless, it is certain that emerging pollutants are getting into the environment every day and that the world is concerned about the effects they might be having on the environment and macro/microorganisms, including humans. After all, most of the emerging pollutants discussed in this book have already been detected in biota, marine life, such as fish, and drinking water supplies.

Even though the risk of exposure to emerging pollutants on our health is still uncertain, tracking the origin, fate and accumulation effects of this new class of pollutants is still a big challenge. Today, there is an increasing market share for bottled drinking water and the associated economic benefits, deflecting considerable attention from wastewater treatment and management. Therefore, major sources of emerging pollutants like domestic, municipal and industrial wastewater escape strict monitoring and discharge regulation. It is evident that millions of micro-, small and medium-scale industries leave large amounts of unmanaged and untreated wastewater from undesignated zones and illegally discharge it into aquifers, rivers, drains and so on; but if managed well, wastewater treatment and the recycling of higher quality reusable water can be a good business prospect. According to one of the United Nations' Sustainable Development Goals 2030, clean water and sanitization is critical to many other development challenges and a more holistic water agenda, including water resources and wastewater management, is needed. It is clear that to achieve these goals, constituents like emerging pollutants in wastewater need to be fully recognized and included within the overall water cycle assessment as one of the important steps forward to enhance sustainable development. With this background, the present book aims to address the issues related to emerging pollutants and their treatment.

Therefore, it is necessary to design extensive emerging pollutant monitoring, identification and treatment programmes. Even though this task is a resource-intensive exercise, it is a necessary one for preparing the human resources and establishing infrastructure for minimizing future risk. Therefore, future research should emphasize filling these knowledge gaps. This can be achieved by providing support to researchers and innovators to develop cost-effective methods and protocols for the effective management of emerging pollutants within the context of an overall risk assessment and evaluation for all water and wastewater related hazards, noting that emerging pollutants concern both drinking water and the larger environmental emergency. In this context global organizations like the WHO, federal agencies and local governments are likely to mobilize resources and continue to support relevant studies to frame guidelines accordingly. Therefore, this book tries to bridge the knowledge gap in the topic.

Before concluding, I take this opportunity to thank CRC Press, Taylor & Francis Group, for accepting my proposal; and Dr Gagandeep Singh and Aditi Mittal of CRC Press for coordinating the drafting of my manuscript. My sincere thanks to Thivya Vasudevan and Matthew Van for their help in proof reading the final manuscript. I express my gratitude to my students at Sustainable Materials and Process Lab-CNMS-JU for their constant queries which contributed to the betterment of the manuscript. Also, my sincere thanks goes to some very important people for making this book happen: my parents (Veeramma, Sharanappa), in-laws (Yashodha, Mallikarjun) and brothers (Girish, Santosh and Harish), my family, especially my wife Ashwini, and my children Anvita and Vrishank for providing me with the right atmosphere, sacrificing their time and allowing me to work on this book without disturbance.

S.K. Nataraj
Kadakolla-Bengaluru

Author Biography

Professor S.K. Nataraj is currently working at the Centre for Nano and Material Sciences (CNMS), Jain University, Bangalore, India. He obtained his PhD on 'Membrane Based Separation Processes for Industrial Effluent Treatment' in 2008 from the Centre of Excellence in Polymer Science (CEPS), Karnatak University, Dharwad, India. Immediately after completion of his PhD, he moved to the Alan G. MacDiarmid Energy Research Institute (AMERI), Chonnam National University, South Korea to pursue a Postdoctoral Associate assignment (2007–2009) to work on energy materials. During 2009–2010, he accepted a second assignment as Postdoctoral Associate at the Institute of Atomic Molecular Sciences (IAMS), Academia Sinica, Taiwan to develop ion-exchange membranes for fuel cell applications. Further, he worked at Qatar University, as a visiting fellow at Cambridge University (2010–2011) and continued as a full time Postdoctoral Research Associate (2011–2013) at Cavendish Laboratory, University of Cambridge, UK. Later, he moved to India to work as a DST-INSPIRE Faculty Fellow (2013–2015) at CSIR-CSMCRI, Bhavnagar, where his main areas of research were to develop sustainable materials and processes for energy and environmental applications, including supercapacitors, fuel cells, membrane processes for wastewater treatment, functional nanomaterial-based devices for wastewater treatment and the value addition of bioresources. Based on his achievements, he was awarded 'best researcher' from Jain University for the year 2017 and has been admitted to the prestigious role of Fellow of the Royal Society of Chemistry (FRSC), London, UK.

He teaches courses in Separation and Purification Technologies, Battery-Fuel Cells and Solid-State Materials at postgraduate level. He has now published more than 90 research articles, 14 US/PCT patents, 15 book chapters, a single authored book and 1 edited book. He has over 4000 citations and serves as an editorial board member for several Journals. His socially relevant works have been highlighted in several magazines and news articles, such as *Nature Asia, Nature India* and *Outlook.*

1 Introduction to Emerging Pollutants and Their Treatment Techniques

1.1 INTRODUCTION

Today, humanity is facing greater water scarcity due to lack of freshwater, unequal distribution of resources and increasing contamination of water resources at various levels. Knowing the importance of water scarcity and creating awareness, the United Nations recognizes March 22 each year as World Water Day. This has followed a series of measures with greater inputs from several studies and reports on very wet and some very dry geographic locations, increased waste disposal from industries, a sharp rise in global freshwater demand and lack of efficient treatment processes for used water. As of 2021, over 800 million people lack access to clean water worldwide. In underdeveloped and developing countries, deprived communities and families without easy access and affordable water have suffered from disease and poverty for generations and are still suffering today [1, 2].

On the other hand, several parts of the world are experiencing water stress as a result of increasing imbalance between water use and available water resources. The water stress indicator in several studies identifies the extent of water withdrawal with respect to total renewable resources, which is exceedingly high. Water stress is a criticality ratio, which implies that water stress depends on the variability of resources. Water stress causes deterioration of freshwater resources in terms of quantity and quality [3]. Water stress in various parts of the world varies, ranging from 20% for basins with highly variable runoff to 60% for moderate zone basins. Overall, the world average of 40% of water stress is with diverse distribution from region to region.

The stress on water sources directly affects the environment, and this effect is multiplying due to increased industrialization, urbanization, and agricultural activities, consequently reducing the availability of clean water. In this process, polluted water is of great concern to aquatic organisms, plants, humans, and climate and indeed alters the ecosystem. Yet, it is estimated that around 75% of India's water pollution is due to municipal waste from cities. Owing to poor infrastructure, the world's cities are able to treat less than 20% of the wastewater, while the remainder is discarded untreated. It is abundantly clear that the amount of waste generated and discharged into the environment today is greater than at any other time in the history of our planet. While the world's population tripled in the 20th century, more than 1

DOI: 10.1201/9781003214786-1

billion people lack access to safe drinking water and more than 2.6 billion people lack adequate sanitation. In addition to this, given ever-increasing water usage, lack of rains and unexpected weather conditions, these figures should be much higher [4].

Although the world's population increased almost threefold in the 20th century, water resources remain the same. However, the use of renewable water resources has gained importance manyfold. It is estimated that within the next 50 years, the world population will increase by another 40% to 50%. This unprecedented population growth and decreasing underground resources, coupled with industrialization and urbanization, will result in increased water stress and demand which is expected to induce serious consequences on the environment. Specially, water bodies including streams, tanks, lakes, rivers, oceans, aquifers and groundwater have already been turned into hazardous reservoirs. Water pollution with a variety of contaminants are released to the natural environment as household water filters are also rejecting huge amounts of used water into drains almost every second.

A large number of chemicals are continuously being released into our environment in the form of wastewater. The toxicological repercussions of exposure to hazardous compounds and their impact on the health of the population and environment in general are a growing concern. It is now widely recognized that the existence of certain chronic diseases is the outcome of a complex interaction between chemical contaminants in environmental and genetic factors. Water pollution is severely affecting the world today more than ever before and requires urgent intervention. In this direction, researchers, innovators and technologists are working on new concepts and ways to monitor and implement plans to derive solutions.

Every day, millions of tonnes of domestic sewage, industrial discards and agricultural waste are discharged into various water resources. Once, the United Nations (UN) estimated that the amount of wastewater produced from different means annually is about 1500 km^3, six times more water than exists in all the rivers of the world [5]. However, the situation is now much worse because most of the freshwater has now been utilized and laid as wastewater in the environment. Nevertheless, lack of adequate sanitation, treatment techniques and recovery steps in place make it worse, creating a chain of events in increasing water pollution. It is estimated that over 2.5 billion people live without proper sanitation worldwide. Due to drastic decrease in water quality, in many parts of the world, more than 50% of native freshwater fish species and nearly one-third of the world's amphibians are at risk of exposure to toxicity and extinction [6].

Wastewater can be simply defined as a natural resource in the wrong place that has been changed in some way, is no longer suitable or has no means for reuse, and requires disposal or discharge. Water can pick up pollutants from daily human activities like bathing, toilet flushing, laundry, and dishwashing. Pollutants may enter water resources from residential and domestic sources. Commercial wastewater comes from non-domestic sources, such as industrial activities, washing, beauty salon, taxidermy, furniture refinishing, and automotive and industrial cleaning. In general, wastewater may contain hazardous materials and emerging contaminants.

According to UN Environment Programme (UNEP) wastewater expert Birguy Lamizana, 'Municipal (domestic wastewater), industrial, and increasingly domestic, wastewater are primary sources of emerging pollutants in the aquatic environment' (https://www.unep.org/news-and-stories/story/evidence-rising-emerging-pollutants-poisoning-our-environment). The emerging pollutants comprise a wide range of chemicals, substances and microbial pollutants collectively referred to as 'emerging pollutants (EPs)' or 'contaminants of emerging concern (CECs)'. Emerging pollutants are broadly classified into active pharmaceuticals ingredients (APIs), personal care products, pesticides, and industrial and household chemicals. Diverse types are present in highly variable concentrations in freshwater resources such as rivers, streams, lakes and groundwater, and lately huge numbers of plastics and organic contaminants have entered oceans. It is estimated that over 2000 chemicals are currently in use which are potentially classified as emerging pollutants. According to reports, already over 700 EPs, their metabolites and transformation products are listed as being present in the European aquatic environment. On the other hand, chemicals, daily care products, APIs and nanomaterials that have only recently been acknowledged as potential hazards to the environment and are not yet broadly regulated by national or international law are known as emerging pollutants. They are classified as 'emerging' not because the contaminants themselves are new, but rather because of the rising level of concern. Nonetheless, EPs are rarely measured or supervised, and further research is needed to assess their impacts on human health and the environment. The potential human health risks of EPs through exposure to drinking water or food chain is a concern [7].

1.2   DEFINITION AND ORIGIN OF EPS

Historically, many contaminants find their ways to ancient times. Lead has been the most prominent contaminant recorded throughout ancient Greek and Roman civilizations. Definition of emerging pollutants or contaminants in this context will have to cover contemporary stock of substances which are creating harmful effects to the environment. Particularly, the term 'emerging' is relative, newly formed or now coming into prominence, and a contaminant emerging a decade or two ago might no longer qualify as an emerging contaminant. Therefore, emerging contaminants are new compounds or molecules that were not formerly known to exist, or information is jostling our understanding with regard to environmental and human health risks. Therefore, emerging pollutants (EPs) are defined as synthetic or naturally occurring chemicals that are not commonly monitored in the environment but which have the potential to enter the environment and cause known or suspected adverse ecological and (or) human health effects [8].

Within a larger perspective, emerging contaminants have increasingly been identified and segregated from conventional pollutants only recently in the environment, for which alarms were raised much later. This list of emerging contaminants can also include traditional contaminants or old contaminants with new facts or evidence which shed a new viewpoint on the concerns of risks. Emerging pollutants have brought the whole focus of forecasting vulnerabilities of substances that are truly new and unexpected to human and environmental health. Most importantly, different substances, compounds and molecules that were previously considered non-hazardous, non-toxic and unharmed are now repeatedly mentioned in the scientific literature. On the other hand, there are evolving concerns which were known to exist but for which the ecological infections and related issues were not fully understood. Also, to tackle developing issues, new information is changing our understanding of environmental and human health risks.

1.3 CLASSIFICATION OF EPS

A large number of emerging pollutants reported in the recent literature are posing a challenge for environmental regulation authorities and agencies. It is a bigger challenge to prioritize and enlist the emerging pollutants based on their severity and the vulnerable risk to the environment and to human health. Therefore, considering the origin of the contaminant, occurrence level and their quality criteria will only shed light on their behaviour in the environment or on their toxic effects on the environment and human health. Nevertheless, emerging pollutants represent relatively newly discovered groups of unregulated contaminants which occur in surface water, groundwater and major portions in industrial and domestic wastewaters.

In ancient times, Lead (Pb) was an emerging contaminant, and its potential toxicity was established due to toxic metal leaching off luxurious items which were generally used and which depended on Pb-based metal containment. In those days, even though Pb was emerging as a contaminant, those who used it extensively were unaware of the risks Pb could cause, and there was no method to quantify the traces in the environment. Similarly, dichlorodiphenyltrichloroethane, commonly known as DDT, was widely used for different applications including disinfecting community areas and for pest and mosquito control. Later, increased awareness of emerging contaminants like Pb and DDT led to their being banned in petroleum products and disinfection use, respectively. These are the few good examples of how ecologists and conservationists rang the alarm bell and then academic research followed up with factual data to uncover the truth and risks involved with a range of similar emerging pollutants [9, 10].

Most of the contaminants listed in these categories are substances or microorganisms that are not frequently monitored but have the potential to enter the environment and cause adverse ecological/human health effects. Among these, several have been in discussion and commonly were in use for decades; however, new findings about their potential risks under each category are posing serious threats to the environment. And lead (Pb), which has been in use and known for its toxicity for more than 5000 years, still finds its place in several commonly used commodities and continues to seep into our environment and life cycle.

Generally, water contaminants have been broadly classified depending on their nature of occurrence, namely:

1. Physical contaminants mainly affect the physical appearance or other physical properties of water. Physical contaminants include colloids, suspended particles and sediment or organic material suspended in the water of streams, lakes, rivers and large water bodies from soil erosion and decay.
2. Chemical contaminants are natural or synthetic elements or compounds. Chemical contaminants include complex organic compounds, humic acids,

nitrogen, bleach, salts, pesticides, metals, toxins produced by bacteria, and human or animal drugs.

3. Biological contaminants are microorganisms in water. They are also referred to as microbes or microbiological contaminants. Examples of biological or microbial contaminants include bacteria, algae, viruses, protozoan, pathogens and parasites.

4. Radiological or radioactive contaminants are chemical elements with an unbalanced number of protons and neutrons resulting in unstable atoms that can emit ionizing radiation. Examples of radiological contaminants include uranium, cesium, plutonium and radium.

5. Odour, taste and appearance. This can be considered in other non-categorized sections. This category includes uncategorized compounds which are emerging and still finding their association with one or the other previously mentioned categories. However, recent studies show the necessity of categorizing these into careful considerations. Among these are taste, smell and appearance and the reasons for such characteristics. Though taste, smell and appearance are not health concerns in themselves, they do affect the suitability and acceptability of drinking water. Suspended carbonates, solids, dust particles or even air can give turbid water a milky look. An off-putting blackish tint can be caused by metal contaminants like manganese, iron or slime bacteria. A yellowish appearance of water can indicate humic or fulvic compounds, iron or bacteria. Dissolved iron, manganese or bacteria can also give a reddish tint, and surface foam could indicate the presence of surfactants and discarded or leached additives species [11].

Brackish, hard or slightly salty water is unacceptable for drinking but are generally used for drinking purposes in many parts of the world, while high hardness or solids can give an alkaline taste which is largely avoided for household applications. Greasy/oily, fishy or perfume-like taste may be caused by surfactants, and the presence of iron, copper, zinc and nuisance bacteria may cause a metallic taste. Grassy, mouldy, earthy, rotten-egg, fishy, herbal or cucumber odours are also unacceptable for drinking water and may be caused by hydrogen sulfide, bacteria, soft water reactions, algal byproducts or surfactants. A greasy or phenolic smell/odour can originate from oil, gasoline or bacterial contamination, and industrial chemicals can also produce an unacceptable smell.

In context to emerging pollutants, currently it is estimated that more than 20 different classes of emerging pollutants are present in the various aquatic environments. These may be chemical substances or microorganisms that are not generally monitored but have the likelihood to enter the environment and cause an adversarial effect on ecological and human health. However, based on their occurrence, emerging pollutants can be majorly categorized in to seven divisions, namely: (1) personal care products and additive compounds; (2) pharmaceutical or active pharmaceutical ingredients (API); (3) industrial chemicals; (4) surfactants; (5) biological contaminants and (7) nuclear or radioactive wastes. A detailed list of EPs with examples is categorized in Figure 1.1 [12].

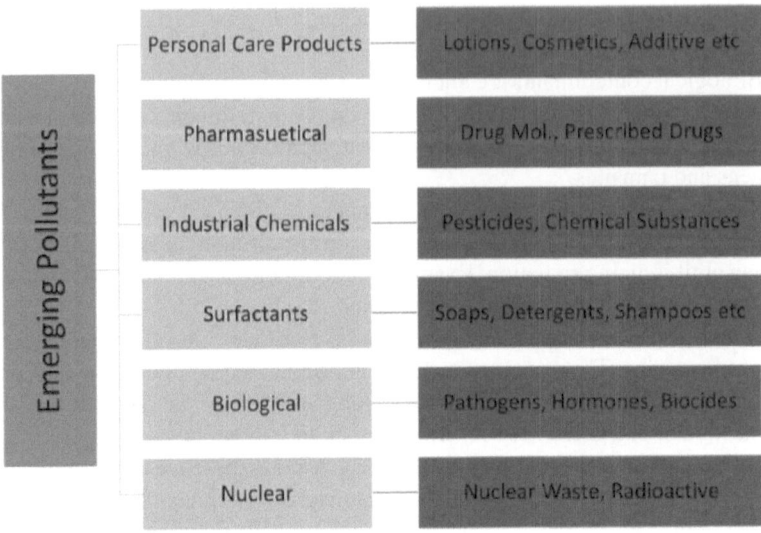

FIGURE 1.1 List of emerging pollutants discharged via domestic and industrial wastewaters found generally in the environment.

With these categories in place, emerging pollutants can be understood in a broad sense, as potentially known or suspected adverse ecological and human health effects which are still to be known are listed in Table 1.1. This list is incomplete and will continue to grow in the future, as regulatory and monetary agencies include new substances category-wise and when potential risks of new substances are assessed. Nevertheless, many EPs are used daily and released continuously into the environment, even in very low quantities, and some may cause prolonged toxicity and endocrine disruption in living organisms and aquatic life [13].

1.4   WHO GUIDELINES ON EPS

According to the United Nations Sustainable Development Goals (UN SDG) adopted in 2015 by all member states as a universal call to action to end poverty, protect the planet Earth and safeguard people's peace and prosperity by 2030. This framework broadly outlines the strategies for developing various stakeholders – including government regulatory and monitoring agencies, educational, academic and research institutions and other non-government organizations – to work on 17 different goals, including drinking water and sanitization. This also considered coordinated effort related sustainable development fields which will support the countries to meet its legal obligations, as well as scientific and technical commitments. These efforts are monitored continuously to address potential challenges arising from the existing multilateral environmental agreements and such emerging efforts in the future [14].

According to Sustainable Development Goal 6, Clean Water and Sanitation, each year millions of people, primarily children, die from diseases associated with inadequate water supply, sanitation and hygiene. It is projected that by 2050, over 50% of

TABLE 1.1
Detailed Category Wise List of Emerging Pollutants

Category 1: Personal Care Products and Additive Compounds

ATII (traseolide)

Bayrepel

Benzaldehyde (phenylmethylene) hydrazone (Eusolex)

Benzophenone

AHDI (phantolide)

alpha-Terpineol

2,4-Dihydroxybenzophenone

4-Methylbenzylidene camphor

4-Oxoisophorone

Acetylcedrene

ADBI (celestolide)

Ethylhexyl methoxycinnamate

Galaxolide

g-Methylionone

Hexamethyldisiloxane (HM or HMDS)

Hexylcinnamaldehyde

Homosalate

Isobornylacetate

Isobutyl paraben

Methylparaben

Methyldihydrojasmonate (methyl
  3-oxo-2-pentylcyclopentaneacetate)

Methyl-iso-propylcyclohexenone

p-t-Bucinal (Lilial)

Biocide compounds

Triclosan

2,6-Di-tert-butylphenol

Butylated hydroxyanisole

2-Ethylthioacetic acid ethylester

2-Methylthioacetic acid ethylester

3-Methylthiopropionic acid

Bromazepam

Butalbital

Carbamazepine

Carazolol

Caffeine

Cefacetrile

Cefapirin

Cefalonium

Cefalexin

Cefazoline

Cefoperazone

Ciprofloxacin

Clofibric acid (metabolite of clofibrate)

Clarithromycin

Citalopram

Clotrimazole

Boisvelone/Iso-E super

Butyl methoxydibenzoylmethane

Cineole

Damascone

Decamethylcyclopentasiloxane (D5)

Boisvelone/Iso-E super

Decamethyltetrasiloxane (MD2M)

Dihydromethyljasmonate

Dodecamethylcyclohexasiloxane (D6)

Dodecamethylpentasiloxane (MD3M)

Drometrizole

Drometrizole trisiloxane (INCI)

Methylsalicylate

Musk ambrette

Musk ketone

Musk xylene

Octamethylcyclotetrasiloxane (D4)

Octamethyltrisiloxane (MDM)

Octocrylene

Oxybenzone

Carvone

Propyl paraben

Tonalide

N,N-Diethyltoluamide

D-Limonene

Triethylcitrate

Butylated hydroxytoluene

Dipropyltrisulfide

Ethylene brassylate

Habanolide

Flumequine

Fenoterol

Flucloxacillin

Fenoprofen calcium salt dihydrate

Fluorouracil

Fluoxetine

Fluvoxamine

Furosemide

Gemfibrozil

Gentamicin

Glibenclamide (glyburide)

Hexobarbital

Hydrocodone

Hydrochlorothiazide

Chlortetracycline

Chlorobutanol

(*Continued*)

TABLE 1.1 (Continued)

Clenbuterol	Chloramphenicol
Crotamiton	Ibuprofen
Cloxacillin	Cholesterol
Codeine	Ifosfamide
Meclofenamic acid	Sertraline
Medazepam	Streptomycin
Metformin	Secobarbital
Mebeverine	Spectinomycin
Mestranol	Spiramycin
Mefenamic acid	Simvastatin
Minocycline	Sulfamerazine
Methylphenobarbital	Sulfadimethoxin
Metoprolol	Sulfadoxin
Mevastatin	Sulfadiazine
Methicillin	Sulfamethazine
Nandrolone	Sulfamethoxazole
Nafcillin	Sulfapyridine
Nadolol	Tetracycline
Naproxen	Temazepam
N-Methylphenacetine	Terbutaline
Neomycin B	Taloxa
Norfloxacin	Tolfenamic acid
Nordiazepam	Tilmicosin
Novobiocin	Timolol
Ofloxacin	Tiamulin
Oxacillin	Tramadol
Omeprazole	Trimethoprim
Oleandomycin	Tylosin
Oxazepam	Verapamil
Oxytetracycline	Valnemulin
Oxprenolol	Zolpidem

Antibiotics and other stimulant APIs

1. Sulfonamides (SAs), sulfadiazine (SDZ), sulfathiazole (STZ), sulfamerazine (SMR), sulfamethizole (SML), sulfamethoxazole (SMX), sulfisoxazole (SFX), sulfamethazine (SMN), and sulfadimethoxine (SDM);
2. Fluoroquinolones (FQs): norfloxacin (NOR), ofloxacin (OLF), ciprofloxacin (CIP), enrofloxacin (ENR), and lomefloxacin (LOM);
3. Tetracyclines (TCs): oxytetracycline (OTC), chlortetracycline (CTC), tetracycline (TCN), and doxycycline (DOX);
4. Macrolides (MLs): clarithromycin (CLA), roxithromycin (ROX), tiamulin (TIA), tylosin (TYL), azithromycin (AZN), and erythromycin (ERY);
5. (v) b-blockers: atenolol (ATE), metoprolol (MET), and propranolol (PROP);
6. lipid regulator: bezafibrate (BF);
7. Antiepileptic: carbamazepine (CBZ);
8. Stimulant: caffeine (CAF); and (ix) dihydrofolate reductase inhibitor: trimethoprim (TMP).

Category 2: Pharmaceutical Compounds

1-Hydroxy ibuprofen	Cyclophosphamide
2-Hydroxy ibuprofen	Desmethylnaproxen
	(metabolite of naproxene)
17-alpha-Estradiol	Danofloxacin

17-alpha-Ethinylestradiol
17-beta-Estradiol
1,1,1-Trichloro-2,2-dihydroxyethane (chloral hydrate)
Acebutolol
Acecarbromal
Aceclofenac
Acemetacin
Acetaminophen (Paracetamol)
Acetazolamide
Acetylsalicylic acid (Aspirin)
Acyclovir
Albuterol
Allobarbital
Alclofenac
Albuterol sulfate
Alprazolam
Amitriptyline
Ampicillin
Amoxicillin
Amobarbital
Anthracene-1,4-dione
Apramycin
Aprobarbital
Atenolol
Azithromycin
Bezafibrate
Baquiloprim
Betamethasone
Beta-sitosterol
Betaxolol
Baclofen
Bisoprolol
Iopamidol
Iminostilbene
Indomethacin
Iohexol
Iomeprol
Imapramine
Iopromide
Ivermectin
Josamycin
Kanamycin sulfate
Lansoprazole
Lamotrigine
Ketoprofen
Levetiracetam
Lidocaine
Lincomycin
Lorazepam
Loratadine
Lithium carbonate
Lovastatin
Marbofloxacin

Dantrolene
Dapsone
Daunorubicin
Diethylstilbestrol
Difloxacin
Diphenhydramine
Domperidone
Doxepine
Doxorubicin
Doxycycline (anhydrous)
Doxycycline (monohydrate)
Dexamethasone
Diatrizoate
Diazepam
Diclofenac
Dicloxacillin
Enoxacin
Epirubicin
Enrofloxacin
Escitalopram
Erythromycin
Esomeprazole
Estriol
Estrone
Estrone sulfate
Ethosuximide
Etofibrate
Fenofibric acid (metabolite of fenofibrate)
Fenfluramine
Fenofibrate
Famotidine
Fenoprofen
Penicillin G
Penicillin V
Pentoxifylline
Pentobarbital
Paroxetine
Phenazone
Phenylbutazone
Phenobarbital
Phenytoin
Pindolol
Prednisolone
Pravastatin
Pipamperon
Primidone
Propranolol
Salbutamol
Ranitidine
Roxithromycin
Propyphenazone
Sarafloxacin
Sotalol

(*Continued*)

TABLE 1.1 (Continued)

Meprobamate

Secobarbital sodium

Category 3: Industrial Chemicals

Trichloropropane (TCP)	Oxadiazon
Perfluorooctanesulfonate (PFOS),	Perfluorooctanoic acid (PFOA)
Dioxane	Methiocarb
Trinitrotoluene (TNT)	2,6-ditert-butyl-4-methylphenol
Dinitrotoluene	Tri-allate
Hexahydro-trinitro-triazane (RDX)	Imidacloprid
Nanomaterials	Thiacloprid
N-nitroso-dimethylamine (NDMA)	Thiamethoxam
Perchlorate	Clothianidin
Perfluoro-octane sulfonate (PFOS) and	Acetamiprid
Perfluorooctanoic acid (PFOA)	
Polybrominated biphenyls (PBBs)	2-Ethylhexyl 4-methoxycinnamate
Polybrominated diphenyl ethers (PBDEs)	Erythromycin Clarithromycin, Azithromycin
Tungsten	Perchlorate
Bisphenol A (BPA)	Dioxins
Perfluoroalkyl and Polyfluoroalkyl Substances (PFAS)	Phthalates
Phytoestrogens	Polychlorinated biphenyls (PCB)
Endocrine Disturpting Compounds	

1. Steroid estrogens: estrone (E1), 17b-estradiol (E2), estriol (E3), 17a-ethynylestradiol (EE2), estrone 3-b-D-glucuronide (E1-3G), b-estradiol 3-b-D-glucuronide (E2-3G), b-estradiol 17-b-D-glucuronide (E2-17G), estrone 3-sulfate (E1-3S), b-estradiol 3-sulfate (E2-3S), and estriol 3-sulfate (E3-3S); and
2. Phenolic estrogenic compounds (PEs) as representative MPs in industrial wastewater which usually constitute a certain proportion in the influent of some WWTPs: bisphenol A (BPA), and 4-nonylphenol (NP).

Polybrominated diphenyl ethers (PBDE)

Triclosan

Category 4: Surfactants

Sodium linear alkylbenzene sulfonate (LABS);	Sodium lauryl sulfate;
Sodium lauryl ether sulfates	Benzalkonium chloride
Stearalkonium chloride	Linear primary alcohol polyethoxylate
Dodecyl dimethylamine oxide	Coco diethanol-amide alcohol ethoxylates
Cocoamphocarboxyglycinate;	Cocamidopropylbetaine
Nonylphenol (NP)	Nonylphenoxy carboxylate (NP1EC)
Nonylphenol ethoxylates (NPE)	Nonylphenoxy ethoxy carboxylate (NP2EC)
Monoalkanolamide	Dialkanolamide
Polyethoxylated monoalkanolamide	Sodium alkene sulfonate
Polyethoxylated dialkanolamide	Sorbitols
Polyglycerol monooleate	Polyglycerol Monoester

Category 5: Biological Contaminants

Analgesics, antibiotics	Hormones
Novel viruses	Microalgae
Medical waste sludge	Anti-neoplastics

Category 6: Nuclear and Radioactive Substances

Uranium	Thorium-230
Radium-226	Radon-222
Polonium-210	Isotopes of radon decay

the world's population will live in water-stressed regions. Nonetheless, it is also true that since 1990, more than 2.5 billion people have gained access to better-quality drinking water sources. Yet, over 600 million people are still waiting to get good quality water. Historically, it is recorded that every day, nearly 1000 children die due to preventable and avoidable water- and sanitation-related diseases. With careful analysis, it is estimated that most of these diseases find their origin due to the contamination of one or the other emerging pollutant.

The UN water programme clearly laid down the guideline with increased priorities for wastewater management, pollution prevention, control and monitoring. It is evident from several studies that more than 80% of wastewater resulting from human and industrial activities is discharged into different water bodies like streams, lakes, rivers or seas without any treatment. UN guidelines on water and wastewater also clearly indicate the need for sustainable management of water resources, control and monitoring of wastewaters. It also emphasizes the need for access to safe water and sanitation as essential for unlocking economic growth and productivity, and it provides significant leverage for existing investments in health and education. The natural environment, including water bodies, forests, soils and wetlands, contributes to management, monitoring and regulation of water availability. It also highlights maintaining of water quality, strengthening the resilience of watersheds and complementing new technologies and innovation in physical infrastructure. Also, several reports and committees recommend establishing institutional and regulatory arrangements for water access, distribution, use and disaster preparedness. Water shortages, clean water and adaptation of wastewater treatment technologies weaken food security and incomes of larger populations that depend on freshwater resources. In this context emerging pollutants may throw new challenges at water and wastewater management; however, the presence of new studies and guidelines may improve the water management aspects, resulting in sustainable economies and making the agriculture and food sectors more resilient to rainfall and weather unpredictability. All these guidelines need to be implemented to meet the threat of emerging pollutants and to be able to fulfil sustainable economic growth to match the growing population.

According to SDG target 6.1, UN member states have set a target of universal and equitable access to safe and affordable drinking water for all by 2030. This further includes (target 6.3) improving the 'water quality by reducing pollution, eliminating dumping and minimizing release of hazardous chemicals and materials, halving the proportion of untreated wastewater and substantially increasing recycling and safe reuse globally'. Target 6.4 describes the 'substantially increased water-use efficiency across all sectors and ensur[ing] sustainable withdrawals and supply of freshwater to address water scarcity and substantially reduce the number of people suffering from water scarcity'. These targets can be fulfilled only by addressing the issues of imminent threat from emerging pollutants both from synthetic or natural origin chemical or microorganism that are not frequently scrutinized or measured in the environment. To achieve this SDG target (6.a) requires clearly laying out the plan of action in which member states need to come together to 'expand international cooperation and capacity-building support to developing countries in water- and sanitation-related activities and programmes, including water harvesting, desalination, water efficiency, wastewater treatment, recycling and reuse technologies' [15].

1.5   HEALTH RISKS AND TOXICITY ASSOCIATED WITH EPS

Recently, researchers have unearthed enormous information about potential risks of several listed and still-to-be-assessed registered emerging pollutants. By now it is clear that emerging pollutants are toxic substances that adversely affect human health, life and the environment at large. According to some studies, EPs can be transported by wind, water and contaminated living things. Most of the EPs have shown dangerous potential to spread across the life cycle which they come in contact with. EPs can persist for long periods of time in the ecosystem and can accumulate, mutate and undergo chemical change and pass from one species to the next through the food chain. As mentioned in the previous section, even though global concerns associated with EPs are being addressed by UN member states via co-operative mechanisms, statistical data on many EPs need to be finalized for their toxicity and potential hazard mechanisms. On the other hand, EPs like Bisphenol-A production and use have already been restricted. However, people and inhabitants near contaminated wastewater resources exposed to EPs can still be at risk from unintentionally discarded resources. Although the UN and its advanced member states recognize the risk of exposing their citizens and inhabitants to EPs-contaminated wastewater, a great number of underdeveloped and developing nations have only recently begun to restrict their use, treatment and discarding mechanisms to the environment [16].

Existence of EPs in domestic and/or industrial wastewater can result from point of use or accidental diffusion or discharge into untreated discharges. In such a situation, and in light of their probable impact on the environment, assessment needs to be made and action needs to be devised. Therefore, this section gives the current state of the art and challenges for monitoring programs and risk assessment tools with respect to emerging pollutants as a base for health risks and toxicity associated with EPs and sustainable water resource management [17].

At present, there is no single method or technique available for sampling, assessment and analysis. In absence of coherent analytical methods and protocols, it is becoming difficult to establish a general method to assess the environmental risk. However, there are typically dedicated techniques available for certain limited EP classes, and for a number of known highly hazardous EPs, detection limits are inadequate to allow proper risk assessment or are still in an initial state. Recent literature suggests that advanced ultra-sensitive instrumental techniques should be used for quantitative determination of prioritized EPs in groundwater as well as wastewater sources. At present, important data on EPs and their intermediates, metabolites and properties that govern their fate in the environment are often not available. One of the main limitations to monitor and assess the effects of EPs on ecosystems is the absence of broader surveys on water quality using different parameters in particular regions or affected areas for water quality assessment and often do not include EPs. A coordinated monitoring of surface water, groundwater and wastewater is not yet realized, but it is urgently required. In addition to the infrastructures, to evaluate the futuristic projection, a specific component integrated into models evaluating the fate of EPs in a multi-compartment ecological approach are missing and must be developed. The main goal of risk assessment is the overall protection of ecological populations in the aquatic and human environments. Also, innovative methods for evaluating the collective risks from combined exposures to several hazardous substances of both chemical and biological origins in a multi-scale approach is required. On the policy end, comprehensive regulations and management measures with respect to use and discharge of EPs into the environment, their occurrence in the environment and eco-system are essential to devising efficient water and wastewater management [18].

Currently, EPs, their metabolites, and their transformation products in the environments are not included in routine monitoring lists. Also, current protocols and regulatory authorities lack information on EPs, and their fate, behaviour and ecotoxicological effects remain little known. However, lately adopted norms and efforts yielding basic information according to which the classes of pharmaceuticals and personal care products (PPCPs) and polar pesticides have been detected predominantly in European aquatic eco-life. PPCPs are found amply in all aquatic compartments at concentrations reaching few nanograms per litre (ng/L) in coastal marine aquatic systems to several micrograms (µm) in wastewater streams of developed countries. Lately, these PPCPs have shown adverse intrinsic properties which may cause biological effects. On the other hand, due to their compatible physico-chemical properties with high water solubility (log $K_{ow} > 3$) PPCPs and pesticide-type EPs undergo constant infusion into the aquatic media, eventually finding their way to the environment and life cycle. Several compounds used in and as ingredients in cosmetic, pharmaceutical, pesticides or food industries, like antimicrobial disinfectants, sanitizers and paraben preservatives, raise concern in particular because of their suspected or proven endocrine disruptive effects on aquatic organisms and human life. Some of the antimicrobial diphenyl ether derivatives, also known as triclosan, that are found in many aquatic ecosystems environments throughout the world need to be investigated in risk assessments. The continuous release of personal care product ingredients and pharmaceutical residues also poses ecological concerns, as they can behave as health-disrupting substances when used as psychotropic drugs, for which worldwide consumption is increasing [19, 20].

Many synthetically prepared substances associated with exert known include anti-epileptics, carbamazepine, and the antidepressant venlafaxine are found in amounts from nanograms to micrograms in aquatic ecosystems of the United States, France and other developed countries. On the other hand, pesticides like glyphosate-based formulations and other herbicides once were often considered relatively nontoxic for living organisms with an environmentally low impact because of its rapid biodegradation and strong adsorption to soil particles. But later studies worldwide showed unacceptable consequences on human life.

Based on recent vast literature, the list of compounds that qualify as emerging pollutants is long and getting longer, including active pharmaceutical ingredients (APIs) used in and as antibiotics, psychiatric drugs, analgesics, flame retardants, industrial additives, anti-inflammatory drugs, steroids, hormones, contraceptives, fragrances, sunscreen agents, insect repellents, polymeric microbeads, microplastics, antiseptics, pesticides, herbicides, surfactants, detergents, and other surfactant metabolites, plasticizers and gasoline additives, among others. These compounds have been categorized as EPs based on their ecotoxicity and potential risk on human life. In particular accumulated molecules like the anxiolytic venlafaxine (VEN) induces loss of appetite, constipation, diarrhoea, nausea or vomiting, or dry mouth;

the antiepileptic carbamazepine (CBZ) is found to cause neurological complications; the preservative methylparaben (MP) is known to add allergic reactions in high concentration; the antimicrobial triclosan (TCS) accumulates in fatty tissues may cause long-term complication in human bodies; and the aminomethylphosphonic acid (AMPA) is known to induce severe metabolic acidosis [21, 22].

Interestingly, today many of the EPs are discharged to the environment out of ignorance. We throw away expired medicines, substances and kitchen or household discards. It is estimated that several kilograms of medicines or pharmaceutical ingredients are thrown away daily. Also, several tonnes of surfactants in the form of soaps, detergents, wet wipes and washing solutions are being discharged. In addition, microplastic constituents in the form of small plastic wrappings, condoms, cotton buds, single use masks and many other products are discarded into the environment in large quantities, adversely affecting the functioning of wastewater treatment plants. However, efforts are being made to adopt new technologies to overcome the limitation of conventional wastewater methods and technologies.

1.6   ENVIRONMENTAL HAZARD ASSOCIATED WITH EPS

In recent times, issues related to environmental pollution and in particular water-related concerns have intensified due to awareness and accessible resources. Many studies involving heavy metals, APIs (CBZ, MP and AMPA), and surfactants and toxicology reports on aquatic ecosystems have revealed alarming data on EPs' potential ecological risks. Specifically, the disposal of toxic heavy metals, evaluation of toxic greenhouse gases, oil spillage, use of non-biodegradable materials, pesticides spreading and APIs are posturing serious accumulation in that particular vegetation. Further, several bioassays using two freshwater and one marine species contaminated with endocrine-disrupting species showed signs of inhibiting or artificially augmenting the function of natural chemical messengers in the body. Further, endocrine-disrupting contaminants in fish and amphibians downstream of contaminated water sources have shown reproductive abnormalities and physical deformities. However, several studies still argue for more research and evidence to prove the possible health effects of low-level endocrine-disrupting chemicals in wastewater and domestic water supplies [23].

On the other hand, common ecological toxicity studies like embryotoxicity and metamorphosis tests on leading aquaculture species like the Pacific oyster (*Crassostrea gigas*) show their vulnerability to a large range of EPs. This is also true with growth inhibition assay studies on the freshwater green-alga Pseudokirchneriella subcapitata, which is a major manufacturer and immobilization test on the freshwater crustacean Daphnia magna, which occupies a key position in the aquatic food chain. Similarly, endocrine disruption can cause adverse effects in fish and other aquatic life through the interruptive development of the brain and nervous system, the growth and function of the reproductive system and the response to stressors in the environment. Endocrine disruptors, also known as hormonally active agents, can cause cancerous tumours, birth defects and other developmental disorders in humans [24].

In addition to this, some chemicals and compounds have only recently been identified, such as PPCPs, additives, surfactants, APIs, perfluorinated compounds (PFCs) and nanomaterials, not because the contaminants themselves are novel, but rather because of the growing level of concern to environment and aquatic life. Chlorofluorocarbons (CFCs) are another important set of chemicals which are widely used as firefighting chemicals and refrigerants; some lubricants and detergents are substances which lay close to environmental exposure and are prone to enter into aquatic ecosystems effortlessly. These compounds have already been proven to be ozone layer-depleting substances as CFCs, and now adverse effects have been detected at low dosage, with tests conducted on lab animals revealing tumour-promoting potential at higher doses [25, 26]. Similar studies have also revealed the close relationship between presence of small concentration of PFCs in blood and occurrence of prostate and breast cancer of the assessed subjects (Figure 1.2). Some of the other commonly detected diseases originating from EPs (pathogens, HMs and organics) in contaminated water sources are typhoid fever, cholera, hepatitis A, diarrhoea and so on.

The last century has witnessed extensive use of water disinfection products like chlorine. However, the effects of disinfection byproducts have hardly been recorded. As chlorination works best in water at higher concentrations to treat organic matter. Organic matter like humic acids interact with chlorine, making it unavailable for disinfection due to which chlorine is added in abundance to disinfect water contaminated with organic matter. This leads to a number of byproducts. During disinfectant many byproducts form as a result of organic matter pollutant interaction with chlorine and other halogens like bromine, resulting in formation of byproducts such as trichloromethanes and other trihalomethane. These unlisted byproducts (EPs) have been evaluated for their carcinogenic tendencies in animals [27, 28].

Recent fascination on nanomaterials has led to the development and use of millions of different classes of nano-sized substances. Nanomaterials are minute-sized materials employed in almost every field of applications, including water filtration,

FIGURE 1.2 Environmental and human risks associated with presence of EPs in drinking and wastewater sources.

sensors, medicine, agriculture and energy. Due to nanomaterials' size and abundance, post-use disposal is a source of great concern. One such example is the fate of titanium dioxide nanoparticles largely used in personal care products, water filtration, catalysis and energy application, and a large portion of them disappears in an environment without a trace. However, several recent studies have discovered a connection of such nanomaterials of different chemical compositions to cancer in humans and aquatic ecosystems.

Cosmetic and pharmaceutical industries run into the billions of dollars in worth. However, the use of tonnes of personal and cosmetic products annually pose a serious threat to the environment. Every household consumes several kilogrammes to tonnes of PPCPs annually in one or the other form, and certainly these used items disappear into water bodies via drainage systems every day. Several recent studies reveal the challenges related to uncontrolled release and disposal of these EPs into the environment. Advanced trace studies also reveal the presence of PPCPs in aquatic systems. Further, PPCPs like hormonal supplements such as estrogens have been found in large amounts in animal manure and urine.

Similarly, several PPCPs and their metabolites have been found in low, moderate and high concentrations in several water/sediment samples. APIs such as paracetamol, ibuprofen, 2-hydroxy ibuprofen, CBZ-diol, Oxazepam, iopromide, and clofibric acid have been found in freshwater lakes. Also, PCPs like triclocarban (TCC), triclosan (TCS) as well as its metabolite methyl-triclosan (M-TCS) were detected in algae in different concentrations. These were further percolated in abundance of aquatic plant biomass. Similarly, across the world, several water bodies have been tested for considerable levels of pharmaceuticals [29, 30].

However, it is rational to assume that in dry areas or during dry seasons, water bodies are more likely to contain higher concentrations of proportions of such EPs. Nevertheless, many unlisted but significant sources of EPs are present in surface and open waters. Therefore, studies are focusing on including such EPs in international agreements with routine monitoring programmes so that their impact on the environment can be well understood.

On the other hand, for the past several years, many research findings reveal the presence of several EPs in wastewaters discarded to surface water and aquatic ecosystems. Every day, manufacturing industries, processing units, agriculture residues and domestic wastewaters are carrying high concentrations of EPs into freshwater resources. Most of these wastewaters are being discharged to freshwaters untreated or undertreated. Once these wastewaters reach lakes, rivers or storage places, they accumulate different physical natures. In this storage period, EP-contaminated water attains favourable conditions to generate a lot of metabolites, derivatives and unknown species, making the treatment process much more complicated.

1.7   NEED OF DETECTION AND QUANTIFICATION OF EPS

It is easy to detect or remove conventional water contaminants like metals, colloidal contaminants, nutrients, bio-contaminants like bacteria, and other disinfection byproducts as a set protocol and removal techniques are available. However, API pollutants like ibuprofen, diclofenac, anti-inflammatories, antidepressants, diazepam, painkillers and remains of contraceptives such as estrogens are plentiful in water sources, with no proper monitoring and tracking protocols in place. This is a subject of great concern. They are affecting the health of many species of fish, algae and certain amphibians, such as frogs. Traces of contraceptives substances lie estrogen hormones are affecting the feminization of the fish population.

On the other hand, morning hygiene routines like using mouthwash leave harmful substances such as triclosan and titanium dioxide into the water down the drain. Teeth whiteners are present in the nanoparticles form (1 to 100 nm) that have made up a new type of EPs whose harmful effects are still unknown. Dangerous substances like perfluorinated compounds are found in many cooking pans, fabrics, rugs or food containers. Also, surfactants, softener used in excess are another source of nanomaterials in domestic wastewater that effects are barely known. The ionic silver particles used to destroy bacteria, fungi, viruses and protozoa in socks to avoid unpleasant odours now has been found in alarming concentrations in water originating from laundry and bathroom discharges [31].

The list goes on. EPs are seeping into wastewater and freshwater resources via several routes and become more concentrated with time. Human wastes; fats, oils and grease; food preparation waste in numerous unknown forms; toxic and poisonous metals from drug, dye and leather processing units; highly acidic or basic compounds; malodorous compounds like formaldehyde, mercaptans, radioactive substances and materials causing high oxygen demand like biological substances.

At present the biggest reluctance among the population makes treatment of wastewater complex and costly, as EPs were simple pollutants prior to legislation. While scientists have recently found their harmful effects, regulatory authorities across the world have not yet implemented any restrictive measures yet. However, studies on several EPs have been unable to determine accurately the level of harmfulness on the environment. This is also because of lack of suitable and affordable technologies, and protocols have not been defined clearly. Until that time, natural freshwater ecosystems will be under severe stress and exposed to harmful pollutants. Nonetheless, efforts need to be focused on theoretical and simulation studies to predict the effects of EPs on aquatic pollution, ecosystem degradation and overexploitation of natural resources. Certainly, several institutions across the globe have gathered adequate tools and resources, but policy makers, action groups and decision-makers need to implement these considerations and act on them.

According to studies, municipal, industrial, and ever more domestic wastewater are primary sources of emerging pollutants in the aquatic environment. Today, the EP list has grown to over 3000 natural and synthetic substances that include a variety of compounds such as antibiotics, analgesics, anti-inflammatory drugs, psychiatric drugs, steroids and hormones, contraceptives, fragrances, sunscreen agents, insect repellents, microbeads, microplastics, antiseptics, pesticides, herbicides, surfactants and surfactant metabolites, flame retardants, industrial additives and chemicals, plasticizers and gasoline additives, among others. While more research is underway and still needs to be accelerated, it is widely recognized that these emerging pollutants are increasingly becoming a hazard. However, to some extent part of this problem could be solved by simply adopting basic waste monitoring and management techniques. However, most of the EPs lack detection and analytical protocols [32].

In this direction, a resolution adopted by the member states of the United Nations of Environment Assembly in March 2019 calling on governments and all other stakeholders, including United Nations agencies, funds and programmes "to support relevant science-policy interface platforms, including input from academia and research to enhance cooperation in the areas of environment and health and consider." This also urges us to find ways of strengthening the science-policy interface, including its relevance for the implementation of multilateral environmental agreements at the national level.

However, in general, citizens cannot be considered generators of EPs; rather, they are transporters. Also, discharge of EPs into the environment is practically impossible to avoid. However, it is important to know that EPs are part of our daily lives, and they are characteristic to our standard of living and our lifestyle. Therefore, EPs are potentially in every product that we consume. Unless the world recognizes the threat, the choice remains that we have to coexist intimately with these substances we consume and end up as EPs in the ecosystem. For now, it is hard and challenging to trace EPs, but they are everywhere.

The monitoring of EPs now needs to start from the source. One of the widely used PPCPs constituent is drug molecules. In such cases, traces of remaining medicines that our bodies have not assimilated goes out as excrete in urine. Currently, several APIs have been found in rivers and lakes. Another set of substances as parabens used in sunscreen creams and lotions which have potential carcinogenic effects. However, it is important to track usage of such substances for their dosage and frequency. Also evident from the several studies is the presence of high concentration of metal traces like mercury and toxins such as hydroquinones which are abundantly used in cosmetic products end up discharging into freshwater sources whose combined effect is still not fully understood.

Amongst the universal regulatory discussions, the UNEP report suggests the preventive principle should guide responses to EPs. Further, it elaborated the need of promoting research, monitoring programmes, reductions in waste, and green chemistry. It should be possible to prevent and mitigate the negative impacts of pharmaceuticals without compromising on their availability, effectiveness, or affordability, particularly in countries where access to important health services is still limited.

1.8   NEED OF REMOVING/SEPARATION/TREATMENT OF EPS

Continuous release of EPs into the environment through various wastewaters can persist in the ecosystem for a very long period, accumulate in living organisms and vegetation, enter the food web and may end up in humans via drinking water. Conventional wastewater treatment processes have been designed to convert wastewater into attaining required water quality standards before being discharged back into the environment. Most of the currently practiced wastewater treatment techniques have proven effective in countering contaminants like bacteria, known bulk chemicals, suspended solids, colloidal contaminants and few toxins. At present, conventional wastewater treatment has been successful in reducing the contaminants to acceptable levels to make the water safe for discharge back into the environment.

Scientists, environmentalists and biologists worldwide are now alarmed that industrial waste production, treatment and disposal can have an impact on the life cycle and the quality pattern and hydrological cycle on the earth, thereby severely affecting surface and groundwater availability. Industrial wastewater production is believed to rise due to increased population, urbanization and industrial growth which may lead to the global temperature at an increasing pace. Temperature increase affects the hydrological cycle by directly increasing evaporation of available surface water and vegetation transpiration [33].

Water scarcity now becomes an important topic in international diplomacy. From the local village to the United Nations, water scarcity is a widely discussed topic in decision making. Nearly 3 billion people in the world suffer from water scarcity. According to World Health Organization (WHO) sources, a combination of rising global population, economic growth and climate change means that by 2050, 5 billion (52%) of the world's projected 9.7 billion people will live in areas where freshwater supply is under pressure. Researchers expect about 1 billion more people to be living in areas where water demand exceeds surface-water supply.

The level of water in the ponds and rivers dwindled and in some cases water bodies dried up completely. Due to increased human and industrial activities, there has been great emphasis on treatment of water and wastewater. Publicly and privately funded research activities are going on to reduce the disposal of EP contaminated wastewater to the environment. Source of water, usage of water, typical use of water and conventional treatment methods are still being used and adopted. However, R&D activities are underway to improve the process efficiencies to produce the high quality of reusable water from industrial sources. Typical contaminant characteristics of major polluting industries have been discussed and enlisted to monitor their hazardous effect on the environment. Now, the time has arrived, and every industry has to adopt efficient wastewater treatment technologies to serve the minimum basic discharge standards. Even after treatment and water reclamation of high reusable quality, water treatment produces high amounts of organic and mineral sludges have been characterized for EPs in filtered water and sedimentation sludge. At present most popular, the ion-exchange process using natural or synthetic resins removes calcium, magnesium and carbonate ions from water, replacing them with hydrogen and hydroxyl ions. However, regeneration of ion-exchange columns with strong acids and alkalis produces a wastewater rich in hardness ions which are readily precipitated out, especially when in admixture with other wastewaters resulting in numerous complex byproducts [34].

1.9   FUNDAMENTALS OF WATER TREATMENT

Wastewater, whether it is domestic, industrial or agricultural residue contaminated, has several undesirable components from the organic, inorganic, biological and hundreds of unknown pollutants that are potentially harmful to the environment and

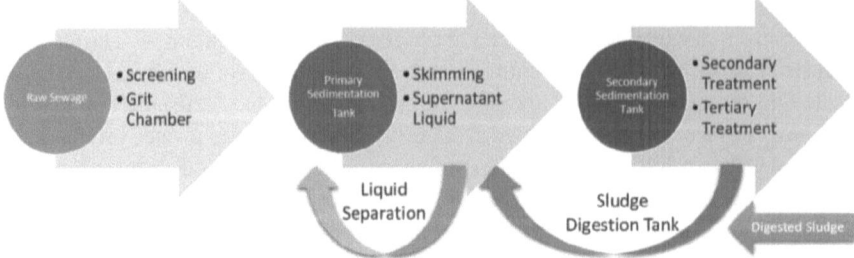

FIGURE 1.3   Schematic showing a conventional three-stage wastewater treatment process, namely, primary, secondary and tertiary water treatment.

human health. The treatment of wastewater and its proper management has become a necessity in order to conserve this vital resource. In addition to these, the unknown but avoidable threats are emerging from a new class of materials and substances that are seeping into freshwater resources via wastewater streams, which are raising concerns in recent times. Research and developmental activities are more focused on these segments of wastewater so that the reclamation of reusable water can be achieved [35].

The main aim of wastewater treatment is the removal of contaminants from water so that the treated water can be reused for beneficial purposes. Currently, the wastewater treatment is carried out in three stages, namely primary, secondary and tertiary or advanced wastewater treatment as shown in Figure 1.3. However, different types of contaminated wastewater necessitate a variety of strategies to remove the pollutant in addition to conventional treatment processes discussed in the previous section. However, many R&D activities are still focused on the individual and hybrid processes which also aim to treat the specific contaminant targeted approach. The following briefly lists conventional methods available and ongoing R&D activities to treat highly polluted wastewaters.

1.9.1   SUSPENDED SOLIDS AND OIL CONTAMINATIONS

Most suspended solids can be removed using simple sedimentation techniques with the solids recovered as slurry or sludge. However, very fine solids (nano-sized) and solids (colloidal) with densities close to the density of water pose special problems. In such a case, filtration or ultrafiltration may be required. Although, flocculation, coagulation and adsorbents are being used with the help of alum salts or the addition of polyelectrolytes with limited success. Many oils can be recovered from open water surfaces by skimming devices. The skimming process is considered a reliable and economical way to remove oil, grease and other hydrocarbons from water, and oil skimmers can sometimes achieve the desired level of water purity. In all oil-contaminated wastewater treatments, skimming is also a cost-efficient method to remove most of the oil before using membrane filters and chemical processes. Skimmers will prevent filters from blinding or fouling membranes prematurely and keep chemical costs down because there is less oil to process [36].

However, skimmers may be ineffective in case of wastewaters contaminated with high viscosity hydrocarbons. In such cases, skimmers must be equipped with heaters powerful enough to keep grease fluid for discharge. If floating grease forms into solid clumps or mats, a spray bar, aerator or mechanical apparatus can be used to facilitate removal. However, hydraulic oils and the majority of oils that have degraded to any extent will also have a soluble or emulsified component that will require further treatment to eliminate. Dissolving or emulsifying oil using surfactants or solvents usually exacerbates the problem rather than solving it, producing wastewater that is more difficult to treat.

Organic contaminants and sludge in the wastewater is generally removed by secondary wastewater treatment processes. The biodegradable organic substances of plant or animal origin is usually possible to treat using extended conventional wastewater treatment processes such as activated sludge or trickling filter. Problems can arise if the wastewater is excessively diluted with domestic wastewater pollutants or is highly concentrated and non-biodegradable pollutants like organics, surfactants, disinfectants, blood and milk. The presence of cleaning agents, disinfectants, pesticides or antibiotics can hinder the effectiveness of the treatment steps, thereby resulting in detrimental impacts on treatment processes upon discharge. Activated sludge is a biochemical treatment process for treating domestic and industrial wastewater in presence and absence of air (or oxygen) which can mitigate the growth of microorganisms by biologically oxidizing organic pollutants, producing a waste sludge (floc). However, treated wastewater will have now contaminated the many oxidized substances of unknown consequences in the form of byproducts. Further, synthetic organic materials including solvents, paints, pharmaceuticals, pesticides, coking products, etc., can be very difficult to treat. These problems necessitate treatment methods that are often specific to the material or substance to be treated. Currently, these byproducts contaminated wastewaters are being treated by such methods as advanced oxidation processing, distillation, adsorption, vitrification, incineration, chemical immobilisation and landfill disposal, with limited success and efficacy.

1.9.2 ACIDS AND ALKALIS TREATMENT

Conventionally, acids and alkalis are being treating using a neutralization process under controlled conditions. However, the neutralization process frequently produces a precipitate that may be toxic in nature and will require further treatment as a solid residue. In some cases, gases may be evolved, requiring treatment for the gas stream. In general neutralized byproducts are of unknown consequences and can be categorized as emerging pollutants. Waste streams rich in hardness ions as from de-ionization processes can readily lose the hardness ions in a build-up of precipitated calcium and magnesium salts. This precipitation process can cause severe furring of pipes. Further, in extreme cases these byproducts may cause the blockage of disposal pipes [37].

1.9.3 HEAVY METALS AND METAL COMPLEXES TOXICITY

Toxic materials including many organic materials, heavy metals such as zinc, silver, cadmium, thallium, etc., and their complex materials of acidic and highly alkaline pH, non-metallic elements such as arsenic or selenium are generally resistant to bio-logical processes unless found in trace quantities. Metals can often be precipitated out by changing the pH or by treatment with other chemicals. Many, however, are resistant to treatment or mitigation and may require concentration followed by land-filling or recycling. However, new emerging concerns of these substances are forcing researchers to find ways to treat, recover and manage metal-based complex emerging pollutants.

1.9.4 Adsorption

Adsorption has been widely used to remove water contaminants due to its low cost, availability of different adsorbents and easy operation. Different adsorbents that have been used include use of magnetic nanoparticles, activated carbon, nanotubes and polymer nanocomposites; these can remove different contaminants, including heavy metals, that are very harmful even at low concentrations. Even though adsorption can remove most water pollutants, it has some limitations such as lack of appropriate adsorbents with high adsorption capacity and low use of these adsorbents commer-cially. Hence, there has been a need for more efficient techniques such as membrane technology.

1.9.5 Membrane Technologies in Industrial Wastewater Treatment

Among other advanced techniques used for industrial wastewater treatment are mem-brane-based separation processes. The membrane separation or treatment process mainly depends on three basic principles, namely adsorption, sieving and electro-static phenomenon. The adsorption mechanism in the membrane separation process is based on the hydrophobic interactions of the membrane and the solute (analyte), generally organic solutes. These interactions normally lead to more rejection because it causes a decrease in the pore size of the membrane.

Membrane technology is known for its simple operational, economical and greener processes compared to conventional methods. This technology follows the physical separation of contaminants from the wastewater sources. The separation of materials through the membrane depends on pore and molecule size. For this reason, various membrane processes with different separation mechanisms have been devel-oped. Recent decades have seen the growth of membrane-based separation processes like reverse osmosis (RO), nanofiltration (NF), ultrafiltration (UF), microfiltration (MF), electrodialysis (ED) where ion-exchange membranes are extensively used, the membrane distillation (MD) process, forward osmosis (FO) and other membrane hybrid process for effective treatment of several EP-category wastewaters, as shown in Figure 1.4 [38].

However, membrane processes such as MF, NF, UF and RO are currently used for reusable water recovery from brackish water, ground water, seawater and wastewater to some extent. These membrane-based separation processes also gain greater impor-tance in treating emerging pollutants with limited success. This is because in mem-brane technology, typically organic polymer-based membranes are widely used separation media. These membranes are composed of hydrophobic polymers like

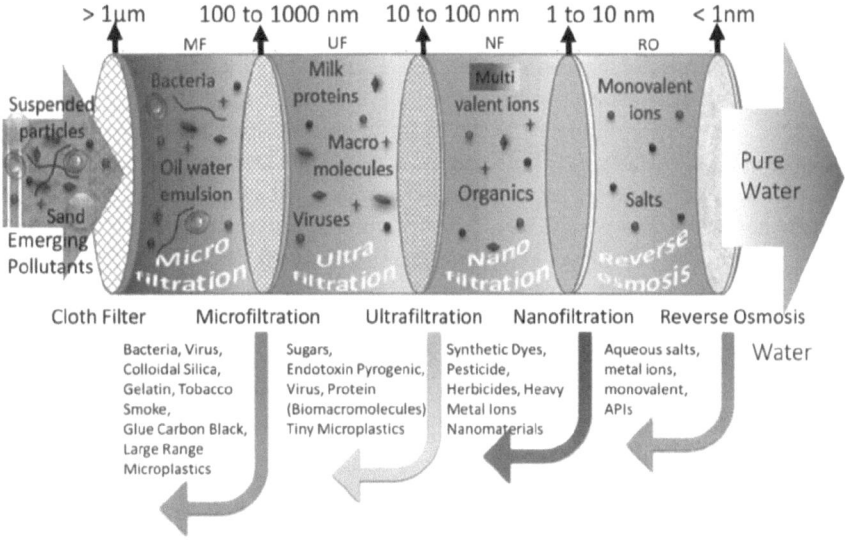

FIGURE 1.4 Membrane-based separation processes with membrane separation spectrum with size exclusive distribution of emerging pollutants.

polysulfone (PSF), polyamide (PA), polyvinyl iodine fluoride (PVDF) and polyethersulfone (PES), which are prone to organic and bio-fouling. This fouling propensity in polymer membranes leads to blockage of membrane pores and decreases membrane performance and life, and it also increases operation cost by demanding an extra cleaning process. This limitation in membrane technology originates from several factors such as deposition of inorganic components onto the surface membrane/solute absorption pore blocking, microorganism and feed chemistry. Therefore, the fouling phenomenon in membrane results in either reversible or irreversible membrane fouling. Reversible fouling is formed by attachment of particles on the membrane surface; irreversible fouling occurs when inorganic particles, organic contaminant interaction and/or strongly biological contamination accumulate on the membrane surface which cannot be removed or cleaned by any of physical means. Recent studies reveal the possible emerging pollutants' accumulation and permeation through membrane media limits the technology for long-term use. As there is a formation of a strong matrix of the fouled layer as a result of solute accumulation, during continuous filtration the process will turn reversible fouling to an irreversible fouling layer.

Therefore, most of the conventional wastewater treatment methods discussed here are designed to treat conventional contaminants at higher concentrations like TDS, pathogens and the loads of the bulk organic chemicals, but generally they are not suitable to remove residues of trace organics and series of EPs. It is clear from the technology survey that an important measure of a successful water treatment technology is the magnitude to which it delivers a public or personal good or service safely, providing safe and quality water for reuse. In this direction, conventional wastewater treatment techniques have proven limited success for EPs treatment, removal and recovery to make wastewater discharge safe and healthy for the environment.

1.10   CHALLENGES IN EPS CHARACTERIZATION

Currently, methods for emerging pollutants recognition, identification, sampling, segregation and analysis are not coordinated. For now, a number of recorded highly hazardous emerging pollutants' detection limits are too high to allow proper risk assessment. However, recently detected other EPs like nanomaterials, microplastics and other polymeric substances, method development, analysis, and protocol establishment has just begun. Several advanced and ultra-sensitive instrumental techniques are being used with several protocols for quantitative analysis of listed EPs in water, suspended substances, contaminated soil and biota. However, data on major portions of EPs and their metabolites properties and characteristics that determine their fate in the ecosystem are often not available. Worldwide surveys on water quality often use various parameters for qualitative and quantitative assessment. Nevertheless, sampling and analysis protocols for each EP in the form of integrated techniques are missing and must be developed [39].

Amongst all, slowly and steadily, more emerging pollutants are identified based on their original substances or possible parental raw material tracing. However, EPs' toxicities, environmental occurrences and features remain unidentified compared to conventional pollutants. Statistical information and records on EPs toxicity mechanisms, environmental occurrences and characteristics need to be comprehensively studied to understand their impact systematically. Recent literature surveys associated with new source identifications, analytical methodology, formation mechanism, and control of EPs have been widely documented. It is apparent from the recent literature that use of novel analytical techniques, including chromatographic techniques that allow detection of polar analytes such as liquid chromatography (LC) and liquid chromatography with tandem mass spectrometry (LC-MS), together with their improved detection limits, has permitted us to detect analytes that were not routinely analysed in the past.

REFERENCES

1. "UN-Water: World Water Day". UN-Water. Retrieved March 10 2020.
2. UN-Sustainable Goal 6, UNDP Oslo Governance Centre, https://www1.undp.org/content/oslo-governance-centre/en/home/sustainable-development-goals/goal-6-clean-water-and-sanitation.html.
3. Dan Wang, Klaus Hubacek, and Yuli Shan, Winnie Gerbens-Leenes and Junguo Liu, Review of water stress and water footprint accounting, *Water*, 2021, 13(2), 201.
4. Christine L. Moe and Richard D. Rheingans, Global challenges in water, sanitation and health, *Journal of Water and Health*, 2006, 4(1), 41–57.
5. The first edition report on *Water for People, Water for Life* was launched on World Water Day, March 22 2003, at the third World Water Forum held in Kyoto, Japan.
6. J.-C. Vié, C. Hilton-Taylor, and S. N. Stuart (eds.) (2009). *Wildlife in a Changing World – An Analysis of the 2008 IUCN Red List of Threatened Species*. Gland, Switzerland: IUCN, p. 180.
7. Evidence rising: the emerging pollutants poisoning our environment, UNEP, https://www.unep.org/events/un-environment-event/emerging-pollutants-wastewater-increasing-threat, UN Environment Programme United Nations Avenue, Gigiri Nairobi, Kenya. Retrieved March 10 2021.

8. Violette Geissen, Hans Mol, Erwin Klumpp, Günter Umlauf, Marti Nadal, Martinevan der Ploeg, Sjoerd E.A.T.M. van de Zee, and Coen J. Ritsema, Emerging pollutants in the environment: a challenge for water resource management, *International Soil and Water Conservation Research*, 2015, 3(1), 57–65.

9. Bieby Voijant Tangahu, Siti Rozaimah Sheikh Abdullah, Hassan Basri, Mushrifah Idris, Nurina Anuar, and Muhammad Mukhlisin, A review on heavy metals (As, Pb, and Hg) uptake by plants through phytoremediation, *International Journal of Chemical Engineering*, 2011, 2011, Article ID 939161, 31.

10. WHO Regional Office for Europe, Air Quality Guidelines, Chapter 6.7, Lead, Copenhagen, Denmark, 2nd edn, 2001. ISBN 9289013583.

11. Francisco G. Calvo-Flores, Joaquin Isac-Garcia, Jose A. Dobado, *Emerging Pollutants: Origin, Structure, and Properties*, ISBN: 978-3-527-33876-4, 2017, p. 528. Weinheim, Germany.

12. Directive 2000/60/EC of the European Parliament and of the Council of 23 October 2000 establishing a framework for Community action in the field of water policy. *Official Journal of the European Union*; L 327, 22.12.2000, 1–73 OJ series: L (Legislation), Number: 327, OJ year: 2000.

13. João C.G. Sousa, Ana R. Ribeiro, Marta O. Barbosa, Cláudia Ribeiro, Maria E. Tiritan, M. Fernando, R. Pereira, Adrián M.T. Silva, Monitoring of the 17 EU Watch List contaminants of emerging concern in the Ave and the Sousa Rivers, *Science of the Total Environment*, 2019, 649, 1083–1095.

14. This UNESCO Project on "Emerging Pollutants in Wastewater Reuse in Developing Countries," fully funded by the Swedish International Development Cooperation Agency (Sida) under the Programme Cooperation Agreement between UNESCO and Sweden for 2014–2017.

15. Sustainable Development Goal 6: Clean Water and Sanitation, United Nations Office for Outer Space Affairs, United Nations Office at Vienna, Vienna International Centre, Wagramerstrasse, 5, A-1220, Vienna, Austria.

16. Wilfried Sanchez and Emilie Egea, Health and environmental risks associated with emerging pollutants and novel green processes, *Environmental Science and Pollution Research*, 2018, 25, 6085–6086.

17. E. Hardell, A. Kärrman, B. van Bavel, J. Bao, M. Carlberg, and L. Hardell, Case-control study on perfluorinated alkyl acids (PFAAs) and the risk of prostate cancer, *Environment International*, 2014, 63, 35–39.

18. Meng Lei, Lun Zhang, Jianjun Lei, Liang Zong, Jiahui Li, Zheng Wu, and Zheng Wang, Overview of emerging contaminants and associated human health effects, *Research International*, 2015, 2015, Article ID 404796, 12.

19. D. R. Sambandan and D. Ratner, Sunscreens: an overview and update, *Journal of the American Academy of Dermatology*, 2011, 64(4), 748–758.

20. E. Gilbert, F. Pirot, V. Bertholle, L. Roussel, F. Falson, and K. Padois, Commonly used UV filter toxicity on biological functions: review of last decade studies, *International Journal of Cosmetic Science*, 2013, 35(3), 208–219.

21. W. C. Li, Occurrence, sources, and fate of pharmaceuticals in aquatic environment and soil, *Environmental Pollution*, 2014, 187, 193–201.

22. Di Poi Carole, Costil Katherine, Bouchart Valérie, Halm-Lemeille Marie-Pierre, Toxicity assessment of five emerging pollutants, alone and in binary or ternary mixtures, towards three aquatic organisms, *Environmental Science and Pollution Research*, 2018, 25(7), 6122–6134.

23. Chao Li, Afruza Begum, Jinkai Xue, Analytical methods to analyze pesticides and herbicides, *Water Environment Research*, 2020, 92(10), 1770–1785.

24. Aline A. Becaro, Claudio M. Jonsson, Fernanda C. Puti, Maria Célia Siqueira, Luiz H.C. Mattoso, Daniel S. Correa, and Marcos D. Ferreira, Toxicity of PVA-stabilized silver nanoparticles to algae and microcrustaceans, *Environmental Nanotechnology, Monitoring and Management*, 2015, 3, 22–29.

25. Elizabeth Phillips, Tetyana Gilevska, Axel Horst, Jesse Manna, Edward Seger, Edward J. Lutz, Scott Norcross, Scott A. Morgan, Kathryn A. West, E. Erin Mack, Sandra Dworatzek, Jennifer Webb, and Barbara Sherwood Lollar, Transformation of chlorofluorocarbons investigated via stable carbon compound-specific isotope analysis, *Environmental Science and Technology*, 2020, 54(2), 870–878.

26. Patrick Höhener, David Werner, Christian Balsiger, and Gabriele Pasteris, Worldwide occurrence and fate of chlorofluorocarbons in groundwater, *Critical Reviews in Environmental Science and Technology*, 2003, 33(1), 1–29.

27. Xing-Fang Li and William A. Mitch, Drinking Water disinfection byproducts (DBPs) and human health effects: multidisciplinary challenges and opportunities, *Environmental Science and Technology*, 2018, 52(4), 1681–1689.

28. Mark J. Nieuwenhuijsen, Mireille B. Toledano, Naomi E. Eaton, John Fawell, and Paul Elliott, Chlorination disinfection byproducts in water and their association with adverse reproductive outcomes: a review, *Occupational and Environmental Medicine*, 2000, 57, 73–85.

29. Anekwe Jennifer Ebele, Mohamed Abou-Elwafa Abdallah, and Stuart Harrad, Pharmaceuticals and personal care products (PPCPs) in the freshwater aquatic environment, *Emerging Contaminants*, 2017, 3(1), 1–16.

30. Michael F. Meyer, Stephen M. Powers, and Stephanie E. Hampton, An evidence synthesis of pharmaceuticals and personal care products (PPCPs) in the environment: imbalances among compounds, sewage treatment techniques, and ecosystem types, *Environmental Science and Technology,* 2019, 53(22), 12961–12973.

31. O.S. Alimi, J. Farner Budarz, L. M. Hernandez, and N. Tufenkji, Microplastics and nanoplastics in aquatic environments: aggregation, deposition, and enhanced contaminant transport. *Environmental Science and Technology*, 2018, 52, 1704–1724.

32. Aliakbar Roudbari and Mashallah Rezakazem, Hormones removal from municipal wastewater using ultrasound, *AMB Express*, 2018, 8, 91.

33. Sophia A. Carmalin and Eder C. Lima, Removal of emerging contaminants from the environment by adsorption, *Ecotoxicology and Environmental Safety,* 2018, 15(150), 1–17.

34. Kailas L. Wasewar, Surinder Singh, and Sushil Kumar Kansal, Chapter 13 – Process intensification of treatment of inorganic water pollutants, *Inorganic Pollutants in Water*, 2020, 245–271.

35. Tomonori Matsuo, Keisuke Hanaki, Satoshi Takizawa, and Hiroyasu Satoh, *Advances in Water and Wastewater Treatment Technology: Molecular Technology, Nutrient Removal, Sludge Reduction and Environmental Health*, ISBN: 978-0-444-50563-7, 2021, Elsevier, Amsterdam, Netherlands.

36. Heinrich Strathmann, *Introduction to Membrane Science and Technology*, ISBN: 978-3-527-32451-4, September 2011. P. 544, John Wiley & Sons, Hoboken, New Jersey, USA.

37. Andrew C. Maizel and Christina K. Remucal, The effect of advanced secondary municipal wastewater treatment on the molecular composition of dissolved organic matter, *Water Research*, 2017, 122, 42–52.

38. Heinrich Strathmann, *Introduction to Membrane Science and Technology*, Weinheim: Wiley-VCH, 2011, 524 pp.

39. Maria Gavrilescu, Kateřina Demnerová, Jens Aamand, Spiros Agathos, and Fabio Fava, Emerging pollutants in the environment: present and future challenges in biomonitoring, ecological risks and bioremediation, *New Biotechnology*, 2015, 32(25), 147–156.

2 Methods to Detect Emerging Pollutants

2.1 INTRODUCTION

The world of chemicals, compounds, and substances prepared by those chemicals is very complex. It is estimated that at any point of time there will be more than 100,000 substances in commercial use worldwide. For decades, there has been no systematic tracking of these synthetic substances disappearing into the environment. However, lately many researchers are following changes in the environmental and climatic changes that are affecting larger ecosystems. With this approach researchers not only monitor well-known environmental contaminants, but more and more EPs are identified. These substances are challenging in terms of quantification due to the low concentrations, complex matrices and wide range of compounds with broad physical-chemical properties. More worrisome is that their toxicities, environmental occurrences, features and characteristics remain less known compared to conventional contaminants. Information and knowledge on their toxicity mechanisms, environmental occurrences and characteristics need to be comprehensively studied and clarified for a better understanding of their impact on environment and human health [1].

The newer approaches are aimed at solving multiple issues on EPs, such as identification and discovery, nature of occurrences, physico-chemical characteristics, the toxicity mechanism and impact on human health. Studies associated with the new source identifications, analytical methodology, formation mechanism, and control of EPs also included setup of new analytical protocols. The majority of regulatory and enforcement agencies responsible for water quality now accept that a small number of well-known substances are responsible for a substantial share of environmental, human health and economic risks. However, the accuracy of this hypothesis is questionable, since the chemicals and substances under observation for their possible threat to the environment by official agencies represent only a tiny fraction. This implies that most potentially hazardous substances are thus not covered by any existing water quality regulations or included in environmental monitoring programs. These substances are challenging in terms of quantification due to the low concentrations, complex matrices and wide range of compounds with broad physical-chemical properties. At present, chemicals and substances under monitoring and analysis are commonly carried out for a known insignificant proportion of organic contaminants, thus overlooking larger important industry-specific, discharge site specific collection points. This is particularly the case with many industrial chemicals which are being discharged, systematically avoiding all monitoring and screening norms [2].

DOI: 10.1201/9781003214786-2

Nevertheless, today waterborne pollutants are perhaps the largest cause of diseases and death in the world. To add to this unknown and mystery class of new contaminants that originated from their post-use attain a number of diverse forms and are difficult to detect and characterize due to their stench behaviour and complex sources of discharge into wastewater. Therefore, the understanding of the method of analysis and health implication of EPs in water is critical to providing a more robust understanding of exposure routes, regulations and mitigation. Here is the list of basic wastewater characterization techniques that are generally classified into physical, chemical and biological natures. Info graphics (Figure 2.1) show the classification of methods and contaminants used to categorize the analytical techniques [3].

Among physical characteristics of Eps in wastewater, consider temperature, solids, odour and colour for categorization. Temperature of an EP at collection point may vary due to more biological activity due to which wastewater will attain a higher temperature. Wastewater at high temperature will affect sampling or accuracy of analysis upon being subjected to analytical tools such as industrial discharge and process wastewater. Fresh sewage is normally brown and yellowish in colour but over time becomes black in colour; this is similar to any other industrial wastewater. Wastewater that includes sewage typically develops a strong odour, and other industrial wastewater contaminated with Eps develop odour with time. Turbidity may originate in Eps contaminated wastewater due to suspended solids in wastewater which will have a higher turbidity, or cloudiness.

Chemical characteristics of wastewater contaminated with various Eps have been reported to generate multi-complex substances at multilevel chemical reactions. Wastewater contains different chemicals in various forms, as mentioned in this

Basic Parameters Monitored in Wastewater for Characterization

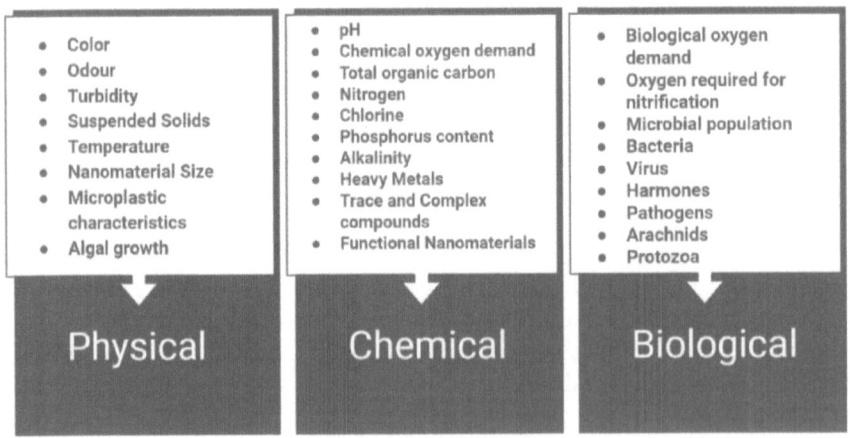

FIGURE 2.1   Infographics showing the basic parameters of EPs monitored in wastewater samples.

chapter. Therefore, the understanding of the method of analysis and health implication of EPs in wastewater or in the ecosystem in general is critical to providing a more robust understanding of exposure routes, regulations and mitigation. EPs in wastewater have been discussed extensively because of their unknown health risks, which may be because newer versions of EPs have following chemical forms. Therefore, chemical characteristics properties originated from these interactions may be listed as COD is a measure of organic materials in wastewater in terms of the oxygen required to oxidize the organic materials. TOC is a measure of carbon within organic materials. Organic nitrogen is the amount of nitrogen present in organic compounds. Organic phosphorus is found in many metabolites of biomacromolecules such as protein and inorganic phosphorus (phosphates, PO_4^-), chlorides (Cl^-), sulfates (SO_4^{-2}) and heavy metals like mercury (Hg), arsenic (As), lead (Pb), zinc (Zn), cadmium (Cd), copper (Cu), nickel (Ni), chromium (Cr), silver (Ag), arsenic, manganese and vanadium. Many or all of these EPs at one place may cause several unknown health risks. More so, the presence of novel solvents and VOCs like acrylonitrile, 1–3-butadiene, chloroform, dichloromethane, ethylene oxides, formaldehyde, toluene, trichloroethylene, 1,4-Dioxane and nanomaterial-based EPs like ultrafine particles, micro- and nano-plastics, engineered nanoparticles, diesel/black carbon and bioaerosols. Therefore, analysing these chemical characteristics in wastewater with accuracy plays a critical role in the wastewater treatment process [4].

In addition to these, it is important to know the biological characteristics of EPs in wastewater and how to regulate them through appropriate treatment methods. These biological characteristics involve BOD, oxygen required for nitrification, microbial population and nitrogenous oxygen demand (NOD). The NOD is the amount of oxygen needed to convert organic and ammonia nitrogen into nitrates by nitrifying bacteria, and microbial life which constantly alters in different conditions due to the presence of EPs in the form of microbes like bacteria, protozoa, fungi, viruses, algae, rotifers and nematodes. In addition to this, presence of oil and grease contamination in wastewater which originate from food waste and petroleum products drastically aggravate the quality of wastewater [5].

2.2 EXISTING ANALYTICAL TOOLS FOR EPS DETECTION AND THEIR THEORETICAL LIMITATIONS

The presence of a wide range of EPs in the aquatic environment and wastewater treatment plant (WWTP) discharges has been confirmed by many studies. In particular, pharmaceuticals and hormones (PPHs), preservatives from personal care products (PCPs), pesticides, phthalates or artificial sweeteners were quantified at ng/L to µg/L levels in many wastewater discharges. Many samples collected from various wastewater following conventional quality parameters were analysed by various laboratories. Some of the monitored parameters were the dissolved organic carbon (DOC), chemical oxygen demand (COD), biological oxygen demand (BOD) that were all characterized by UV-Vis spectroscopy at absorbance max of 254 nm. Total Kjeldahl Nitrogen (TKN, named after the Danish chemist Johan Kjeldahl) is the total concentration of organic nitrogen and ammonia, and

TKN in wastewater as NH_4^+, NO_2^-, NO_3^-, total phosphorus (TP), PO_4^{3-} and total suspended solids (TSS) were quantitatively analysed. All the limits of quantification (LQ) and the analytical standards for using various techniques were followed for further analysis. In addition, UV and DOC were used to calculate the SUVA (specific UV absorbance = 100 × 8 UV-254/DOC) for every sample to assess the evolution of the organic matter aromaticity. Similarly, a wide variety of EPs are now being monitored, sampled and analysed in the dissolved phase using the following methods. Most of the organic origin micropollutants are being both quantitatively and qualitatively analysed, that include analgesics, antibiotics, beta blockers, diuretic, hormones, hypolipemiants, psychoactive drugs, alkylphenols, hypolipemiants, perfluorinated acids, antibiotics, analgesics/anti-inflammatory using LC-MSMS techniques with improvised analysis protocols. Chlorinated solvents, perfluorinated acids (PFAs) and pesticides are being analysed using GC-MS, and pesticides are being quantified using advanced LC-MS/MS GC-MS techniques [6–8].

Lately, an advanced high resolution mass spectrometry (HRMS) has been in use in support of several established analytical tools to support data processing. This helps in analysing all suspected EPs, non-specific detection, data acquisition and screening methods. In many known cases, there is no need for reference standards or markers until the protocol reaches the final confirmation stage. This approach leads to the time and resource saving and allows the inclusion of extensive unknown substances to the list. However, in the case of EPs, forecasting the impending danger of EPs is a complex process. Therefore, future approaches need specific research questions to achieve a better understanding of applicable tools [9].

Nevertheless, various strategies for monitoring EPs in environmental samples have already been developed for various registered pesticides and their metabolites, APIs and their predicted metabolites with the help of all pre-existing relevant information is considered. This approach and information can be extremely significant in identification of suspect substances and compounds to focus upon devising proper sampling and analysis protocol. In some of the tested cases, these hypotheses and basic database information have helped in tracking the suspected EPs and have increased the chances of substances to accuracy.

However, for most of these emerging contaminants, occurrence, risk assessment and ecotoxicological data are not available, so it is difficult to draw suitable analytical protocol and technique for their quantitative and qualitative accuracy. The larger research community still believes that all the EPs, antibiotics, polar compounds and their metabolites present in various forms and mixtures are of greatest concern and pose great challenges to toxicologists. On the other hand, analytical methodology for these diverse groups of EPs is still lacking, and while methods exist for analysing each of these substances independently, analytical chemists try to find answers to the key issues by developing multi-residue methods in which different compound classes can be determined by a single short analysis. But this is a paradoxical situation, since these compounds are not regulated, so there is no urgent necessity to monitor them in the ecosystem; but researchers are after documenting the fact to serve the future purposes [10, 11].

2.3 INSTRUMENTATION AND THEIR LIMITATIONS FOR DETECTING EPS

Source identification, sampling, detection and quantification of EPs and their metabolites and complex products in the various ecosystems is vital to gaining knowledge on their occurrence and fate. The EPs monitoring is highly challenging for several reasons, the number of presently known potential EPs and their metabolites and complex products. Majorly, these EPs continuously attain their form and chemical compositions over time. This takes a journey from the point of their origin or manufacturing site, during usage and disposal with and without treatment, which defines their fate and hazards. This is true with many EPs such as hormones, pyrethroids, APIs, surfactants and certain organophosphorus pesticides that affect the ecosystem at extremely low concentrations and need analytical methods with correspondingly low detection limits [12].

Unlike known pollutants, EPs possess many unknown properties with widely varying physico-chemical properties. These properties in organic substances can be subdivided in PBTs as persistent bioaccumulative and complex formation abilities. More polar substances like pesticides, pharmaceuticals, industrial chemicals and APIs retain endocrine disrupting properties. Trace metals and metal-organic complexes possess toxic and particulate and nanomaterial contaminants originating from microplastics, polymers, inorganics and composites known to induce numerous health risks which are unknown for now.

At present, state-of-the-art methods for EPs sampling and analysis procedure vary amongst various monitoring and analytical agencies and laboratories. Currently, each testing agency and laboratories survey their own dedicated and typical class of EPs list and omit the full range of EPs of potential concern. Furthermore, for the number of known highly hazardous EPs that are monitored, the detection limits are inadequate to allow appropriate risk assessment. Furthermore, nanotoxicity originated from extensive use of nanomaterials, microplastics and novel solvents like ionic liquids, methods of sampling and environmental analysis are in the initial state or virtually non-existent [13].

Presently practiced conventional methods target a specific EP. Advanced ultrasensitive instrumental techniques like GC-APCI-MS/MS, and LC-MS/MS have been widely adopted to characterize possibilities of qualitative and quantitative predictable monitoring. The application of such technologies would streamline sample preparation and offer improved possibilities for simultaneous detection and analysis of multiple EPs. Now the sensitivity of analytical techniques also improve detection limits significantly for EPs with very low predicted no effect concentration (PNEC) of a substance. At present these techniques may be cost intensive; however in the long term such instruments would be highly valuable for quantitative detection and evaluation of prioritized EPs in water, sludge, environment and biota [14].

Currently used conventional methods have been extended to measure only known and theoretically predicted EPs. Standard chemical screening methods based on chromatography with full scan mass spectrometry have become increasingly popular to supplement or replace existing targeted approaches. These specific techniques consider sample analysis through a comprehensive approach, and after careful evaluation of sample physico-chemical characteristics, specific analytical tools can be selected for each EP. This is an effective method of screening samples for very high numbers of EPs and an economical way to pick up unanticipated EPs for which application of quantitative and specific targeted approaches with all associated analytical quality control is not reasonable. In addition to this, another important advantage of this approach is the ability of retrospective detection of other EPs without the need for re-sampling/re-analysis of the unknown samples. An obstacle still to be taken to fully take advantage of the potential of this technology in routine practice is the availability of adequate data processing software and associated databases. This advancement in recent analytical instrumentation eases the process of EPs detection and quantification accurately and reliably.

On the other hand, behaviour and unknown properties of EPs indulge in snooping accurate data interpretation. EPs in different contaminated sources, whether from domestic or industrial wastewater, contain a variety of forms. This is because the transport of EPs from occurrence and their diffusion to waste streams strongly depend on the EPs' properties such as polarity, hydrophilic/hydrophobic nature, volatility, adsorption properties and persistence and surface functional properties of the interacting compartments. EPs detected in domestic wastewater and industrial wastewater that have been directly discharged into rivers where their environmental fate is of concern. These EPs in the environment may undergo partial degradation, adsorb into different living species, may find a place in regimented sludge, may transport into soil and finally may seep into underground water. The large number of EPs can undergo significant biodegradation and transformation in contaminated surface waters, soil and groundwater [15, 16].

However, EPs biodegradation depends on the presence of domestic wastewater pollutants such as microorganisms that are capable of transforming the pollutants' form and the severity through metabolic activities. Once these pollutants attain complexity, EPs become susceptible for increased and long-term toxicity in the environment. These metabolites and complex substances especially in sediments and soil become difficult samples. The natural biodegradation of EPs can vary significantly between compounds and has not been specifically studied for many pollutants like hormones, surfactants, nanomaterials and APIs. These complexities associated with EPs are being studied through realistic biodegradation simulation approaches for possible analytical methods. Further, intermediates of EPs and their complex products may undergo complete or partial photodegradation and that form exhibit properties which appreciably affect the standard detection and analytical protocols [17–19].

In addition to this, properties such as adsorption behavior of organic contaminants, APIs and surfactants can differ widely in different water and wastewater samples because they occur in both ionized and unionized form, which affects their interaction sample preparation reagents and thereby interfering in accurate detection. The presence of EPs in domestic wastewater sludge and contaminants in fertilizer adulterated agricultural soils can affect not only the sorption behavior as a result of interaction with adjacent constituents leading to unknown by-product formation materials and their detection protocols. Even though adverse ecological behavior of increasingly found microplastics and engineered nanomaterials is largely unknown. Transformations of microplastics and functional nanomaterial before and after entry to the ecosystem when they interact with humic acid, common ionic species and other organics in the wastewater samples may demand advanced or multiple analytical techniques and require further dedicated research. Nanomaterials toxicity is also known to cause damages to tissue and cell cultures resulting in increased oxidative stress, inflammatory cytokine production and cell death in humans and living organisms depending on their composition and concentration. However, unlike conventional substances, microplastics nanomaterial and their modified substances possess varying properties. Therefore, new paradigms and areas specific research will be needed for microplastics, and nanomaterials identification, sampling and analysis [20, 21].

2.4 ADVANCEMENTS IN APIS DETECTION AND QUANTIFICATION

The active pharmaceutical ingredient (API) is the substance used in a finished pharmaceutical product and formulations. The API is the part of a pharmaceutical drug that produces the anticipated effects in humans. These APIs are intended to furnish pharmacological activity or to otherwise have direct effect in the diagnosis, treatment, cure, disease mitigation and prevention of disease, or to have direct effect in restoring, correcting or modifying effects in human bodies. Some drugs in its final formulation as individual active species or as combinations of several ingredients as therapies to induce multiple actions on affected portions treat different symptoms or act in different ways [22, 23].

Discovery and large-scale production of APIs follow long and tedious synthetic and evaluation procedures. The APIs are conventionally prepared using organic, inorganic, natural and in some cases synthetic polymers. Recent nanotechnology and drug discovery innovations have also enabled the use of new classes of functional and nanomaterials. The major concern with pharmaceutical waste in water today originates from their manufacturing and transportation point itself, as in both cases regulations fail to track the pollutants. This is because, in recent years many pharmaceutical companies and corporations have opted to send APIs manufacturing overseas to cut costs. This has caused significant changes to how these drugs, APIs and their waste are regulated, with more rigorous guidelines and inspections put into place [24].

The process of knowing side effects of APIs post usage become much more complicated as it is the biologically active component of a drug product. In ideal conditions, activities of APIs are well documented in combination with other components of formulations such as tablet, capsule, cream, injectable, etc., that produce the intended effects. Therefore, for decades APIs in high-quality drugs have been successfully used to treat diseases pertaining to oncology, cardiology, CNS and neurology, orthopaedic, gastroenterology, nephrology, pulmonology, ophthalmology and endocrinology. With well-regulated and monitored protocols, APIs can potentially create a more sustainable healthcare system by introducing more innovative products.

Pharmaceutical drugs come in different forms and formulations, since the solubility and dissolution rate of drug molecules and other constituents may depend on the formulation and, moreover, on the manufacturing methods. Therefore, it can be of interest to characterize an active pharmaceutical ingredient thoroughly for their physico-chemical properties. Many of the drug molecules and formulations are manufactured in powder form following specified preparation procedures. Therefore, understanding the physico-chemical properties of APIs or ingredient solid materials is key to successful drug detection and establishing analytical protocols. These physio-chemical properties during and after use of drugs can have an impact on the drug's bulk properties, product performance, processability, stability and product disposal in wastewater. When APIs are actively performing their activity in humans, the melting point, chemical reactivity, apparent solubility, dissolution rate, optical and mechanical properties, vapour pressure, and density can all be affected. On the other hand, all the metabolites and by-products generated post disposal into wastewater will also have varying physico-chemical properties. To characterize such pollutants GLP-, GCP- or cGMP-compliant physical characterization serves best to meet the regulatory norms [25].

In standard quality control protocols all manufactured APIs in powder form are primarily characterized by its particle size distribution determined by fractional sieving and other various advanced techniques. In addition to this, the dissolution profiles of each particle size fraction generally measured using particulate size, distribution and shape, surface property like zeta potential, surface area and porosity, powder flow characteristics, polymorph analysis, microstructure properties like encapsulation capacity and release, thermal properties and morphological properties like scanning electron microscope (SEM), transmission electron microscope (TEM) and luminescence microscopy (LM). Post usage of APIs in wastewater needs to undergo basic compositional characterization using x-ray powder diffraction (XRPD), thermal analysis (DSC and TGA), particle size technology and all other physico-chemical properties mentioned earlier [26–28] (Table 2.1).

TABLE 2.1

List of Various Physico-Chemical Property Analysis Tools Used for Characterization of APIs in Various Sources

Molecular Level	Particulate Level	Bulk Level
FTIR or RAMAN microscopy	Particle size	Powder X-ray Diffraction
Light microscopy	Light and electron microscopy	(XRPD)
NMR spectroscopy	(SEM, TEM)	Thermal analysis (DSC, TGA)
FTIR or RAMAN spec.	Particle morphology particle	Surface area (BET N2),
UV-vis or Fluores spec.	size analysis	porosity, porosimetry
Interferometry	SEM-EDX for elemental	Powder flow characteristics
MALLS and SEC-MALL	particle size in suspension	(bulk density, pourability,
	Aggregation studies	dustability)
	Zeta potential	Physico-chemical properties
		(viscosity, LogP, vapour
		pressure and more)

2.5 ADVANCEMENTS IN HEAVY METAL ION DETECTION AND QUANTIFICATION

The most common earlier known cationic contaminants and still listed as some of the dangerous emerging contaminants in wastewater include heavy metals (HMs), used in various applications, and their complex substances, formed as a result of interaction with other constituents in wastewaters. The most common toxic heavy metals found in wastewater include arsenic, lead, chromium, copper, mercury, nickel, silver, cadmium and zinc. However, newer contaminants as EPs have been added to the list owing to their potential threat, such as aluminium, antimony, barium, boron, cobalt, manganese, selenium, strontium, tin and organic tin. It is also important to know that HMs such as cadmium, chromium and lead are natural minerals found in the earth's crust and are typically present in water at various concentration levels [29].

The discharge of even low concentration of heavy metals into water bodies creates serious ecological problems and may lead to an increased wastewater treatment cost. Heavy metals, also known as trace metals, are the most persistent pollutants in wastewater. The discharge of heavy metals into wastewater can occur through a variety of routes, including domestic, industrial and pesticides used in agriculture residues. Some adverse impacts of heavy metals to aquatic ecosystems include habitat destruction from sedimentation, death of aquatic life, algal blooms, debris, increased water flow and other short- and long-term toxicity from chemical contaminants. Copious amounts of HMs present in the leftover residues, wastewater and resulting sludge drastically reduce the productivity of a plant species, inhibit the nutrients uptake capacity and alter metabolic processes. Even trace amounts of HM in wastewater and ecosystem severe effects on animals may include reduced growth and development. Many of the HMs in water resources are carcinogenic in nature, which can damage organs and the nervous system, and exposure to high doses can even cause death. In humans, HM toxicity can severely affect the nervous system, causing mental illness, lower energy levels and altered blood composition to kidneys, liver, lungs and other vital organs. Long-term exposure to HM results in physical, muscular and neurological degenerative processes that mimic Parkinson's disease, Alzheimer's disease, muscular dystrophy and multiple sclerosis. Several HMs cause allergic reactions [30–32].

Currently, several methods and techniques are being followed to mitigate the adverse effects of HMs in human health, animals and the environment. The detection of HM in various pollutants is performed using spectroscopic and chromatographic techniques that require tedious sample preparation and skilful handling techniques. Conventionally, a large number of HM cations are analysed using inductively coupled plasma mass spectrometry (ICP-MS) which proved to be distinctively advantageous because it can determine several cations simultaneously. This technique is advantageous more so because ICP-MS can detect HMs like sodium and calcium as high as 10 mg/L or higher, and traces like arsenic and lead as low as 10 µg/L concentration. The ICPMS spectrometric is particularly useful to determine several HMs like Ca, Cr, Cu, Fe, K, Mg, Mn, Mo, Na, P, Se, Al, As, Cs, Hg, Pb, Sb, Sn and Zn in various concentrations [33, 34].

Further, literature reveals several existing techniques useful for HM detection in water and waste samples. Some focus on an interesting hypothesis and methodologies that addresses many of the existing challenges in the detection of HMs. Many of the HM detection and analysis follow standard protocols that are sensitive, rapid, simple and selective methods for the detection of HM cations, in order to achieve a reliable solution to improve their trace-level monitoring in water and wastewater samples. In this direction, the popular other conventional techniques for the detection of HM ions are high-performance liquid chromatography (HPLC) coupled with electrochemical or UV-Vis-detectors, atomic absorption spectroscopy (AAS), electrothermal atomic absorption spectrometry (EAAS), flame atomic absorption spectrometry (FAAS), wet chemical methods such as colorimetry, titrimetry and gravimetry, and electrochemical techniques [35–38].

However, despite the high accuracy and sensitivity reached in spectrometric methods, most of them require expensive, complex and sophisticated equipment and are capital and operational intensive. These instrumentation techniques also require highly trained and skilful technicians, making them difficult to use in on-site measurements for portable and easy-to-use detection. In such circumstances, newer inexpensive but fewer HM-specific methods follow electrochemical methods for high sensitivity, fast response, low power cost, simpler approach and ease of adaptability in order to be integrated into portable, disposable devices for the in-situ multi-element analysis of HMs.

For detection of HMs, several nanomaterials-based systems use an interesting strategy of enhancing the sensitivity of the sensors due to the possible interaction of HMs with functional groups in the nanomaterial electrodes with high electrochemical activity enabled using an electron-transfer process derived from unique electronic, physical and chemical properties. Several electrodes have also been prepared on a screen-printed electrodes form modified with mercury, bismuth, gold, selenium, tin, carbon and conducting and functional polymers in voltametric systems have been used for the simultaneous detection of HMs at residual levels, displaying a low limit of detection (LOD) and a large linearity range. Similarly, glassy carbon electrodes (GCE) have also been successfully used for HM detection following chemically modified routes with bismuth nanoparticles in assistance with square-wave anodic stripping voltammetry (SWASV) for the simultaneous detection of Pb^{2+} and Cd^{2+} in water and wastewater samples with detection limits of 0.2 gL^{-1} and 0.6 gL^{-1} for Pb^{2+} and Cd^{2+}, respectively. Nanocomposites-based electrodes such as graphene/bismuth nanocomposite films further found to be ultra-sensitive for sensing several HMs Zn^{2+}, Cd^{2+}, and Pb^{2+}, simultaneously. Several new advancements in thin-film technology use polymer-based chips on silver as reusable HM sensors with varying sensitivity. This technique is extendable to on-site detection of toxic HM like lead with the help of microfluidic device technology. Despite achieving ultra-high sensitivity from polymer, nanomaterial and composite electrodes, they suffer from reproducibility, and new strategies have also been devised to overcome sensitivity and reproducibility issues with HM detection. Even with several advancements, simultaneous detection of multiple HM in water samples is still a difficult task, unlike ICP-MS [39–41].

To improve the ion selectivity and sensitivity of HM detection, a series of ion-selective chalcogenide glass sensors were developed and tested for detection of HMs in water samples. These unique chalcogenide-based electrodes were successfully used for detection and quantification of various HMs like Ni^{2+}, Cu^{2+}, Zn^{2+}, Cd^{2+} and Pb^{2+} in a variety of concentrations. Further, recent advances also explored the use of electronic tongue (e-tongue) systems following a multisensory array based on non-specific or low-selective sensing devices. This method works by taking advantage of HMs cross-sensitivity to identify a liquid medium, mostly using electro-analytical methods. Nonetheless, reliable and sustainable HM characterization techniques usually required substantial data sets to compare and evaluate reference techniques [42–44].

Arsenic and high concentration of its metabolites in wastewater are a major concern for aquatic as well as humans due to arsenic's high toxicity. The arsenic-containing wastewater mainly originates from smelting copper and other non-ferrous metals. On the other hand, the waste acid mainly comes from circulating cooling water and domestic wastewater. Inorganic species of arsenite and arsenate have a bigger toxicological relevance than the organic species on water quality degradation. Along with arsenic, chromium in hexavalent species form discharged from various industries and food packaging units was recognized as a human carcinogen widely analysed using the combination of HPLC and ICP-MS.

Recent studies prove an increased concentration level of antimony (Sb) in several mineral waters, soft drinks, fruit beverages and industrial wastewaters. On the other hand, the Sb-based complex substances are gaining alarming attention because of their toxicity and harm to the environment. Some studies trace the origin of Sb and antimony trioxide (Sb_2O_3) in drinking water from their packaging material PET bottles (polyethylene terephthalate) which has been used as a catalyst in PET production. Sb-based metal complexes and Sb are qualitatively and quantitatively measured using AAS with high sensitivity. In addition to several instrumentation or device-based techniques, there are polymer and paper-based free-standing film chemical-sensors that have been used for HM detection and quantification [45, 46].

Over the past decade, paper-based analytical devices (PADs) have gained increasing focus due to their futuristic potential, including simple preparation techniques, the ability to be used for on-site detection, improved sensitivity and multiple analysis detection. One such film was natural-polymer lignin-based films deposited onto gold inter-digitised electrodes through the Langmuir–Blodgett technique. These thin-film and free-standing sensing strips are being used for the detection of HM ions in water and wastewater samples. To achieve higher accuracy, these sensing films are used in combination with Fourier transform infrared spectroscopy (FTIR) and Langmuir isotherms to explore the physical interactions formed between lignin phenyl groups. Through this method, EPs like Cu^{2+} ions have been detected in several water samples with the help of impedance measurements with accuracy. Advancements in the field of carbonaceous nanomaterials such as multi-walled carbon nanotubes (MWCNT), graphene and graphene oxide, in combination with several reactive polymers like polylactic acid (PLA) nanofiber nanocomposites, are being explored in HM ion detection and analysis. Among different carbonaceous nanomaterials, CNTs have been extensively used for HM ion detection with the help of integrated flow injection analysis system. These CNTs-based probes are successfully being used for the analysis of water samples contaminated with trace levels of HMs including Ba^{2+}, Cd^{2+},

Cu^{2+}, Pb^{2+}, Ni^{2+}, Mn^{2+} and Fe^{3+}. Also, electrospun nylon-based ternary nanocomposites and cellulose nano-whiskers are being used for silver nanoparticles sensing in water samples. Further, paper strips modified with gold nanoparticles in device form have been used for detection of cadmium (Cd) nickel (Ni), ferric (Fe), Cu and Cr through colorimetric changes. Additionally, similar paper strips in combination of electrochemical tools have been used to colorimetric detection of Pb and Cd [47–49].

Even though thin-film and free-standing paper-based analytical tools have shown high sensitivity for HMs, they fail to quantitatively detect HMs for sustainable and reliable usage. Despite the availability of various efficient analytical techniques in use, there are still challenges to overcome in order to improve HM detection. Analytical tools for specific HM detection also majorly depend on the limit of detection, quantitation, accuracy, precision and selectivity of the instrument. Another desired quality of the successful analytical tool is reproducibility and stability of probes in complex wastewater samples and retaining sensitivity in presence of counter ions interference. Furthermore, detecting multiple HMs at once is a hard task, as in most cases the sensor and probes in almost every instrument and analytical tool presents specific sensitivity to only one or two heavy metals. Other major limitations of the HM analytical tools currently in use are the affordability and point of use to HMs in affected areas. Therefore, research and innovation activities need to focus on devising HM detection and quantification tools which are sophisticated, sensitive, accurate and affordable all at once for multiples of HM in wastewater sample analysis.

2.6 NEW PROTOCOLS TO DETECT AND QUANTIFICATION OF ORGANIC POLLUTANTS

Emerging organic pollutants are chemical substances of natural or synthetic origin that occur in the ecosystem and environment, resist bio-degradation and bio-accumulate through the food web which may adversely affect human health. Organic substances, chemicals or their complex variations present in various forms and concentrations in products like finished pharmaceutical drugs, surfactants, personal care products, dye, humic substances, phenolic compounds, petroleum, surfactants, pesticides, hormones, food additives, pesticides, plasticizers, wood preservatives, laundry detergents, disinfectants, flame retardants and other organic compounds in water generated mainly by human activities [50].

On the other hand, all emerging organic pollutants also undergo further classification based on their varying chemical properties, fate after use, ecological impact and transportation of traces, and also the diagnostic methods for their quantitative and qualitative detection, as shown in Table 2.2. In each category organic compounds are still considered very harmful.

National and international regulatory agencies categorize organic pollutants based on their volatility. Very volatile organic compounds (VVOCs) have the lowest boiling points (< 0 to 50–100 °C), and they readily mix into the air through outdoor emissions from reactive organic compounds and their interactions. Due to stringent

TABLE 2.2

General Classification of Organic Emerging Pollutants in Water and Wastewaters

Description	Abbreviation	Boiling Point (°C)	Example
Very volatile organic Compounds (Gaseous)	VVOCs	< 0 to 50–100	CO, CO_2, formaldehyde
Volatile organic compounds	VOCs	50–100 to ~250	Aliphatic and aromatic solvents, terpenes
Semi-volatile organic compounds	SVOCs	~250 to ~380	Pesticides, plasticizers (Ex: phthalates)
Dissolved organic matters	DOMs	>380	Dyes, APIs, pesticides,
Particulate organic matters	POMs	>380	Pesticides, polycyclic aromatic hydrocarbons

monitoring and regulatory norms, manufacturers using chemicals like propane, butane, methyl chloride and solvents that produce VVOCs need to be equipped with devices which can sense, monitor and detect emissions to the environment. On the other hand, volatile organic compounds (VOCs) can be found as both indoor and outdoor pollutants. VOCs have a boiling point range of 50–100 to 240–260 °C. VOCs such as formaldehyde, d-limonene, toluene, acetone, ethanol (ethyl alcohol), 2-propanol (isopropyl alcohol) and Hexanal, at room temperature, can contaminate water and are readily found in industrial wastewaters.

On the other hand, semi-volatile organic compounds (SVOCs) are a class of organic compounds having moderate to high boiling points in the range of 240–260 to 380–400 °C. Although these chemicals are relatively less volatile, they generally include a myriad of industrial chemicals such as industrial chemicals and petrochemicals, DDT, chlordane, plasticizers (phthalates), phenols, carbonyl compounds, fire retardants (PCBs, PBB), ethers, aliphatic and aromatic esters, anilines, pyridines and many others. It is pertinent to note that SVOCs exhibit a definite degree of volatility and hydrophilicity due to which SVOCs can readily pollute water resources, and they are abundantly found in industrial wastewaters. SVOCs are extensively used in several industries, including pharmaceutical, plastic and package, polymer products, etc. Being copious and resistant to degradation, SVOCs can be transported with particulate matter, water and wastewater. The trace of SVOCs are found in several segments of ecosystems such as soil, water, air and solid wastes. Because of their definite or known harmful effects as individual pollutants, SVOCs can form several complex transformations when they come in contact with other pollutants like heavy metals. Therefore, with time and persistent interaction with other variables, SVOCs attain highly stable emerging contaminant status. These SVOCs in the environment further retain bioaccumulation capability and capability of long-range transportability, and finally induce toxicity to the environment. Therefore, it is important for manufacturers to monitor and take into account the type and measure of controlling, mitigating, detecting and neutralizing SVOCs being released in the form of wastewater [51].

Recently, a new class of organic emerging pollutants such as dissolved organic matter (DOMs) are increasingly found in water resources, their origin generally traced from extensive industrial activities. Naturally occurring DOMs consist of soluble organic materials derived from the natural decomposition plants, soil matters and other microbial residues. DOCs found in water resources are a heterogeneous class of water-soluble compounds, substances and chemicals that contain reduced organic carbon and come from a variety of biological and geological sources with a wide range of chemical reactivity. DOMs found in water and wastewater reservoirs are complex mixtures of organic molecules made up of carbon, hydrogen, oxygen, heavy metal complexes and heteroatoms nitrogen, phosphorus and sulphur. Recently found DOMs in water and wastewater resources are characterized by contaminants that are industrial substances like dye, humic substances, phenolic compounds, petroleum, surfactants, pesticides and pharmaceuticals. The presence of DOMs and their persistent variants in water may in the long run produce chemicals that are toxic to the environment, and to humans upon consumption.

On the other hand, hydrophobic organic pollutants (HOPs), such as oil, grease, liquid petroleum products and other polyaromatic hydrocarbons, come primarily from automobile discharges ubiquitously found in the aquatic environment. In general, HOPs limit their known harmful effect on the environment as a result of association between DOM and HOPs, which limits the bioavailability of free HOPs. However, DOMs readily undergo chemical transformation during several artificial and natural disinfection processes, leaving behind several emerging pollutant by-products of unknown consequences.

Therefore, it is very important to monitor and devise efficient analytical tools for VOCs, SVOCs, DOMs and POMs that are both free and bound forms as transformative, metabolites and by-products. As new classes of organic emerging pollutants are increasingly characterized for their harmful effect in wastewater, an efficient analytical tool is necessary to better understand and develop effective treatment processes. This section further explores the analytical tool available for detection and quantification of emerging organic pollutants in wastewater and treated water from water treatment plants.

In general, COD and BOD tests are used to indirectly measure the amount of organic compounds in water and wastewater. Particularly, the current international standard BOD measurement protocol is followed to monitor and measure organic pollutants in water and wastewater samples using five-day BOD assay (BOD_5). The other important methodologies used for characterizing DOM and the hazardous potential of organic pollutants are total concentration of organic carbon by dissolved organic carbon (DOC), aromaticity by UV absorbance and specific UV absorbance (SUVA), and size distribution by ultrafiltration (UF), polarity by polarity rapid assessment method (PRAM), UV and fluorescent chemical components by UV spectrometer and fluorescence excitation-emission matrix (EEM) and fluorescence regional integration (FRI) [52].

Further, dissolved carbons and their variants are being analysed for definitive forms and structure using qualitative spectrometry techniques. Conventional mass spectrometry (MS) is widely used to detect an ubiquitous amount of organic pollutants in various samples. Several LC-MS techniques have been used for organic EPs identification, such as quadrupole, ion trap and time-of-flight mass analysers, and MS sources such as atmospheric pressure photoionization, electrospray ionization, atmospheric pressure chemical ionization and direct electron ionization. On the other hand, to understand and assess the quantitative impact, emerging organic pollutants on the environment are being analysed using quantitative analytical techniques like gas chromatography (GC) and liquid chromatography (LC). GC provides an outstanding advantage for compatible analytes because of the high-resolution separations that are intrinsic to the method to predict pollutants based on their quantity. For the analysis of polar or thermally labile organic substances, compounds and dissolved chemicals, LC is best suited. The advanced LC techniques in combination with enormous databases are being used to detect and quantify organic compounds like PAHs, organochlorine compounds and perfluorinated acids in various samples [53].

In addition to this, advancements in nanomaterial sensing properties include CuO nanosheet decorated glassy carbon electrodes, used to sense ethanol in water samples. Organohalides-based pesticides like acetamiprid are analysed for both qualitative and quantitative measures using an aptamer-based colorimetric approach [54]. Advances in nanomaterial-based sensing probes have been widely applied for detection of various pollutants with the help of spectrometric techniques. One such probe uses gold nanoparticle (AuNPs)-based reagents with the ability to undergo catalytic activity to oxidize a pollutant. In this approach contaminated water samples undergo colour change which is further evaluated using a UV-VIS (visible) absorption spectrometer thereby detecting EPs to lowest possible concentrations (1 µg/L). Similarly, a sensor with PtNPs-coated surface allows the specific detection of acetamiprid and atrazine. The polychlorinated biphenyls (PCBs) and organophosphorus pesticide (OP) like malathion, parathion and methyl parathion are detected using a colorimetric detection scheme employing Au nanoprobes and their aggregations. Furthermore, other well-studied OPs for sensitive detection in environmental samples are phenol and its derivatives, such as aminophenol or nitrophenol in water sources detected using GC-MS or HPLC-based methods.

Except for advanced nanomaterials-based sensing probes, conventional analytical techniques proved to be expensive, laborious and time-consuming, illustrating the requirement for new approaches for on-site applications. However, economic benefits still need to be evaluated for nanomaterials-based probes for their potential real application scenarios in spiked water samples. In a larger context, use of affordable techniques to detect emerging organic pollutants such as VSOCs and DOMs needs further theoretical and large database understanding for realizing accuracy and sustainable protocol set-up. In this direction researchers have come up with a promising perylene-based probe to be able to evaluate complex water samples on a site-specific basis, since DOM is site-specific.

However, it is clear that nanomaterials-based chemical and electrochemical probes will govern the future water analytical tools market. These sensing probes are increasingly proving their potential for range of organic pollutant detection with reliable accuracy. Nevertheless, described sensing strategies and their preparation routes are also cost effective to be applied in environmental monitoring and on-site screening as a fast and easy-to-prepare pre-scan method. Nevertheless, these advanced sensing probes indeed overcome the high-concentration detection barriers. Once a too high and upper limiting concentration of an organic pollutant is detected on site by integrated electrochemical, colorimetric or optical methods, the verification could be performed by the well-established conventional lab-based GC-MS or HPLC-based techniques [55].

Because several industrial, domestic and agricultural wastewaters have left a range of organic pollutants during the past several decades, it is difficult to predict their present fate in soil and water. However, taking futuristic risk into account, the UNEP organization has developed a complete analysis procedure for various emerging pollutants, including emerging pollutants. The standard procedure includes either several matrices or individual matrices for developing an analytical protocol. Furthermore, different novel techniques for specific matrices are readily available, as evident from several study reports and case studies that show there is no complete or step-by-step analytical procedure suggested for specific emerging organic pollutants. As a result, performance-based methods are used by analytical laboratories with contaminant-specific procedures. Such workability allows analytical chemists to choose the most suitable technique depending on their applications and drawbacks. Nevertheless, organic pollutants comprise a range of molecular weight and properties from different chemical classes, which necessitate different detection protocols, procedure, schemes and hybrid techniques for accurate detection of pollutants. As a consequence, for now no holistic approach is available to monitor all emerging organic pollutants with the same nanomaterial-based sensor platform. To address this limitation, future research needs to be focused on developing suitable combinations of different sensing schemes.

2.7 DETECTION AND QUANTIFICATION OF ENDOCRINE DISRUPTING CHEMICALS

It is now well established that pharmaceuticals and human hormones which are generally classified as endocrine disruptors (EDs) are ubiquitous contaminants of wastewater effluents. Endocrine toxicity in humans and animals results when ED chemicals interfere with the synthesis, secretion, metabolism, binding action, transport or elimination of hormones necessary for endocrine functions resulting in loss of normal tissue function, development, growth or reproduction. Endocrine disruptors are the class of organic, metal-organic pharmaceutically active substances which are generally referred to as hormonally active agents. Recent studies around the world are reporting the trace levels of pharmaceuticals and human hormones in domestic and industrial wastewater treatment plant (WWTP) effluents. Even though human hormones contaminants have been recorded in wastewater samples as early as 1965, the studies on adverse effects of such steroids hormones on human health have only recently been recorded. Steroid hormones have been extensively detected in domestic wastewater samples and were not completely eliminated during wastewater treatment with advanced treatment technologies. It is also evident from the studies that endocrine-disrupting chemicals, or endocrine-disrupting compounds, can interfere with endocrine or hormonal systems inducing severe toxicity to living beings. Consumption endocrine disruptors contaminated water can cause cancerous tumours, birth defects and other developmental disorders in humans. Some of the examples of endocrine disruptors include polychlorinated biphenyls (PCBs), polybrominated biphenyls (PBBs), dixons, bisphenol A (BPA) from plastics, dichlorodiphenyltrichloroethane (DDT) from pesticides, vinclozolin from fungicides and diethylstilbestrol (DES) from pharmaceutical agents. More often, endocrine-disrupting substances are found in milligram per litre concentrations; however, even at trace concentrations, these compounds are known to be harmful to the environment [56].

In addition to this, PPCPs and APIs, and other EDs, are an extremely diverse group of compounds that interfere with the functioning of natural hormones in living organisms. Every known emerging pollutant is known to cause endocrine disruption in humans; therefore, it is difficult to determine which chemical or substance should or should not be classified as endocrine disruptors. Several studies reveal that worldwide manufacturing of ED contaminants has increased from 1 million (M) to 500 M tons per year. Nevertheless, EDs of both natural and synthetic origin have been associated to exert known or suspected adverse consequences on human health and the environment, as shown in Figure 2.2. For instance, APIs, which are specially prepared to produce a biological response in a target organism in humans to diagnose specific disease, may also produce a similar response in non-target entities following chronic exposure to even trace concentrations of these compounds. Similarly, some EDCs have been analysed for their potential to interfere in the human body's endocrine function and thus resulting in adverse reproductive, oxidative distress, hypertension, neurological, anaemia, developmental disorders and immune effects in

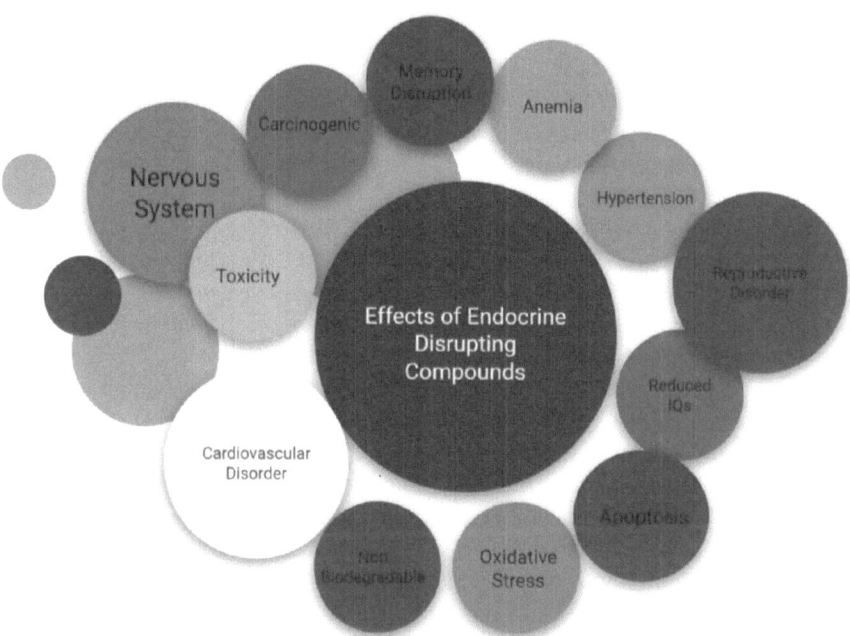

FIGURE 2.2 Infographics showing the effects and suspected health risks of endocrine disrupting substances and their complexes in humans.

humans and wildlife. Today, EDs are found in many water resources as several daily used products leach many contaminants into the environment but are mainly found in domestic wastewater effluents.

Environmental protection and monitoring agencies throughout the world increased efforts in regulating discharge, management and treatment. Even though these concerted efforts make no difference to existing toxicity in the environment with respect to the technical design, implementation or understanding of research in this area, new efforts can dramatically impact the communication of the work to non-experts, future treatment techniques and on-site detection to protect future impending disasters. Among several EDCs, APIs and drugs, only a small percentage of APIs or produced drugs such as hormone receptor agonists/antagonists have serious disruptive potential. However, each EP has an environment in high concentration or bioaccumulation, and thus all pharmaceuticals can be embraced in a broad category that can be dangerous to the ecosystem. This necessitates the procedure and protocols of multitude effects to remove all emerging ECDs from the immediate endocrine system.

EDs can occur in various ways both from natural and synthetic origin chemicals, and some of these mimic a natural hormone, fooling the body into overresponding to the stimulus. Many synthetic chemicals are widely considered to be EDs, including certain APIs, pesticides, drugs and hormones excreted by animals and humans. In this scenario, detection of EDs throughout the ecosystem, water and wastewater has become a high environmental concern, and it is of great significance to increase the efficacy of wastewater treatment technologies to overcome environmental pollution. Also, several regulatory and monitoring authorities worldwide, including the World Health Organization (WHO), consider detection and monitoring of EDs in wastewater. There are several methods and analytical tools used to detect EDs qualitatively and quantitatively. Most of the EDs are water soluble and easily undergo solubility in different conditions because of their chemical structures. The identification and assessment of these EDCs requires highly sensitive analytical techniques for detection at nanograms per liter scales [57].

The most commonly used approach to APIs, pesticide and other EDs analysis involves the use of chromatographic methods. The GC-MS with electron ionization (EI) and the combination of liquid chromatography with tandem mass spectrometry (LC-MS/MS) were identified as techniques most often applied in multi-residue methods for APIs and pesticides by several researchers. However, GC is still preferred over LC methods for VOCs and SVOCs due to higher-resolution and lower detection limits. The use of chromatographic methods are still the most efficient tools in combination with mass-spectrometry to detect EDs with excellent sensitivity and precision. GC-MS remains the most useful and sensitive method for the detection of trace levels of SVOCs of endocrine disruptive concerns. The most often used MS detector for EDs is the quadrupole MS with EI; however, tandem MS is widely used for detection of ultra-low concentrations. On the other hand, the powerful technique of gas chromatography time-of-flight mass spectrometry (GC-TOF-MS) provides the possibility of performing untargeted fast GC of pesticides in complex samples of wastewater samples. Nevertheless, a combination of GC-MS and LC-MS/MS techniques still prove to be the best approach to multi-compound (EDs class) analysis.

Liquid chromatography, atmospheric pressure chemical ionization (APCI) and electrospray ionization (ESI) as ionization techniques are widely used to detect the EDs in wastewater of more polar, less volatile (SVOCs) degradable pesticides residues. On the other hand, GC-MS allows cost-effective and easier analytical operation than LC-MS/MS, GC/MS and GC/tandem MS generally have lower LODs when compared to LC-MS/MS. Therefore, for reliable identification and quantification of compounds at ultra-trace level, chromatographic methods are required. GC-MS and GC-tandem MS are used for SVOC analytes, while LC-MS/MS is still an effective tool to detect more polar and less volatile compounds.

On the other hand, recent years have seen innovative methods for identifying chemical interactions with a molecular target, such as a hormone receptor or enzyme, chemical sensors and electrochemical sensing probes that can sense the target molecule, and these methods have become increasingly cost effective. These sensors and probes with target-specific interactions might initiate a sequence of downstream biological effects that lead to adverse outcomes, yet for now molecular effects and adverse responses are not usually evaluated in the same test. Therefore, it necessitates urgent research to develop fast, robust analytical methods for efficient monitoring and determination of wide-ranging EDs in water and wastewater samples. It is also interesting to realize that there are many other chemicals for which limited evidence is available on their possible endocrine activity or for which the evidence of endocrine activity is insignificant and debatable. An even greater number of chemicals have not yet been screened and tested for potential endocrine activity using any of the available methods. Most of the organics, HMs, APIs and SVOCs have still been defined as emerging pollutants of endocrine disruptive concern.

In this direction, the United States and European nations have established programmes to increase the efforts in managing EDCs emitting from different sources. The Environmental Protection Agency (USEPA) started the Endocrine Disruptor Screening Program (EDSP) to identify screening methods, and the European Organization for Economic Co-operation and Development (OECD) initiated programs to address EDCs toxicity, testing strategies that can be used to determine whether chemicals are endocrine disruptors, but this process is still in progress. These efforts need to be supplemented with the effective database on EDCs which is lacking at present, and all efforts within the scientific community are in progress for a strategy to definitively determine whether a chemical is an endocrine disruptor, and definitions of the term vary.

2.8 DETECTION AND QUANTIFICATION OF PATHOGENS AND OTHER MICROPOLLUTANTS

Pathogenic or microbial contamination in wastewater could come from either animal or human faecal wastes and contain different kinds of protozoa, viruses and bacteria, capable of causing diseases in humans and animals and harm to the environment at large. The harmful effects originating from pathogenic contamination in wastewater range from mild acute illness, through chronic severe sickness, to fatality. The WHO reports that millions of people died of diarrhoeal diseases globally every year. This situation escalates when pathogenic contaminants in wastewater disturb the ecosystem, causing infectious hepatitis, cholera, amoeboid giardiasis, typhoid, bacillary dysentery and bilharzia are some of the more common diseases [58].

Biological emerging pollutant-contaminated wastewater originate from human, animal feedstock's waste, dairy farms, ship sewage discharge, oil spills from commercial ships and biochemical industries as shown in Figure 2.3. The biological wastewater formation process occurs at several levels. In surface water, microbial and pathogenic pollutants originate via the natural process of animal and plant decomposition. However, biological wastewater formation is a complex process involving an intersection between biology and biochemistry which is yet to be understood completely. Severity of biological wastewater treatments rely on the quantitative presence of bacteria, nematodes or other small organisms which continuously break down organic wastes using normal cellular processes. Biological wastewater typically contains a buffet of organic matter, pathogenic organisms, HMs and toxins, as well as their amplifiers such as garbage and partially digested foods.

The most common transmission of pathogens takes an oral consumption route to spread in humans and animals. Although many pathogens can live for a short time outside the human body, waterborne and water-transmitted pathogens are resilient (bacterial cysts and oocysts together) as a key infection mechanism. This unrelenting pace of pathogenic contaminants transmission in water, soil and treated sludge forces greater attention for their mitigation and management. In addition to primary wastewater treatment, biological wastewater treatment often is used as a secondary treatment process to remove material remaining to remove BOD, DOMs and other organic contaminants using dissolved air flotation (DAF) and aerobic and anaerobic treatment techniques. However, the primary water treatment process leaves large amounts of solid wastes, sediments and substances such as oil to be removed from the wastewater. Therefore, detection and quantification of pathogenic contaminants in wastewater becomes very important.

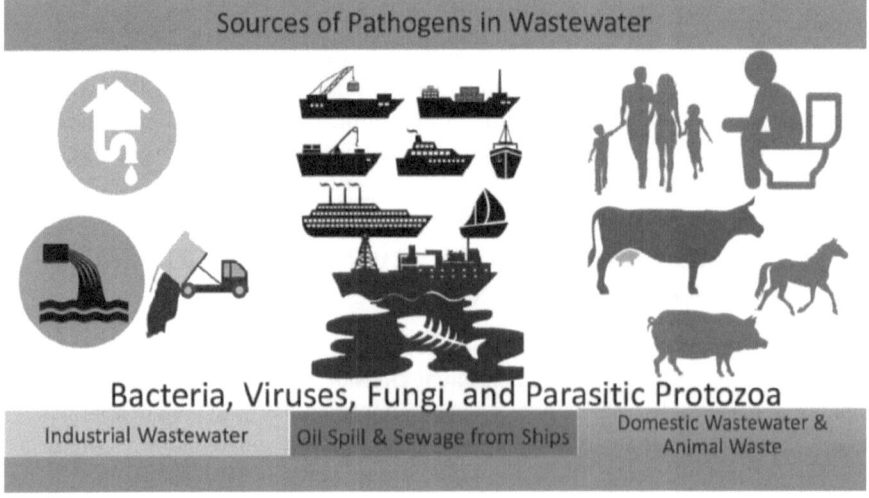

FIGURE 2.3   Infographics showing the origin of various pathogenic EPs in the environment.

Although the environmental and human health risks associated with pathogens have been well documented, the understanding and information on microbial communities present in WWTPs remains limited. Bacteria are major contributors to the degradation of complex organic compounds in wastewater treatment systems and are essential for the optimal operation and preservation of biological treatment systems. However, the emergence of new combinations of pathogen diversity or metabolism of such communities can affect the whole wastewater treatment process. Therefore, understanding new biological contaminants and their functioning requires advanced detection and analytical tools. In the past, conventional bacterial diversity studies were performed using culture-dependent methods. However, these culture-based methods only permit the study of those fraction of bacterial communities which can be cultivable. Some recent advanced culture-independent molecular techniques such as automated ribosomal intergenic spacer analysis (ARISA), denaturing gradient and temporal temperature gel electrophoresis (DGGE/TTGE), 16S rRNA gene clone libraries and fluorescent in situ hybridization (FISH) have also proved to have limited capabilities in detecting and quantification of specific pathogen communities in wastewater [59].

It is important to identify the different morphological characteristics, namely colony size, elevation and pigmentation to detect and analyse bacterial colonies in wastewater samples. In the recent past, colonies were isolated from different media plates and sub-cultured to obtain pure bacterial cultures, and then to maintain the culture maintenance, LB was used. Several studies have been reported for bacterial diversity identification and detection using both culture-dependent and culture-independent analysis as complementary methods. This starts with basic analysis of sample collection following characteristics determined for COD, BOD, FOG, TOC, TS, TSS, total nitrogen (N) and total phosphorus (P), recorded using a conventional technique from wastewater samples, followed by microbial analysis using polyacrylamide gel electrophoresis (PAGE). Further, total DNA from wastewater samples and the cultivable fraction are extracted using the Ultra Clean™ Soil DNA Isolation Kit and the Microbial DNA Isolation Kit. Temporal temperature gel electrophoresis analysis is carried out using polyacrylamide gels, urea and TAE buffer. To identify the specific bacterial colony, DNA sequence and phylogenetic analysis, sequence accession numbers, bacterial community analysis, detection and quantification of lipase activity and detection of trans-esterification activity will be recorded following standard analytical protocols for each step.

Despite several measures by environmental monitoring agencies, today thousands of chemical compounds are found in natural waters, most of them in very low concentrations (pg to μg L^{-1}). Many of these minute pollutants are classified as micropollutants originated from domestic chemical substances used in homes, workplaces or in the urban environment in various forms and concentrations. Several micropollutants trace back their parent substances in daily used products like detergents and their additives, food, plastic additives and flame retardants used in speciality textile, PCPs, pharmaceuticals and their metabolites that are excreted in urine and faeces. Micropollutants are also made up of a large portion of non-domestic wastewater pollutants like HMs, pesticides or hydrocarbons, leached during rain runoff from roads, buildings, and urban parks and gardens

The EPs list contains a mixture of different types of contaminants which are not commonly monitored and measured with a specific criteria, as a mixture of contaminants EMPs in micron and submicron level impose adverse effects on human health and the aquatic world. The occurrence of organic EMPs in wastewater treatment plants (WWTPs) have received considerable attention due to limitations in detection due to their size that allows them to escape most treatment techniques. Nevertheless, intensive sampling and analysis efforts worldwide are being made to improve understanding of the occurrence, behaviour and fate of EMPs in wastewater and WWTPs. Although many reported studies highlight the fact that environmental concentrations of the EMPs are very low, pg/L to mg/L in some cases is considered as below the toxicity threshold to induce an acute effect. However, their higher concentration and long-term cumulative impacts on human health and ecosystem safety remain largely unknown. Therefore, to understand the behaviour, transport and fate of EMPs in the soil, water and WWTPs, it is necessary to develop reliable and versatile analytical methods for rapid detection of multiple-class EMPs in water, wastewater and sludge.

At present several analytical techniques, like ultra-high performance liquid chromatography in combination with tandem mass spectrometry (UPLC-MS/MS), solvent extraction (USE) and solid phase extraction (SPE), are being commonly used to analyse APIs and pharmaceuticals at trace concentrations. Organic EMPs are analysed using high-resolution mass spectrometry (HRMS), and recent developments in effect-based water analysis used multidimensional chromatography and ion mobility techniques.

Presently, practiced conventional analytical tools have shown significant improvements in predicting EMPs qualitatively. APIs and a number of pharmaceuticals are monitored precisely using high chromatographic separation techniques for efficiency, low detection limits and reliable quantification with the help of four isotopically labelled internal standards. This method was validated and then successfully applied for analysis of several residual APIs, pharmaceuticals and organic MPs in real wastewater. Nevertheless, the primary challenge in detection and analysis of EMPs lies in how to significantly and precisely extract and quantify EMPs for both qualitative and quantitative analysis. This process requires advanced research, as EMPs involve multiple-class contaminants covering a wide range of pK_a and log K_{ow} values, polarities and stabilities that are largely pH dependent. Therefore, futuristic aims should focus on developing a selective, sensitive and reliable analytical method, based on affordable and reliable techniques for fast detection of multiple-class EMPs in both water and wastewater samples.

2.9 DETECTION AND QUANTIFICATION OF MICROPLASTICS

Humans produce an average of 380 M tons of plastic every year (> 380 M in 2019), which is a slight decrease compared to earlier years (407 M in 2015) due to increased awareness. However, several reports indicate that nearly 50% of the annually produced plastic is for single-use purposes. Most of the single-use plastic ends up as waste in the environment after a few days of its production and few moments of use, as depicted in Figure 2.4. Synthetic plastics have invaded all aspects of society. Plastics are everywhere in the average household, from toothbrushes to eateries,

FIGURE 2.4 Infographics predicting the journey of plastic waste from polymer extraction from petroleum well to formation of microplastics via natural disintegration.

utensils, packing foods, containers, toys, entertainment devices like musical keyboards, furniture and at last to sleeping plastic-filled pillows. Synthetic polymers are an incredibly versatile group of materials that can be used in almost every product manufactured for human use, and the production of polymeric materials has been growing continuously year after year. Plastics have reached the status of an ever-growing market commodity [60].

However, plastic does not biodegrade, and it cannot be disposed safely into the environment; instead, it breaks down into progressively smaller pieces of plastic. This means that in our environment, there are pieces of plastic measurable in both metres and nanometres. Even with several measures in place to collect used plastics, a major portion of the non-degradable plastic ends up in soil, and in the aquatic environment in disintegrated forms. From the point of disposal into the environment, tiny pieces of plastic, also known as "microplastics," end up everywhere, including water bodies like lakes, rivers and oceans.

Microplastics have their origin from discarded polymeric products that come from a variety of sources. The main source of microplastic is larger plastic debris that degrades into smaller and smaller pieces. In addition, microbeads, a type of microplastic, are very tiny pieces of manufactured polyethylene, polypropylene, polystyrene and PET plastic that are added as exfoliants to health and beauty products. Several variants of polyethylene are also used as cleansers and toothpastes. It is of great concern now that a major portion of microplastics easily pass through in conventional water filtration systems and end up in the lakes, rivers and oceans, posing an existential threat to aquatic life. A recent WHO report with the support of over 50 studies indicates that by 2050 there will be more plastics in oceans and other aquatic life than fishes. Another report reveals the presence of microplastics in over 90% of packaged drinking water bottles produced using advanced filtration techniques by every commercially successful brand collected the world over for testing. Many of these studies conducted on tracing microplastics in water counted thousands of microplastic particles in every litre of drinking water. Further, scientists worryingly reported that 'Theoretically, if a person consumes microplastic, some of them are small enough to pass through the gut wall and enter the circulatory system.' In any case, the presence of microplastics show increasingly worrying trends; however, the impacts of microplastic consumption on human health remains unknown.

Microplastics comprise a wide range of natural and synthetic polymer materials composed of different substances, with different densities, nature, solubilities, chemical compositions, shapes and sizes. There is no scientifically agreed definition of microplastics; however, new classes of emerging pollutants are classified frequently based on their size as plastic particles less than 5 mm (millimetre) in length. However, this is a rather arbitrary definition and is of limited value in the context of drinking water, since particles at the upper and lower end of the size range are unlikely to be found in treated drinking water. A subset of microplastics less than 1 μm (micron) in length are often referred to as nanoplastics.

Recently, and increasingly highlighting the impending threat by the increased plastic waste in the environment, microplastics have now been classified as emerging pollutants. There is some evidence related to the occurrence of microplastics in the water cycle which are increasingly found in both tap and bottled drinking water. Studies have also pointed towards the potential health impacts from microplastic exposure and the removal of microplastics during wastewater and drinking-water treatment. Owing to their potential health risks and prolonged non-degradability nature, recommendations have been made by several regulatory agencies with respect to monitoring and management of microplastics and plastics in the environment. Several research reports have made their preliminary assessment on microplastics' impact on human health and bridge the gaps in understanding key concerns. Several studies suggest possible entry routes of microplastics into drinking water sources, including surface runoff, both treated and untreated wastewater, combined municipal waste and sewer overflows, industrial effluent, disintegrated plastic waste and atmospheric deposition. Surface runoff carrying discarded and disintegrated plastic waste and wastewater effluent are recognized as the two main sources. Nevertheless, dedicated databases and information sets are required to quantify the sources and associate them with more specific plastic waste streams. Each bottle and cap used for packing drinking water could be a source of microplastic. It is also suspected that plastic bottles and caps used in packed bottled water may also be a source of microplastics in drinking water.

The WHO report reveals the three possible routes by which microplastics could enter humans and impact health. Namely, microplastics can enter the body physically and damage internal structures. Polymer softening and finished plastic manufacturing process use various additives in the form of plasticizers. Therefore, microplastics can enter humans via chemical interactions, additive leaching and intermediate formation upon exposure to different external variables. Another possibility is the formation of biofilms as a result of microorganisms attaching to microplastics and forming colonies, which could cause harm upon exposure to such EPs.

Currently, size-exclusive sieving and concentration of sample methods are used to detect microplastics in water. In freshwater studies, microplastic analysis follows a physical counting method under microscopic observation which ranges from around 0 to 1000 particles per litre. Only limited studies have so far identified measured water sample microplastic counts from 0 to 10,000 particles per litre depending on the source of the sample; however, more studies are needed to determine average frequency and site-specific speculations. Therefore, a comparison of the data between freshwater, bottled water, treated wastewater and oceanic samples that lacks accuracy should not be made.

However, studies reveal that a wide variety of microplastic particles with varying shapes have been found in freshwater samples while the polymers most frequently detected roughly correlate with plastic waste dumped in the environment. In drinking water samples, microplastics are predominantly found in fragments, fabrils and fibre shapes mainly composed of polyethylene terephthalate (PET) and polypropylene (PP) polymers. So far, it is believed that plastic polymers (thermoplastics) which are widely used as commodity polymers, covering major segments of synthetic macromolecules to prepare a variety of end products, are generally considered to be of low toxicity. However, finished plastic products may have microplastics that can contain unbound monomers and additives which are vulnerable to leaching and physical separation. These additive and leached microplastics may cause impacts in the body, depending on a range of physico-chemical properties of the polymer, particle size, surface area and shape. The fate, transport and health impacts of microplastics following ingestion in humans and living beings are not well studied. However, it is hypothesized that hydrophobic chemicals in the environment may absorb organic pollutants, thereby turning them into a complicated class of EPs to trace precisely.

According to the research reports, microplastics larger than 150 micrometres may not enter the human body, but smaller micro-particles may easily get in without a trace. But at present it is estimated that microplastic uptake is limited. Almost all the emerging microplastic pollutants are transparent and inert to most of the detection tools; therefore, nanosized microplastics may easily enter the human body. However, based on limited study reports, it is suggested that microplastics in drinking water may not pose a health risk at the current level. Yet, the microplastic toxicity in water is a function of both hazard and exposure. Higher doses of nanosized microplastic and its accumulation via ingestion, inhalation or injection still pose a health risk. Therefore, dedicated research programs need to be adopted to detect, monitor and remove microplastics from water and wastewater sources, because plastics are ubiquitous in the environment and will remain in use for decades to come.

Nonetheless, a number of research gaps need to be filled to better assess the risk of microplastics in wastewater, treated water and drinking water. This will allow creation of systematic data sets, microplastic detection, analysis and monitoring mechanisms. To establish a reliable information bank on emerging microplastic pollutants, well-designed and quality-controlled investigative studies need to be undertaken. As microplastics are also found in bottled drinking-water bottles, research is also needed to understand the significance of efficient wastewater treatment techniques and their limitations.

2.10   SUMMARY ON DETECTION AND QUANTIFICATION OF EPS

There are so many numerous, diverse and ubiquitous emerging pollutants present in the environment having sources from domestic, industrial and sometimes rarely natural origins. EPs are frequently lumped into categories that describe their origin, occurrence, use and post-usage characteristics. EPs can be understood in a broad sense as any synthetic or naturally occurring chemical or any microorganism and their derivatives, intermediates and metabolites that are not commonly monitored or regulated in the environment. Presently most of the EPs escape monitoring and screening mechanisms because of lack of advanced techniques. However, researchers and regulating agencies used several existing and conventional tools, as summarized in Table 2.3, to characterize EPs to some extent that include APIs, pesticides, surfactants, PPCPs, SVOCs, etc., that are consistently being found in wastewaters.

TABLE 2.3

Summary of Various Techniques used for Qualitative and Quantitative Detection and Analysis of Various Emerging Pollutants

Emerging Pollutants	Source	Analytical/Instrument Technique
Organic EPs, DOMs, SVOCs	Water and wastewater	GC-TOF-MS, GC-MS, LC-MS/MS, APCI and ESI
Pathogens	Wastewater	PAGE, ARISA, DGGE/TTGE), 16S rRNA gene clone libraries, FISH, thin film probes, nanomaterials-based sensors
Pharmaceuticals	Wastewater	LC-Orbitrap-MS
Pesticides, pharmaceutical product, PCPs, industrial products, lifestyle products	Surface and wastewater	LC-MS/MS
Pharmaceuticals, lifestyle products, drugs of abuse, pesticides, nitrosamines, flame retardants, plasticizers, perfluorinated compounds	Freshwater and wastewater	LC-TOFMS
Endocrine active compounds	Wastewater	LC-QTOF/MS
PCPs, pharmaceuticals, illicit drugs	Wastewater, final effluent and river water	UHPLC-MS/MS
Pesticides, drug residue, APIs	Pharmaceutical wastewater	UHPLC-Q-Exactive Orbitrap MS
Pharmaceutical, herbicides, stimulant, illicit drugs, preservative agents	Surface water, wastewater, suspended particulate matter, sediments	UHPLC-HRMS
Pharmaceutical and personal care products, drugs of abuse, pharmaceuticals	Wastewater samples and WWTPs, drinking water	LC-MS
Antibiotics, APIs	Aquifer sediments	UHPLC-Q-orbitrap

Pharmaceuticals, transformation products, pesticides	Treated wastewater	UHPLC-QqLIT-MS/MS
Artificial sweeteners, flame retardants, fungicides, herbicides, industrial chemicals, insecticides, pharmaceuticals, plasticizers.	Lake, river	LC-HRMS/MS
Parabens, hormones, anti-inflammatory drugs, triclosan, bisphenol A	Wastewater	GC-MS with derivatization
Polycyclic and nitro-aromatic musks, brominated, chlorinated flame retardants, methyl triclosan, chlorobenzenes, organochlorine pesticides, polychlorinated biphenyls	Wastewater and seawater	GC-MS/MS
Pharmaceuticals, APIs	Wastewater	FT-NIR
Heavy metal contaminants	Wastewater	SP-ICP-MS, ICP-OES, chemical sensors, ion selective probes
Trace metals	Wastewater	AAS, ICP-OES
Mercury	wastewater	HPLC-ICP-MS
Flame Retardants	Wastewater samples	GC-ICP-MS
Nanomaterials	Wastewater	SEM, TEM, EDX, XRD, XPS, ICP-OES, Raman
Microplastics	Water and wastewater	OM, SEM, MS, GPC

In addition to chromatographic techniques, spectrometric methods and all other conventional microscopic techniques like TEM, SEM and SEM-EDAX have also been used for EPs characterization. Thus far, these services were specifically used for conventional materials characterization in particulate or bulk materials forms for their particle size, morphology and count studies. Light microscopy techniques like IR, fluorescence, polarised light, DIC mode, Confocal Raman microscopy and inter-ferometry are all also being used to assist chemo-sensors and ion selective probes. Generally known organic compounds in wastewater are still being detected using GC, while non-volatile, polar and thermolabile ECs are analysed using LC.

In addition to conventional organic, pharmaceutical, oil/grease, pathogenic and heavy metal contaminants, nanomaterials and plastic pollutants are posing serious threats to human and environmental health. However, XRD, SEM, TEM, XPS and other spectroscopic techniques are being extensively applied for the identification and characterization of solid form nanomaterials, plastics and pharmaceutical material in terms of the structural order or disorder of solid pollutants. XRPD, XPS and other spectroscopic techniques are efficiently applied to detect and quantify EPs for their characteristics like polymorphic structure, crystallographic changes, and quantify active ingredients in the final dosage form. These techniques can also be used to monitor and control the quality of many EPs and their raw materials, excipients and finished products. This is important because any change in the morphology of additive materials and nanomaterials, or in the crystalline state of active ingredients in the final product, resulting from the manufacturing process, can influence a water quality and wastewater fate which directly affects the environment upon bioavailability.

Therefore, future innovations in analytical tool design and their applications should consider the specific parameters of a pollutant and their references as a basis for qualitative and quantitative characterization of EP-contaminated wastewater accurately and precisely. These procedures may consider the protocol and steps including: (1) identification of existing forms of the EPs constituent and their parent raw material; (2) identification of the type of EPs and their nature; (3) identification of the solid forms and sizes; (4) physical and chemical interaction and stability; (5) distinguishing EPs for their purity, impurities and composites in wastewater; and (6) quantification of EP.

There are other ways to handle some pollutants like polymers. The PET bottle, used for packing material for mineral water, is one of the most recycled items in the world. Recycled PETs (polyesters) are used for clothing, apparels and shoe making. But this rule does not apply to other polymer pollutants. Even if microplastics are removed from water resources completely, at this rate of production and discharge, they will continue to persist in the environment in millions of tons every year. This condition applies to nanomaterial-based EPs as well.

Nevertheless, the existence of various chemical substances will continue to persist, and increased population, industrialization and consumption of commodity end products will continue to increase their presence in the environment. To make this condition worse, the new class of pollutants (EPs) discussed in this chapter deceive all existing techniques, or the majority of the EPs are not responsive to conventional analytical techniques. Therefore, to make a safer future, the ubiquitous EPs need to be removed from wastewater and environment. The EPs will continue to be an enduring and curious target of the scientific community due to ever-increasing synthesis and usage. Therefore, to reduce the impact of EPs on the environment, many regulatory agencies including WHO push for efficient EPs removal and recovery technologies. They also advocate for continued efforts to minimize EPs production, uncontrolled usage and unregulated discharge. In this direction, many water treatment technologies have been developed for EPs removal, and they can be categorized broadly into natural attenuation, conventional water treatment techniques and advanced treatment processes.

REFERENCES

1. *Identifying and Reducing Environmental Health Risks of Chemicals in Our Society: Workshop Summary: The Challenge: Chemicals in Today's Society*, ISBN 978-0-309-30115-2, 2014, p. 179, Washington, DC: National Academies Press (US).

2. Violette Geissen, Hans Mol, Erwin Klumpp, Günter Umlauf, Marti Nadal, Martinevan der Ploeg, Sjoerd E.A.T.M. van de Zee, and Coen J. Ritsema, Emerging pollutants in the environment: a challenge for water resource management, *International Soil and Water Conservation Research*, 2015, 3(1), 57–65.

3. Sara König, Hans-Jörg Vogel, Hauke Harms, and Anja Worrich, Physical, chemical and biological effects on soil bacterial dynamics in microscale models, *Frontiers in Ecology and Evolution*, 2020, 8, 53.

4. Milind S. Ladaniya, *Citrus Fruit: Biology, Technology and Evaluation*, ISBN: 978-0-12-374130-1, 2008, Amsterdam, Netherlands: Elsevier Inc.

5. Hassan Khorsandi, Rahimeh Alizadeh, Horiyeh Tosinejad, and Hadi Porghaffar, Analysis of nitrogenous and algal oxygen demand in effluent from a system of aerated lagoons followed by polishing pond, *Water Science and Technology*, 2014, 70(1), 95–101.

6. Susan D. Richardson and Susana Y. Kimura, Water analysis: emerging contaminants and current issues, *Analytical Chemistry,* 2020, 92(1), 473–505.

7. Devika Vashisht, Amit Kumar, Surinder Kumar Mehta, and AlexIbhadon, Analysis of emerging contaminants: a case study of the underground and drinking water samples in Chandigarh, India, *Environmental Advances*, 2020, 1, 100002.

8. David Alvarez, and Tammy Jones-Lepp, Chapter: Laboratory qualifications for water quality monitoring, in *Book Water Quality Concepts, Sampling, and Analyses*, Yuncong Li, Kati Migliaccio (eds)., 2010, pp. 199–226, eBook ISBN: 9780429189265, Boca Raton, FL: CRC Press.

9. Yanna Liu, Lisa A. D'Agostino, Guangbo Qu, Guibin Jiang, and Jonathan W. Martin, High-resolution mass spectrometry (HRMS) methods for nontarget discovery and characterization of poly- and per-fluoroalkyl substances (PFASs) in environmental and human samples, *TrAC: Trends in Analytical Chemistry*, 2019, 121, 115420.

10. I. Ross, J. McDonough, J. Miles, P. Storch, P. Thelakkat Kochunarayanan, E. Kalve, J. Hurst, S.S. Dasgupta, and J. Burdick, A review of emerging technologies for remediation of PFASs, *Remediation Journal,* 2018, 28, 101–126.

11. R. Cumeras, E. Figueras, C.E. Davis, J.I. Baumbach, and I. Gracia, Review on ion mobility spectrometry. Part 1: current instrumentation, *Analyst*, 2015, 140, 1376–1390.

12. Torsten C. Schmidt, Recent trends in water analysis triggering future monitoring of organic micropollutants, *Analytical and Bioanalytical Chemistry*, 2018, 410, 3933–3941.

13. Syahidah Nurani Zulkifli, Herlina Abdul Rahim, and Woei-Jye Lau, Detection of contaminants in water supply: a review on state-of-the-art monitoring technologies and their applications, *Sensors and Actuators B: Chemical,* 2018, 255, 2657–2689.

14. Abdel-Mohsen O. Mohamed, Evan K. Paleologos, and Fares M. Howari, *Pollution Assessment for Sustainable Practices in Applied Sciences and Engineering*, 2021, ISBN: 978-0-12-809582-9, Amsterdam, Netherlands: Elsevier Inc.

15. Fabio Gosetti, Eleonora Mazzucco, Maria Carla Gennaro, and Emilio Marengo, Contaminants in water: non-target UHPLC/MS analysis, *Environmental Chemistry Letters*, 2016, 16, 51–65.

16. Aurea C. Chiaia-Hernandez, Martin Krauss, and Juliane Hollender, Screening of lake sediments for emerging contaminants by liquid chromatography atmospheric pressure photoionization and electrospray ionization coupled to high resolution mass spectrometry, *Environmental Science and Technology*, 2013, 47(2), 976–986.

17. Olalekan C. Olatunde, Alex T. Kuvarega, and Damian C. Onwudiwe, Photo enhanced degradation of contaminants of emerging concern in waste water, *Emerging Contaminants*, 2020, 6, 283–302.

18. Debora Fabbri, María José López-Muñoz, Alessandro Daniele, Claudio Medana, and Paola Calza, Photocatalytic abatement of emerging pollutants in pure water and wastewater effluent by TiO_2 and Ce-ZnO: degradation kinetics and assessment of transformation products, *Photochemical & Photobiological Sciences*, 2019, 18, 845–852.

19. H. Barndõk, M. Peláez, C. Han, W. E. Platten III, P. Campo, D. Hermosilla, A. Blanco, and D. D. Dionysiou, Photocatalytic degradation of contaminants of concern with composite NF-TiO_2 films under visible and solar light, *Environmental Science and Pollution*, 2013, 20, 3582–3591.

20. Jing Sun, Xiaohu Dai, Qilin Wang, Mark C. M. van Loosdrecht, and Bing-JieNi, Microplastics in wastewater treatment plants: detection, occurrence and removal, *Water Research,* 2019, 152, 21–37.

21. Paul U. Iyare, Sabeha K. Ouki, and Tom Bond, Microplastics removal in wastewater treatment plants: a critical review, *Environmental Science: Water Research and Technology*, 2020, 6, 2664–2675

22. Pierre Oesterle, Richard H Lindberg, Jerker Fick, and Stina Jansson, Extraction of active pharmaceutical ingredients from simulated spent activated carbonaceous adsorbents, *Environmental Science and Pollution Research*, 2020, 27, 25572–25581.

23. E. Kristiansson, J. Fick, A. Janzon, R. Grabic, C. Rutgersson, B. Weijdegård, H. Söderström, D.G.J. Larsson, Pyrosequencing of antibiotic-contaminated river sediments reveals high levels of resistance and gene transfer elements. *PLoS ONE*, 2011, 6, e17038.

24. Lisa (Song) Liu, Ariel Mouallem, Kang Ping Xiao, and Jerry Meisel, Assay of active pharmaceutical ingredients in drug products based on relative response factors: Instrumentation insights and practical considerations, *Journal of Pharmaceutical and Biomedical Analysis*, 2021, 194, 113760.

25. Gerlinde F. Ploger, Martin A. Hofsass, and Jennifer B. Dressman, Solubility determination of active pharmaceutical ingredients which have been recently added to the list of essential medicines in the context of the biopharmaceutics classification systeme biowaiver, *Journal of Pharmaceutical Sciences*, 2018, 107, 1478–1488.

26. Yasmin Vieira, Eder C. Lima, Edson Luiz Foletto, and Guilherme Luiz Dotto, Microplastics physicochemical properties, specific adsorption modeling and their interaction with pharmaceuticals and other emerging contaminants, *Science of the Total Environment*, 2021, 753, 141981.

27. Fella Hamaidi-Chergui and Mohamed Brahim Errahmani, Water quality and physicochemical parameters of outgoing waters in a pharmaceutical plant, *Applied Water Science,* 2019, 9, 165.

28. Marta Llamas, Iñaki Vadillo-Pérez, Lucila Candela, Pablo Jiménez-Gavilán, Carmen Corada-Fernández, and Antonio F. Castro-Gámez, Screening and distribution of contaminants of emerging concern and regulated organic pollutants in the heavily modified Guadalhorce River Basin, Southern Spain. *Water*, 2020, 12, 3012.

29. Geoffrey K. Kinuthia, Veronica Ngure, Dunstone Beti, Reuben Lugalia, Agnes Wangila, and Luna Kamau, Levels of heavy metals in wastewater and soil samples from open drainage channels in Nairobi, Kenya: community health implication, *Scientific Reports*, 2020, 10, 8434.

30. Jessica Briffa, Emmanuel Sinagra, and Renald Blundell, Heavy metal pollution in the environment and their toxicological effects on humans, *Heliyon*, 2020, 6, e046913.

31. Mohamed Lamine Sall, Abdou Karim Diagne Diaw, Diariatou Gningue-Sall, Snezana Efremova Aaron, and Jean-Jacques Aaron, Toxic heavy metals: impact on the environment and human health, and treatment with conducting organic polymers, a review, *Environmental Science and Pollution Research*, 2020, 27, 29927–29942.

32. P. W. W. Kirk, and J. N. Lester, Significance and behaviour of heavy metals in wastewater treatment processes IV. Water quality standards and criteria, *Science of the Total Environment*, 1984, 40(1), 1–44.

33. Sung Hwa Choi, Ji Yeon Kim, Eun Mi Choi, Min Young Lee, Ji Yeon Yang, Gae Ho Lee, Kyong Su Kim, Jung-Seok Yang, Richard E. Russo, Jong Hyun Yoo, Gil-Jin Kang, and Kyung Su Park, Heavy metal determination by inductively coupled plasma – mass spectrometry (ICP-MS) and direct mercury analysis (DMA) and arsenic mapping by femtosecond (fs) – laser ablation (LA) ICP-MS in cereals, *Analytical Letters*, 2019, 52(3), 496–510.

34. Nancy Lewen, Shyla Mathew, Martha Schenkenberger, and Thomas Raglione, A rapid ICP-MS screen for heavy metals in pharmaceutical compounds, *Journal of Pharmaceutical and Biomedical Analysis*, 2004, 35(4), 739–752.

35. Fahad N. Assubaie, Assessment of the levels of some heavy metals in water in Alahsa Oasis farms, Saudi Arabia, with analysis by atomic absorption spectrophotometry, *Arabian Journal of Chemistry*, 2015, 8(2), 240–245.

36. Graeme E. Batley, and Jaroslav P. Matousek, Determination of heavy metals in sea water by atomic absorption spectrometry after electrodeposition on pyrolytic graphite-coated tubes, *Analytical Chemistry*, 1977, 49(13), 2031–2035.

37. Su Li, Chencheng Zhang, Shengnan Wang, Qing Liu, Huanhuan Feng, Xing Ma, and Jinhong Guo, Electrochemical microfluidics techniques for heavy metal ion detection, *Analyst*, 2018, 143, 4230–4246.

38. Baban Kumar Bansod, Tejinder Kumar, RitulaThakur, Shakshi Rana, and Inderbir Singh, A review on various electrochemical techniques for heavy metal ions detection with different sensing platforms, *Biosensors and Bioelectronics*, 2017, 94, 443–455.

39. Zou Li, Li Chen, Fang He, Lijuan Bu, Xiaoli Qin, Qingji Xie Shouzhuo Yao, Xinman Tu, Xubiao Luo, and Shenglian Luo, Square wave anodic stripping voltammetric determination of Cd^{2+} and Pb^{2+} at bismuth-film electrode modified with electroreduced graphene oxide-supported thiolated thionine, *Talanta*, 2014, 122, 285–292.

40. Lian Zhu, Lili Xu, Baozhen Huang, Ningming Jia, Liang Tan, and Shouzhuo Yao, Simultaneous determination of Cd(II) and Pb(II) using square wave anodic stripping voltammetry at a gold nanoparticle-graphene-cysteine composite modified bismuth film electrode, *Electrochimica Acta*, 2014, 115, 471–477.

41. Nisar Ullah, Muhammad Mansha, Ibrahim Khan, and Ahsanulhaq Qurashi, Nanomaterial-based optical chemical sensors for the detection of heavy metals in water: recent advances and challenges, *TrAC Trends in Analytical Chemistry*, 2018, 100, 155–166.

42. Lei Li, Haiwei Yin, Yang Wang, Jianhua Zheng, Huidan Zeng, and Guorong Chen, A chalcohalide glass/alloy based Ag^+ ion – selective electrode with nanomolar detection limit, *Scientific Reports*, 2017, 7, 16752.

43. Andrey V. Legin, Yuri G. Vlasov, Alisa M. Rudnitskaya, and Evgeni A. Bychkov, Cross-sensitivity of chalcogenide glass sensors in solutions of heavy metal ions, *Sensors and Actuators B: Chemical*, 2016, 34(1–3), 456–461.

44. L. Eddaif, A. Shaban, and J. Telegdi, Sensitive detection of heavy metals ions based on the calixarene derivatives-modified piezoelectric resonators: a review, *International Journal of Environmental Analytical Chemistry*, 2019, 99(9), 824–853.

45. Przemysław Niedzielski, and Marcin Siepak, Analytical methods for determining arsenic, antimony and selenium in environmental samples, *Polish Journal of Environmental Studies*, 2003, 12(6), 653–667.

46. Montserrat Filella, Antimony and PET bottles: Checking facts, *Chemosphere*, 2020, 216, 127732.

47. Yanfang Guan, and Baichuan Sun, Detection and extraction of heavy metal ions using paper-based analytical devices fabricated via atom stamp printing, *Microsystems and Nanoengineering*, 2020, 6, 14.

48. Ruiyu Ding, Yi Heng Cheong, Ashiq Ahamed, and Grzegorz Lisak, Heavy metals detection with paper-based electrochemical sensors, *Analytical Chemistry*, 2021, 93(4), 1880–1888.

49. H. Manisha, P. D. Priya Shwetha, and K. S. Prasad, Low-cost paper analytical devices for environmental and biomedical sensing applications. In: S. Bhattacharya, A. Agarwal, N. Chanda, A. Pandey, A. Sen (eds.). *Environmental, Chemical and Medical Sensors. Energy, Environment, and Sustainability*, 2018, Singapore: Springer.

50. J. V. Tarazona, A. Castaño, and B. Gallego, Detection of organic toxic pollutants in water and waste-water by liquid chromatography and in vitro cytotoxicity tests, *Analytica Chimica Acta*, 1990, 234, 93–197.

51. Urszula Kotowska, Maciej Zalikowski, and Valery A. Isidorov, HS-SPME/GC–MS analysis of volatile and semi-volatile organic compounds emitted from municipal sewage sludge, *Environmental Monitoring and Assessment*, 2012, 184, 2893–2907.

52. Fernando L. Rosario-Ortiz, Shane Snyder, and I. H. (Mel) Suffet, Characterization of the polarity of natural organic matter under ambient conditions by the polarity rapid assessment method (PRAM), *Environmental Science and Technology*, 2007, 41(14), 4895–4900.

53. Hanan Abd El-Gawad, Validation method of organochlorine pesticides residues in water using gas chromatography–quadruple mass, *Water Science*, 2016, 30(2), 96–107.

54. Cheng Luo, Wei Wen, Fenge Lin, Xiuhua Zhang, Haoshuang Gua, and Shengfu Wang, Simplified aptamer-based colorimetric method using unmodified gold nanoparticles for the detection of carcinoma embryonic antigen, *RSC Advances*, 2015, **5**, 10994–10999.

55. Chengzhou Zhu, Guohai Yang, He Li, Dan Du, and Yuehe Lin, Electrochemical sensors and biosensors based on nanomaterials and nanostructures, *Analytical Chemistry*, 2015, 87(1), 230–249.

56. M. A. La Merrill, L. N. Vandenberg, M. T. Smith et al. Consensus on the key characteristics of endocrine-disrupting chemicals as a basis for hazard identification. *Nature Reviews Endocrinology*, 2020, 16, 45–57.

57. Bethany M. DeCourten, Joshua P. Forbes, Hunter K. Roark, Nathan P. Burns, Kaley M. Major, J. Wilson White, Jie Li, Alvine C. Mehinto, Richard E. Connon, and Susanne M. Brander, Multigenerational and transgenerational effects of environmentally relevant concentrations of endocrine disruptors in an estuarine fish model, *Environmental Science and Technology*, 2020, 54(21), 13849–13860.

58. Dae-Young Lee, Kelly Shannon Lee, and A. Beaudette, Detection of bacterial pathogens in municipal wastewater using an oligonucleotide microarray and real-time quantitative PCR, *Journal of Microbiological Methods*, 2006, 65(3), 453–467.

59. Lone Høj, Rolf A. Olsen, and Vigdis L. Torsvik, Effects of temperature on the diversity and community structure of known methanogenic groups and other archaea in high Arctic peat, *The ISME Journal*, 2008, 2, 37–48.

60. T. K. Dey, M. E. Uddin, and M. Jamal, Detection and removal of microplastics in wastewater: evolution and impact. *Environmental Science and Pollution Research*, 2021, 28, 16925–16947.

3 Limitations of Existing Treatment Techniques in Removing Emerging Pollutants

3.1 INTRODUCTION

Emerging pollutants are primarily synthetic chemicals and substances that have been widely studied and included in academic programs due to their forecasted threat to natural environments. EPs have their origin in many conventionally used commodity substances and chemicals; however, their effect on the environment is largely known. Knowingly and unknowingly, EPs are discharged or discarded into the environment in the form of wastewater. EPs originate from organic substances, biological waste, hormone-mimicking synthetic macromolecules, nanomaterials and disintegrated plastics are widely detected in domestic wastewater as well as industrial effluents. EPs in wastewater also exist in various shape, size, physical nature and morphologies. EPs concentration in wastewater may range from a few nanograms per litre (ng. L^{-1}) to a few hundred grams per litre (g L^{-1}). Such diverse class, concentrations and physical nature of EPs in the aquatic environment may cause ecological risk such as interference with various functioning and activities of humans and living beings such as endocrine system, organ malfunction, microbiological resistance and so on [1–4].

Even though there are limited analytical tools in place, increasing and continuous detection of EPs in various wastewater has brought new challenges to water pollution control. As in the previous chapter, emerging detail reiterated the need for greater research interest in EPs' detection and accurate quantification. Every year, people around the world consume an unprecedented amount of PPCPs, surfactants, nanomaterial-induced locations, toothpaste and sunscreen. Most of them end up in surface water and wastewater through human excretion and direct discharge of domestic wastewater. Usually, almost all EPs have good solubility in water and other common solvents used for domestic applications, but upon discharge as effluent, they are difficult to degrade and transform. The situation turns worse when most EPs produce metabolites or transformation products with similar or even higher toxicity, resulting in highly contaminated water pollution which can potentially cause deleterious effects in aquatic and human life in various concentrations [5].

At present, there is no legislation regarding monitoring and efficient management of EPs, and therefore their removal from wastewater at various treatment plants is not

DOI: 10.1201/9781003214786-3

mandatory. However, as new reports emerge on the health issues as a result of exposure to EPs, this concern is gaining in importance to adopt suitable assessment and treatment technologies. At present, most of the EPs are not considered as priority pollutants as per the European Union and WHO, but they have been categorized under a list of pollutants with permissible concentration of pollutants in surface waters. Therefore, due to lack of suitable EP treatment and removal techniques in the long term these substances may cause eco-toxicological changes like feminization of organisms, several disorder diseases in human body functioning and microbiological resistance, among others.

On the other hand, novel EPs like nanomaterial waste and microplastics are posing great concerns to the environment. In these new classes of EPs, quality-assured toxicological data are needed on the most common forms of plastic particles. Further, little information is available on the understanding of nanoplastics and microplastic behaviour and their risk at various levels in the environment. Finally, to avoid the direct exposure of various EPs to humans and living beings via air, water and soil, a better understanding of overall effects of exposure to EPs from the broader environment is needed. Nonetheless, it is clear from the data available that EPs cannot be completely removed from the environment using conventional wastewater treatment processes.

At present most concerns regarding EPs have focused on the water and wastewater samples in the environment owing to the tendency of domestic and industrial wastewater. At present, EPs are looked at as a new class of pollutants, with much attention having been paid in current research to their occurrences, sources, distributions and abundances, and less focus as detection, identification and separation methods. However, several studies are underway to understand the mechanisms of conventional methods like adsorption and desorption mechanisms, eco-toxicological effects, effective separation techniques, etc. However, EPs have been increasingly detected in drinking water, freshwater and wastewater. So far, conventional physical and physico-chemical methods like adsorption, coagulation, oxidation and membrane-based processes are mainly tried and tested for EPs removal in water and wastewater samples. However, little attention has been paid to the EPs' interaction and their interference with individual separation and removal behaviour with techniques used which are relevant to human health [6–8].

On the other hand, sewage wastewater is the largest source of biological contaminants and lately domestic wastewater is increasingly characterized for their EPs content. However, special attention now must be paid to EPs-contaminated wastewater disinfection processes in order to fulfil the legislation requirements. Major waterborne diseases originate from the presence of bacteria like *E. coli* in domestic wastewater. Therefore, treatment of EPs is essential to water to be reused with proper disinfection processes in presence. In this direction, conventional disinfection processes such as chlorination or ozonation are known to produce disinfection by-products (DBPs), also known as secondary pollutants, with potential carcinogenic effects. Conventional disinfection methods include thermal, physical and a combination of chemical processes. Boiling, sunlight exposure, pot filtration, chlorination, ozone, ultraviolet light and chloramines are primary methods for disinfection (Figure 3.1). Traditionally, boiling, sunlight exposure and pot filtrations have been used for

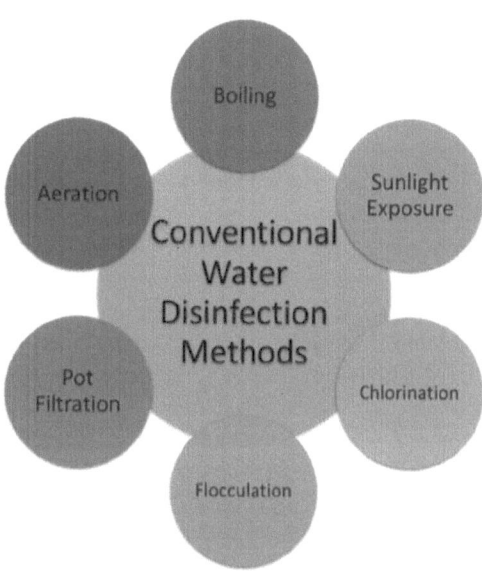

FIGURE 3.1 Conventional water disinfection methods adopted for disinfecting contaminated surface and ground water.

disinfecting of biological contaminants and removal of suspended pollutants. Lately, advanced chemical disinfectants in combination with chlorination and flocculation are being used to decontaminate wastewaters. In chemical disinfection methods, potassium permanganate, chlorine dioxide and photocatalytic disinfection are popularly used. Organic pollutants like humic acid and other polyaromatic hydrocarbons are removed or degraded using oxidation process and nanofiltration, respectively. Several other studies have shown the inefficiency of conventional treatments to remove several EPs from wastewater. Nonetheless, recent advanced oxidation processes (AOPs) provide a valuable alternative as they generate highly reactive radicals that have the potential to simultaneously oxidize chemical EPs and inactivate biological contaminants [9].

In addition to conventional pollutants like organics, biological pollutants, HMs and pharmaceuticals, two other major groups of contaminants like DBPs and their metabolites escape all conventional treatment techniques. DBPs produced as a result of interaction between disinfection reagents with target contaminants that are incomplete and by-products such as haloacetic acids or trihalomethanes from chlorination are generated that have potential toxicity. Treatment of such DBPs are complex processes, and most of the conventional or combination of treatment processes fail to address the issue. To avoid by-product formation, the ozonation process is often used as an alternative or hybrid treatment that could potentially reduce or eliminate DBPs. However, ozone is also known to produce a suite of by-products due to its harsh pollutant degradation mechanism. However, given the effectiveness of the ozonation process, studies are underway to improve the use of ozone processes with minimum risk to the environment.

Some of the advanced wastewater treatment techniques in practice are hybrid chemical oxidation process, ion-exchange techniques, electrodialysis and membrane-based separation processes. For the removal of most of the soluble organics, commonly used methods are adsorption using activated carbon, functional polymers, active nanomaterials and composite nanomaterials. Chemical coagulation and advanced oxidation techniques are now commonly used for surface water and wastewater treatment applications. However, these methods use added harsh chemicals like aluminium and iron metal salts to assist the pollutant separation in treatment plants. Currently, membrane-based separation processes are the most efficient and cleaner techniques, namely MF, UF, NF, RO, FO and electrodialysis; most of these processes follow size-exclusive or molecular-sieving mechanisms to separate pollutants. These processes are widely considered sustainable and will continue to be in use for wastewater treatment in the next few decades to come, owing to the excellent water quality production and pollutant separation [10].

However, membrane-based separation processes are capital- and operational-intensive techniques. Another major issue with these processes is membranes susceptible to fouling while treating highly contaminated wastewater, as most of the commercially successful membranes are made of polymeric material which possesses relatively high hydrophobicity. Organic and bio-fouling is difficult to remove and can greatly restrict the permeation of product water, pollutant rejection rates and overall productivity of the membrane process. Membrane fouling is the single most important hurdle in the widespread use of membrane separation processes, necessitating high operational and maintenance costs.

Some of the other most effective conventional water treatment methods that have stood the test of time are simple physical separation techniques like screening, settling, sedimentation and chemical methods such as precipitation, adsorption, coagulation, filtration with coagulation, ion exchange and advanced oxidation processes have been used for the removal of several EPs from polluted wastewater. One of the main disadvantages of these methods is that all conventional methods follow a multi-step treatment path following some or all of the steps listed here, namely collection, straining, chemical addition, coagulation, floc formation, sedimentation, clarification, filtration and finally disinfection. Nevertheless, within the realm of water treatment, future studies must emphasize the low-input technology such as physical and physico-chemical methods that certainly deserve better recognition for their potential for low-cost removal of EPs in wastewater [11, 12].

3.2 PROPERTIES OF DIFFERENT EPS AND THEIR INFLUENCE ON ANALYSIS

It is evident from the several reports that the methods followed currently for sampling, and analysis are not harmonized. Current methods do not follow any general procedure or protocols, being typically focused on a specific class of EPs. EPs have their origin in many commonly used chemicals and substances but were recently characterized as highly hazardous EPs even within a permissible limit. For many recently categorized EPs such as microplastics and nanomaterials, detection and analytical method development is still at an early stage. Overview of all existing analytical techniques suggests inherent limitations of an individual method to serve as common-category EPs. In such a situation, advanced ultra-sensitive instrumental techniques should be used in combination with a standard database for quantitative determination of EPs in wastewater. On the other hand, there is a need for a library involving lists of pollutant properties in their nano-size and their metabolites that determine their fate in the environment. Generally, several surveys on water quality contaminated with conventional pollutants often use different parameters for water quality assessment and often are not completely applicable for EPs. Therefore, a harmonized, comprehensive and data-based monitoring of groundwater, surface water and wastewater contaminated with a series of EPs is not yet achieved but urgently required. Another, major disadvantage that current approaches and analytical tools are missing are specific components integrated into EPs-assessing models through a comprehensive environment (air and water sampling), and new policies must be formulated to include such approaches [13].

The major risk assessment approaches need to outline the specific as well as general set of guidelines for detection, identification and quantification procedures for EPs. This will help in effectively managing the overall protection strategy of ecological communities in the aquatic environment and in human health. New methods also need to consider properties based on individual as well as cumulative risks assessment for individual as well as a group of EPs. This approach also helps in developing strategies to handle analytical protocols from combined exposures to several mixtures of EPs. However, these tasks are highly challenging for several reasons: (1) currently, there are already very high numbers (estimated at > 5000) of known EPs and their relevant transformations and metabolites existing in the environment; (2) EPs and their transformative products continuously change their form and quantity over time due to changes in production, use and disposal as well as external conditions; (3) certain pharmaceuticals, PPCPs and biological EPs like hormones, pyrethroids and organophosphorus pesticides affect the aquatic ecosystem even at extremely low concentrations which are difficult to detect at present; and (4) absence of highly sensitive analytical tools, probes and sensors make this overall process tedious and prolonged [14].

In a new strategy, EPs can be monitored from the point of occurrence and in transport of EPs from diffuse points at sources, and this phenomenon strongly depends on the nature and properties of EPs. EPs' presence and form in the environment can be directly characterized by tracing possible parent sources and their properties such as volatility, adsorption properties, polarity and persistence and their functional as well as reactive group site. It is evident that EPs from domestic or industrial wastewater water sources are different relatively from groundwater and surface water sources. This is because EPs treatment plants continuously may undergo degradation, sorption at the sediment and transport in the aqueous phase under the influence of external atmospheric stresses. Whereas ground and surface water EPs persist in an isolated system, which is less susceptible for transformations. EPs in wastewater discharges can undergo significant biodegradation and transformation depending on the presence of a community of organisms, metabolic networks and the bioavailability of contaminants, however, this phenomenon can later influence the quality of surface and underground water as EPs leach or persist in environment, especially in sediments and soil. Also, the natural degradation of an EP can be directly influenced by the individual or collective properties of the substance and can vary significantly between compounds, and this has not been specifically studied for many EPs. On the other hand, many EPs like long chain hydrocarbons, polymers and composite nano-material do not readily undergo natural degradation. In such a situation, the most desirable approach for realistic biodegradation of such EPs is to carry out simulation tests using experimental systems that approximate real environmental conditions as much as possible. However, the most uncertain situation will still exist with property prediction of the intermediates, by-products and end-products of the disinfected, photo-degraded or chemically transformed EPs [15].

Another obstacle for efficient monitoring of EPs is insignificant or lack of modelling frameworks available or developed for EPs. The major challenge is devising the strategy to study the EPs model in various possible sources of groundwater, surface water, wastewater, sediments and soils and transport model study data with relevant monitoring and regulatory authorities. All the strategies discussed regarding understanding of properties of EPs will be useful to design the proper wastewater treatment and waste management techniques. EPs can be transported by different processes depending on their properties, such as by runoff, erosion or leaching, and enter into groundwater or surface water. However, during the transportation or management process EPs may be intercepted by the soil through adsorption or can be degraded during the transport and never reach the water bodies. Once EPs reach the water bodies, further they transport downstream and settle in solution or suspended in water bodies in various forms.

However, very little experimental evidence is available in the literature on EPs transport, partitioning and distribution in catchments. The existing simulation models on EPs transportation and distribution have not been parameterized except for pesticides and few pharmaceuticals. In pharmaceuticals, properties such as adsorption behaviour can vary widely in different soils and waters as they occur in both ionized and unionized form, which affects their interaction with different compounds in the soil during runoff. The presence of biological EPs in domestic and agriculture waste sources like manure or sludge can affect not only the sorption behaviour of these materials but also their persistence due to their surface charge. This is also true with the environmental behaviour of engineered nanomaterials (ENM), which is largely unknown for now. Transformations of nano-sized and reactive nanomaterials before and after entry to the environment many times attain unpredictable status. This is because nanomaterial EPs, through their unique surface properties and reactivities, undergo surface modification by humic acid, interactions with common cations and dissolution under natural conditions. These properties necessitate further research to understand properties on their transportation and distribution. Nanomaterial-based EPs behave differently from that of colloidal and non-particulate contaminants. So new paradigms will be needed for nanomaterial-based EPs in water, soil, sediment and biota. Research also needs to focus on engineered nanomaterials to generate sufficient data to develop, refine and calibrate simulation models for studying EPs [16].

For futuristic modelling on EPs, study extensively depends on size and extent of wastewater catchment areas. The model needs to consider the characteristics of catchment such as the transfer of water, particles and chemicals across interfaces between the soil, groundwater, and surface water compartments for accurate prediction. Transport of EPs may cross the interfaces such as groundwater-level, hyporheic zones which are physically, chemically and biologically known as highly complex zones. These interfacial factors need to be recognized and investigation methods and resulting protocols need to be simplified. Presently, several models are available to investigate the reactive transport at different scales for calibration, validation and process understanding, particularly in the context of pesticide fate screening. For other EPs, similar hypotheses can be adopted as one dimensional unsaturated water flow in soil, where the EP is likely to behave as an adsorbing and degrading solute.

On the other hand, a new class of EPs, plastics, are synthetic organic polymers that have been produced, used and discarded in millions of tons worldwide. There are two types of plastics: the thermoplastic which melts upon heating, and thermosets that disintegrate at heating. Crystalline and amorphous plastics behave differently when they are subject to heat, pressure and other external stresses. Crystals plastic become liquids at a specific temperature (T_m). Amorphous plastics have a wide range of temperature stability and often gradually undergo melting when they are subjected to heat. Viscoelastic parameters of plastics serve as alternative criteria for plastics which prevent direct melting of plastics, instead undergo transition under applied stress.

Plastic in residual form are lightweight and low-density materials that can be easily transported long distances via hydrodynamic processes. Plastics in various forms, shape and size, upon accumulation in water streams, undergo disintegration as a result of rigorous flows, temperature and ocean currents which turns plastic into microplastics. Further, owing to continued mechanical actions like flow and currents, plasticizer leaching, biodegradation, and photo-oxidative degradation over time, microplastics gradually turn into nanoplastics with a particle size below 1 um. Several studies reveal that micro/nanoplastics are stable in water, potentially for thousands of years, because of their chemical stability. Within the aquatic and marine ecosystem, micro/nanoplastics gradually attain complex pollutant compositions. Even though microplastics are chemically stable and inert to many chemicals, they gather different forms and compositions. First, they provide excellent incubating atmosphere to organic pollutants and microbial colonies to grow, as microplastics are excellent carriers of toxic organic chemicals, owing to their large specific surface area and strong hydrophobicity. Some studies suggest microplastics provide excellent conditions to grow or accumulate polycyclic aromatic hydrocarbons found in marine ecosystems. Second, microplastics can absorb numerous heavy metals such as Zn, Cu, Pb, Ag, Cr, and nano-scaled pollutants like TiO_2 as well. And third, these microplastics distributed deep into the aquatic reservoirs, where light transportation is at a minimum, provide a conducive atmosphere for pathologic accumulation and growth on surface [17].

Microplastics by size, appearance and nature are often mistaken for food by marine organisms. On the other hand, organic or HM adsorbed microplastic EPs play an important role in the toxic effects to marine organisms. This also leads to mechanical damage to the organism and decreases their feed efficiency. It was found that the enzymatic activity in marine organisms such as acetylcholinesterase and isocitrate dehydrogenase decreases owing to a mixture of microplastics and polycyclic aromatic hydrocarbon pollutants like pyrene. Recent studies also reveal that the HM and organic EPs adsorbed microplastics alter the gene expression in fishes and other marine organisms. These contaminated fish and marine food stock, after entering the food chain, can seriously threaten human life and other living beings [18].

Microplastics consist of carbon and hydrogen atoms bound together in polymer chains and their product processing several additives are being added to facilitate their processability. Most of the chemical additives such as plasticizers and colourants, typically present in plastics or microplastics, leach out of the plastics after entering the environment. Therefore, for pragmatic reasons, five major properties need to be included in a microplastics definition: (1) chemical composition, (2) physical state, (3) particle size, (4) solubility, or dispersibility in water, and (5) degradability. These new recommendations would help to quantify limit or threshold values that are derived as much as possible from existing guidelines for microplastic EPs.

On the other hand, inorganic and hybrid polymeric systems possess similar properties to organic polymers. Polysiloxane, made of silicone, is a well-known example of inorganic polymer widely used in many household and speciality applications. This inorganic polymer is composed of a silicon-oxygen backbone with organic hydrocarbon side groups, which gives these widely used polymers plastic-like characteristics. Many inorganic and hybrid plastics are flexible, malleable or rubbery, and these may also undergo easy disintegration under extreme stress conditions to result in microplastics. As with many other EPs, the half-lifetime of a plastic pollutant depends not only on the properties and concentrations of the polymer, molecular weight and processing technique, but also on environmental characteristics such as temperature, pH and salinity. Therefore, the International Union for Pure and Applied Chemistry (IUPAC) has set the upper boundaries for microplastics at 100 μm. These proposed size range limits have been studied for their effects in the environment. Therefore, proposed size ranges have been aimed at harmonization of terminology between polymer scientists, medical therapists, biologists and ecologists. This strategy also helps in bridging different disciplines, preferably using the same terminology to reflect similar properties, phenomena and mechanisms of microparticle formation and evaluation.

On the contrary, the information and knowledge gap existing on nanotoxicity effects can be attributed to the lack of effective and accurate analytical methodologies for nanopollutants. Nanopollutants derive their name and properties from physical characteristics like size, shape, aggregation state, chemical composition and structure. Chromatographic, microscopic and spectroscopic techniques serve limited purposes in detection and quantitative estimation of nanopollutants in different forms and compositions. In the absence of specific technique and methodology, often a combination of conventional techniques are used for separation, purification and accurate detection. To track toxicity of nanopollutants, monitoring their initial steps of interaction with the environmental matrices is important, such as sources and sampling points. During this journey, surface properties of nanopollutants play a role that also determines their dynamics of agglomeration and decantation that may interfere in accurate analysis.

This strategy is already being implemented with reasonable efficiency for a sustainable and cost-effective technological development. All these concerted efforts with innovative research allow us to study and manipulate the properties of nanoscale materials systematically for various safe applications and discharge within regulatory norms that were previously unknown.

3.3   LIMITATIONS OF EXISTING ANALYTICAL TOOL IN DETECTING EPS

The main objective of risk assessment is not only to create overall protection of ecological communities in the aquatic environment, but also to guard people living in contact, either directly or indirectly. Earlier, the conventional methods and procedures followed for evaluating the environmental risk of a substance or chemical consisted fundamentally of comparing its concentrations in ecological compartments. These ecological levels are known as predicted environmental concentration (PEC) with concentrations below which unacceptable/adverse effects on a living being will most likely not occur named as predicted no effect concentration (PNEC). But, these methodologies and procedure pose a serious hurdle in identifying different PNEC values reported for a particular class of pollutants. Also, human health risk assessment using such protocols need to be slightly modified including tracking pollutants in which human exposure through different pathways is projected before comparison with references. In this procedure, risk estimate usually depends on the robustness and quality of databases or reference data set. Hence, for an accurate evaluation of environmental and health risks, including anthropological exposure, uncertainty and variability aspects, must be deliberated as important parameters.

There is a wide disagreement between the numbers of EPs that potentially exist in the environment and the number of pollutants identified, listed and monitored on a priority basis. The major problem in listing EPs and assessing them for their potential threat is that multiple contaminants in nature originate from multiple sources in multiple chemical classes. This problem intensifies when multi-class pollutants mix with both low concentration levels and considerable time-based and spatial variability. In addition to this, analysing multi-class pollutants at trace or ultra-trace concentrations such as parts per billion (ppb) or parts for trillion (ppt) requires the use of different sampling techniques, apparatus and protocols, and each protocol involves different sample preparation methods. Developing and identifying unknown trace contaminants is highly time consuming and costly. It evidently for now that chemical monitoring of all potential EPS in every section is an incredibly difficult task. Further, the target analysis of pre-selected or identified sets of potentially EPs often misses site-specific toxicity assessment, which is unknown to the regulatory at present. Therefore, this pattern of analytical assessment route is not able to explain ecotoxicity effects of complex environmental samples [19, 20, 21].

Nevertheless, detailed chemical and compositional analysis of wastewater is a prerequisite for assuring the safety of the environment from EPs. Current limitations in identifying contaminants in water also limit the analyst's ability to predict or formulate the necessary future measures to assess environmental impacts. Recent developments in analytical techniques discussed previously have widened the analytical window; that is, the EPs' class range that is amenable to a specific identification method. Using these upgraded tools, much has been documented and learned about the occurrence, fate and transport of EPs in the wastewater.

For analytical purposes, a conventional sampling technique follows the analysis of samples within the analytical window for which information on specific structure and concentration is referenced. However, in such cases, a specific compound in multi-class of mixtures will have minimal quantity. This process may miss the larger derivative forms or remaining residues. It also may be true that the specifically identified compounds accounted for a smaller percentage of different wastewater samples. Therefore, the remaining major portion of the total content typically remains uncharacterized or only in aggregate form such as an average functional group content, or their distribution. For example, even a reasonable method using a good mass spectra technique can extract information on the majority of the contaminants that are detected in a sample; most still remain unidentified and/or unquantified because of a lack of reference spectra and/or reference compounds. The suggestion that can be inferred from the mass spectra may be sufficient to recommend structures using spectral findings. However, verification and quantification of the proposed structures are often impossible because of a lack of reference spectra.

Meanwhile, the majority of conventional organic carbon present in various environmental settings in the form of SVOC is not agreeable to specific identification, but in that circumstance, aggregate properties are often used as indicators for pollutant detection. In such a situation, other physico-chemical properties are used to predict EPs such as elemental composition, functionality, acidity, ultraviolet absorption, metal binding properties, total halogen content, fluorescence properties and molecular size distribution. To deal with these concerns, monitoring and regulatory agencies have structured as systematic protocols by creating maximum contaminant or concentration limits (MCLs) for a new class of pollutants known to have significant potential to contaminate environment such as EPs.

For future challenges, new analytical procedures need to include several interdependent procedures carefully tuned to provide maximum efficiency. Some specific routes of the analytical operations that may apply for EPs' tracking, sampling, separation and detection are shown in Figure 3.2. In each case, the sensitivity of the step and method defines the lower limit of the sample size that must be processed. Sample size and occurrence are among the most significant considerations in devising analytical protocols. If the required sample is too dilute or contaminated with mixtures of pollutants, direct use of sample as analytes is not feasible. In such situation, the sample extraction, transportation and storage logistic add to the cost of the sample analysis. Hence, in complex samples like wastewater, the sample is handled carefully to preserve, recover, isolate and/or concentrate the analytes to yield analysable extracts that are compatible with a suitable instrumental tool. This careful consideration leads to a final effluent or desired pollutant concentrate known as the sample preparation step that can be analysed using the instrumental methods technique. On the other hand, mixtures or multi-component samples undergo a derivatization step, which is a micro-analytical procedure that serves one or several of the following purposes: (1) increase recovery from the aquatic matrix, (2) facilitate separation from other organic constituents and/or (3) improve identification or detection. This is very

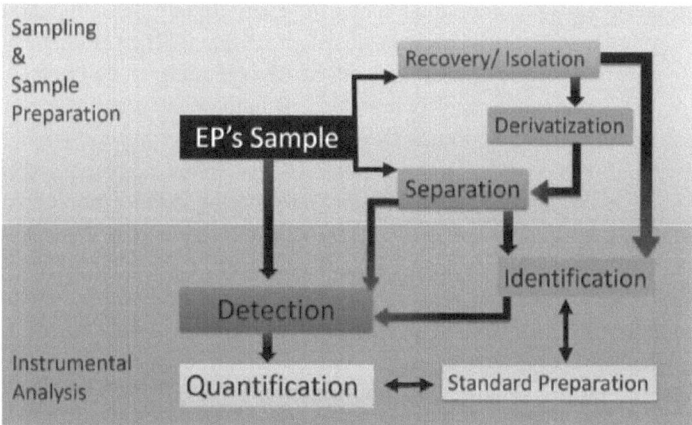

FIGURE 3.2   General analytical protocol for the analysis of organic trace compound.

significant as reference compounds for most of the EPs are not available, data comparison with source tracing is often necessary for structure verification and the development. This will also help in setting up rapid detection and quantification procedures. This general procedures are used to characterize aggregate pollutant parameters like DOC to reflect the physico-chemical properties originating from the heterogeneous mixture of organic molecules in wastewater. This protocol is also followed to analyse molecular size distribution of dissolved macromolecules using selective membranes or GPC, organic pollutant parameters like the absorbance and fluorescence of UV-Vis light, the bonding configurations of 13Carbon, 1H-NMR, IR and potentiometric titrations for acidity level measurements [22].

On the other hand, the possibility of using a range of ionization techniques provide flexibility to ensure detection of most molecules present in a real wastewater sample. These techniques follow the route of ionization, detection and quantification of contaminants. However, identification of molecular structure using ionization techniques in the environment or wastewater is a difficult task, since a target sample may contain complex mixtures containing varying concentrations of organic compounds. Another major limitation of the ionization technique is the isolation of specific molecules and interference of trace contaminants during measurement of a specific molecule. Nevertheless, the ionization technique sensitivities have been greatly improved with the help of advanced probes and supporting methodologies to address samples with concentration levels below detection limits.

Further, integrated techniques in identifying mixed or passive samples may allow rewarding benefits both for the low-concentration pollutant and to overcome the limitations of conventional pollutant detection. Thus, it is likely that combination technique development strategies based on passive sampling in wastewater will provide protection from possible environmental damage. This methodology will also minimize instrument operational costs and improve characteristic and reliable EPs detection. In addition, passive sampling techniques in combination with suitable probing methodologies are likely to provide robust information on EPs. Thus, the structural categorization of active compounds in wastewater containing complex matrices remains a key point to improve ecological risk assessment. Thus, in the case where any specific physico-chemical, chemical and biological analyses do not lead to identification of risky substances and compounds of activity at specific locations, the combination of techniques approach can be applied to interpret unknown EPs and their combined effects.

Therefore, to avoid errors in analysing and interpreting EPS information in wastewater which may be due to human interference or instrumental dependency on a database, recent advancements have adopted intelligent testing strategies (ITS). In recent times, ITS have been increasingly used by researchers, technologists and the scientific community as viable tools for studying chemical substances and their structures, which is cost effective for EPs detection and in rare cases reduces possibilities of animal testing situations for confirmation.

3.4   LIMITATIONS OF EXISTING TECHNIQUES IN REMOVING EPS FROM A CONTAMINATED SOURCE

Conventional wastewater treatment techniques fail to address EPs' complete removal from wastewater sources due to the two-front unsuccessful strategy. On the one side, discharged toxic untreated wastewater of various origin EPs entering into clean water bodies lose the track of their percolation, which is a grave cause of concern. Another reason is that the absence of proper wastewater treatment technologies at the point of wastewater generation and discharge with minimum discharge criteria has led to a drastic increase in EPs-related problems. According to a recent United Nations World Water Development report (2017), nearly 80% of the global industrial and domestic wastewater is discharged without appropriate treatment. To make it worse, this percentage varies within various income group countries, yet overall it is estimated that more than 90% of industrial and domestic wastewater is discharged into various water bodies and the aquatic environment as untreated or partially treated.

To overcome such an enormous problem is an overwhelming task. Nevertheless, various physico-chemical, adsorption, oxidation and biological treatment technologies have been adopted and are being used to treat highly polluted effluents before being discharged into the environment. The presence of EPs such as pesticides, pharmaceuticals, surfactants, surfactant metabolites, perfluorinated compounds, drugs of abuse, hormones, other endocrine disruptors industrial additives and agents, and personal care products in the effluents of wastewater and their treatment plants is a matter of well-documented concern. Concerns related to emerging pollutants mainly arise because many of the chemical compounds are still unregulated at present. These chemical substances could be included as potential candidates under regulatory observation in the near future depending on the study results on their potential health effects and occurrence. The main concern often related to EPs is the very nature of unregulated and partially treated discharges from WWTPs due to the lack of efficient treatment technology. As a result of lack of suitable treatment, technology and continued widespread use of many of potential EPs increases the risk. However, most of the conventional technologies like ozonation of adsorption and their combinations in use today in WWTPs fail to remove all contaminants efficiently. In fact, the current regulations and norms related to wastewater treatment so far does not include most of the chemicals and substances appearing on the EPs list.

Therefore, conventional wastewater treatment techniques used in WWTPs are not specifically aimed at eliminating EPs. As a result, several have found that WWTPs only partially remove several EPs, such as pharmaceuticals including carbamazepine or diclofenac with less than 25% efficiency, and these partially treated pharmaceutical wastewaters continuously discharge into many aquatic systems at potentially hazardous levels that could achieve chronic levels over a period of time.

Discharge of incorrectly treated industrial and domestic wastewaters into the environment results in widespread contamination of drinking-water resources or water reserved for domestic purpose with a wide range of contaminants. Despite the use of relatively advanced oxidation process, PPCPs have been found in trace amounts in treated water samples. Regardless of the method followed or the presence of relatively low concentrations in natural waters, EPs pose a potential threat to human health and aquatic ecosystems upon accumulation over a period of time. The diverse industrial substances like PFASs are a group of fluorinated organic chemicals used in the production of grease repellent, firefighting foams and wetting agents, and there is a lack of suitable technique to remove PFAS residues from wastewater, as these are highly resistant to degradation by chemical and biological processes. PFASs possess unique physico-chemical properties, such as saturated carbon–fluorine bonds and low vapour pressure, which prevent the breaking or degradation of these highly stable organic substances during treatment processes. Efficient treatment and removal of both PFASs and PPCPs by existing biological wastewater treatment is often limited, resulting in the occurrence of trace to high-level concentrations of these EPs in downstream water supplies.

Conventional domestic wastewater treatment techniques, including screening, grit removal, primary sedimentation and biological treatment, are considered as ineffective for complete removal of EPs from wastewaters. These conventional methods are relatively effective against suspended solids, colloids and degradable compounds. In most of the cases, all these contaminants either end up in sludge or convert into different intermediates. Then again, degradation, oxidation and disinfection techniques leave behind debris of by-products, residues and intermediates. Some of these techniques also form sediments which need to be addressed specifically before disposal into the environment [23, 24].

In this context, more progressive and industrially advanced countries have adopted advanced tertiary wastewater treatment techniques to increase the quality of drinking water. This general trend is increasing worldwide and in recent years has reduced the possibility of EP traces intruding in post-treated reusable or drinking water supplies. This strategy is followed in many European nations, where presently more than 90% of urban wastewater receives tertiary treatment; however, treated water was characterized for many traces of EPs. Therefore, many studies concluded the need for adopting advanced tertiary treatment techniques to adequately avoid intrusion of potentially hazardous EPs. In this direction, membrane filtration processes and advanced oxidation processes (AOPs) have been widely implemented for the removal of multi-class EPs such as micropollutants, SVOCs, dyes, HMs, etc., from wastewater. In this direction, the most advanced and hybrid processes, which have been used with reasonable success for EPs removal, are ozonation, advanced photo-catalysis using $UV/H_2O_2/Fe^{3+}$, electrochemical reactions, membrane bioreactors, other pressure-driven membrane processes and adsorption. Other physico-chemical methods such as Fenton using hydrogen peroxide oxidation, or others activated by UV and ultrasound, have also proved significant in organic EPs degradation and sep-aration. However, several limitations on use of these techniques remain pertinent, such as secondary pollutants, technical viability, cost-effectiveness ratio, pertaining degraded residues and sustainability of industrial-scale implantation, are sceptical.

On the other hand, membrane-based separation processes are considered as the most environmentally benign and sustainable among other water treatment technolo-gies. Membrane separation science and technology follows the engineering approaches for the separation, removal and transport of solutes/substances based on their size and nature with the help of semi-permeable membranes. The well-known benefits of membrane separation technologies over conventional methods are simple operation, greener process footprint, easy maintenance, high separation efficiency and easy cleaning steps that make them an exceptional choice for wastewater treat-ment. Another major advantage with these technologies is that membrane processes can be adopted to operate either alone or in combination as a hybrid process for multiple EPs' removal of a different nature. Membrane processes like MF are effec-tive for suspended solids and bacterial and colloidal pollutants removal. Among other processes, ultrafiltration (UF) and nanofiltration (NF) are being widely used due to their pore size and pore size distribution characteristics as well as easy operability. These processes also provide good adaptation possibilities with a wide range of pore size available to efficiently treat effluent of EPs' nature from WWTP. Conventionally, membrane-based separation processes have been widely used in applications such as oil-contaminated wastewater reclamation, tannery wastewater treatment, selective separation of dye and salt in textile wastewater. A combination of membrane pro-cesses (hybrid) have been adopted to simultaneous treatment of multiple wastewater and resource recovery or value addition. These processes provide several operational and application benefits when compared to all other conventional wastewater treat-ment techniques, such as: (1) simultaneous product permeation and recovery possi-bility during removal of EPs in the treatment plant; (2) low operational and cleaning cost (~30% economical); (3) lower carbon footprint; (4) continuous product recovery and EPs removal; (5) availability of different plant size to suit the treatment volume and space and (6) processes that can easily be modulated to suit the space. With these advantages, membrane processes can effectively remove large-size EPs which can-not be easily treated in the secondary treatment step of WWTPs. Suitable selection of the UF membrane with their appropriate molecular weight cut-off (MWCO) yield favourable and efficient removal of biological contaminants which appear frequently in domestic wastewater.

Likewise, ultrafiltration can remove some of the EPs like harmones and harmones-type synthetic drugs and macromolecular contaminants with relative efficiency. On the other hand, the nanofiltration process can remove most of the dissolved organics, dye molecules and multi-valent ions to close to 90% rejection. However, nanofiltra-tion is ineffective towards molecular weight below 200 Da, which limits its universal application for low-molecular-weight EPs. Among all membrane processes, reverse osmosis can eliminate the majority of divalent ions and organic molecules above 50 Da. However, many of the membrane-based separation processes are capital intensive and require regular operational maintenance. In addition to this, the pressure-driven membrane processes like MF, UF, NF and RO suffer from severe organic and bio-fouling. At low operating conditions and with a hydrophilic membrane, fouling is reversible; however, fouling in NF and RO processes is irreversible, and this condition worsens as the concentration polarization proceeds with solute cake formation at higher operating pressure.

Several studies have demonstrated that EPs such as PPCPs, surfactants, dyes, HMs and bio-macromolecular contaminants can be effectively removed using the NF process via size (steric) exclusion and electrostatic interactions. As most EPs have relatively high molecular weight, larger than the MWCO of typical NF membranes, removal by the size-exclusion mechanism can be effective. Removal of EPs can be further enhanced via electrostatic repulsion by tailoring the NF membrane surface and pore charge, because charged pollutants and a large suite of PPCP chemicals with a low pKa are negatively charged at the pH of natural waters [24].

Compared to UF membranes, low MWCO NF membranes give relatively high contaminant retention rates due to their low pore size range. However, the permeation or the flux in saline feed waters or charged pollutants limits NF performances. NF membranes are always interrupted by the presence of charged molecules of ionic salt species which are constrained by limited NF membrane surface selectivity for the removal of EPs over the ionic passage. However, high salinity feed solutions or wastewater effluents contain high levels of magnesium, calcium and dissolved silica, resulting in inorganic pollutant accumulation on the NF membrane surface, also known as inorganic scaling. During desalination, typical NF membranes show high rejection of ionic spices; however, upon increased concentration and accumulation, membrane scale-forming occurs. A major limitation of the NF process is that inorganic scaling on the surface of membrane restricts at high water recovery rates, which necessitates constant cleaning of the membrane. At high operating pressure (10 to 20 bar), high-concentration solutes accumulate, leading to scaling on the membrane surface and resulting in decline in water flux. To overcome these disadvantages, several novel NF membranes have been developed for providing selective separation of EPs while permitting the passage of scale-forming species. This has been achieved in membrane using surface modification techniques, composite membrane formation and use of an extensive nanomaterials combination to induce anti-fouling properties in NF membranes.

The NF membranes are generally prepared using interfacial polymerization (IP), a state-of-the-art technique for the fabrication of commercial NF and RO membranes. The IP membrane preparation technique gives limited control over membrane surface properties and also involves relatively toxic monomers and solvents during the selective layer preparation of NF. However, over the years, several researchers have modified the molecular-level design of the IP process to improve NF membrane selectivity through surface charge and porosity-controlled separation of emerging pollutants. On the other hand, new strategies of preparing NF membranes via layer-by-layer (LbL) assembly coating were adopted to improve the separation performance of NF membranes. With this approach, this ionic combination of both synthetic and natural polymers has been used to coat NF selectively, which also offers better water permeation efficiency with environmentally benign fabrication. Unlike the IP process, the LbL method utilizes aqueous-based polyelectrolyte solutions as selective layer modification precursor in a direct, simple and systematic membrane coating the NF membrane. This allows the uniform coating of polyelectrolyte solution via a systematic LbL coating sequence in several optimized cycles to yield a multilayer, polyelectrolyte-modified NF membrane. The LbL coating technique has several advantages in tuning surface properties of NF membranes in the treatment of wastewaters containing a wide range of EPs. Polyelectrolyte-coated multilayer NF membranes have succeeded in selective and effective removal of some of the EPs in saline wastewaters. However, these polyelectrolyte multilayer NF membranes have still suffered from high scaling potential.

On the other hand, the pressureless membrane process forward osmosis (FO) has received much attention due to its significant potential in recovering reusable water from wastewater sources. FO process has been now considered as evolving alternatives to other pressure-driven membrane processes such as RO, NF, UF and MF processes. Several studies reveal that the FO process has potential to treat wastewater contaminated with EPs. A major limitation of the process is the absence of compatible membrane and draw solution which can be used sustainably. However, recent evidence shows increased efforts among researchers resulting in preparation-suitable FO membranes. Advances on reusable and recoverable draw solutes/solution are also gaining pace to make the FO process viable.

To overcome demerits of the membrane process in treating high EPs in contaminated wastewater, a futuristic approach needs to be adopted in addressing shortcomings such as membrane fouling, recovery of hazardous pollutants and high operational cost. This approach would lead to development and operational optimization to demonstrate the large-scale potential of these technologies for EPs-contaminated wastewater treatment, considering the sustainability criteria. Also, efforts are vigorously being pursued to achieve a better understanding on the dynamics of the membrane separation process in addressing issues associated with fouling and other operational parameters. The focus is on altering the membrane interface to effectively reject and remove EPs from wastewater sources with emphasis on higher removal and permeation rates. To improve the FO process, efforts are focused on draw solution recovery and decreasing energy consumption for both viable water and draw solute/solution recovery. In contrast to pressure-driven processes where unidirectional drift governs the mass transport, FO is a complex-membranes process having exposed to feed and draw solution on either side, experiences significantly different in active layer parameters. Other important parameters which can influence the mass transport and solute retention efficiency of the FO membranes include the nature of the membrane, thickness, wettability membrane morphology, pore characteristics and the solubility factors. Also, the FO process depends greatly on its operating parameters, such as feed and draw solution flow rate, high flux, low fouling propensity and good rejection rates with low concentration polarization. Membranes for FO need to be specially prepared considering the dual nature of the active layer function. Using highly contaminated wastewater at one side and a highly concentrated draw solution at the other side, the FO membrane needs to be optimized for its wettability and pore characteristics to obtain an appropriate semi-permeable barrier. In many organic pollutants, rich wastewater poses a serious threat to membrane stability.

Nevertheless, membranes in most wastewater treatment processes suffer from severe fouling problems. Regardless of process and nature of wastewater and contaminant separation, designing or selection of suitable membranes is a very significant step. The selected membrane should address the issue of severe solute scaling, as well as internal concentration polarization (ICP), which is expected in almost all the membrane processes while applying for long-term wastewater treatment conditions. In addition to the membrane properties process, efficient removal of EPs from wastewater depends on the process design and operational parameters. Therefore, designing an efficient membrane treatment configuration is of utmost importance in satisfying the long-term sustainability criteria. The process selection and operational condition optimization should consider the nature of effluent or wastewater, concentration levels, wastewater volume, environmental and post-treatment solid waste disposal aspects. However, membrane-based technologies in suitable combination will still be effective in treating EPs compared to other novel technologies. The membrane processes, including FO, are in the forefront and prove relatively advantageous in EPs' removal compare to existing physical, physico-chemical, biological and other advanced chemical treatment processes. Nonetheless, development of an effective overall policy and strategy for the removal of EPs from complex wastewater media is urgently needed.

3.5   PHYSICAL AND CHEMICAL NATURE OF EPS

Environmental pollution and its degradation has deep roots in industrialization, urbanization, overexploitation of resources, irresponsible discharge of waste and natural degradation of discarded waste in the ecological system. The major challenge today for policy makers is the tracing and managing of a new class of emerging pollutants. Diverse groups of pollutants of metal-based inorganic, organic substances, microplastics, biological toxins and negligently discarded nanomaterials are a serious problem to the environment worldwide; subsequently, they can affect both flora and fauna, human health and larger living beings [25].

Every water resource on the earth's surface faces the most vulnerable ecological compartments of potential wastewater discharge. Consequently, water pollution has become a topic of utmost concern and distress worldwide due to the nature of novel pollutants that are entering the environment. As discussed in previous sections, the current state-of-the-art technologies have failed to track and mitigate the occurrence, fate, risks and removal of EPs in water. The challenges are mounting ahead for improving existing technologies to track and remove EPs urgently as part of a sustainable water resources management strategy. A series of studies emphasize categorizing different pollutants based on their properties in the aquatic environment that are characterized by their ever-changing dynamics over a period of time. It is, therefore, essential to strengthen research efforts and resources invested in tracking, identifying and detecting in all possibilities to reduce the impacts and risks created by EPs on water bodies. Nevertheless, recent efforts in upgraded detection techniques of some EPs in the aquatic environment led to the conclusion that the tracking parent materials and their possible properties are key for monitoring the new class of pollutants.

It is also important to understand all the possible mechanisms of EPs' release and their accommodation. New efforts are underway to gather specific information on the dynamics of EP transformation in water to avoid risks for water quality, flora and fauna in river basins. For decades to come, the issue of EPs will be an enduring challenge due to the increased use of a large number of chemical substances in various forms and compositions. At present, the cost of pollutant monitoring, tracking, identification and detection is relatively high. This restricts the number of EPs under scanner as substances of ecological threat. For that reason, creating an extensive databases on EPs is a need of the hour. This approach would offer data on the properties of EPs and their metabolites, which helps in water quality surveys. The main task of the survey should be to include the systematic listing of properties, including physical, chemical, physico-chemical, biological and nano-toxicological. These methodologies would be rudiments for the development of suitable detection and sustainable water treatment methods for the removal of EPs.

A critical analysis is needed of the physical, chemical and physico-chemical parameters on occurrence point, the state of their intermediate forms and properties of end EPs. This process needs to follow a series of laboratory and field studies in aquatic environments. Simultaneously, it is necessary to identify the possible limitations and gaps in the implementation of EPs removal technologies. Initial experimental assessments in coordination with database on understanding physico-chemical properties may contribute towards removal of the unfavourable affects.

The production and use of chemicals, also known as contaminants of emerging concern, were first documented as potential EPs in the aquatic environment in the early 1800s based on their physico-chemical characteristics. These were basically listed out from the wide range of substances produced by humans and discarded after use as a result of indispensable needs for the modern society. As society adopted new social standards, the use of surfactants, personal care products, pharmaceuticals, steroid hormones and sunscreens found their way to household application and the aquatic environment after use. Even though EPs existed in the environment for several decades, comprehensive data on their toxicity are not available in a systematic form. Then they have been gradually understood for their qualitative and quantitative occurrence via various available analytical tools, which are hazardous for ecosystems. This is due to poor evidence, triggered by the complex characteristics of the EPs in the environment related to their physico-chemical properties, causing unexpected behaviour in water.

The EPs are mainly characterized by their physico-chemical properties as organics with polarity such as pharmaceuticals, industrial chemicals, pesticides, inorganic heavy metal and their complexes, biological, particulate pollutants like nanomaterials and microplastics. Once they reach the environment via domestic and industrial waste streams, EPs become more polar, acidic and alkaline than natural chemicals, as shown in Figure 3.3. Some inorganic EPs are not particularly toxic but are catastrophic to the environment when they accumulate in high concentration or are used extensively. These include fertilizer components, such as nitrates and phosphates.

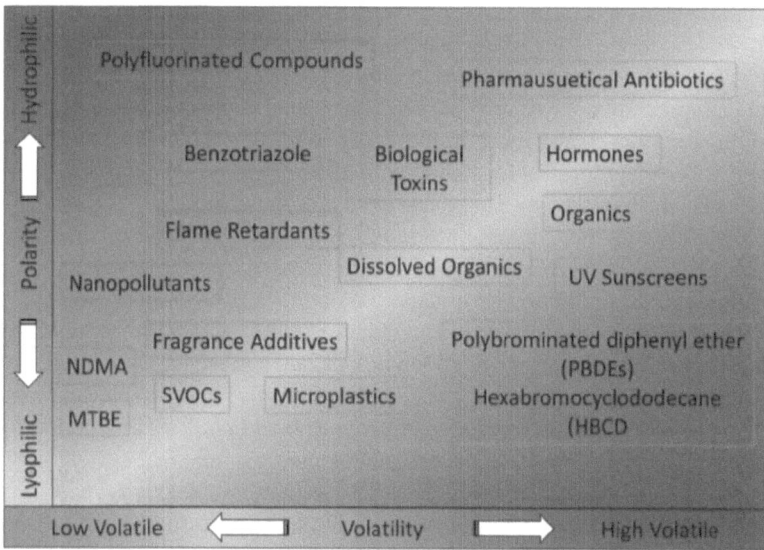

FIGURE 3.3 Chart of EPs listed based on their physico-chemical properties such as volatility and polarity.

Nitrates and phosphates cause algal colonization in surface water, which results in decrease in the oxygen level of the water (COD/BOD) and blue baby syndrome in drinking water. These change in physico-chemical transformation, making them dangerous at certain concentrations. Many of the EPs are hydrophobic in nature and can easily seep into the food chain and accumulate in the lipid-rich tissues. These physico-chemical phenomena can influence the endocrine system in animals and humans through direct or indirect exposure.

A group of organic substances that was recently included in the list of EPs is characterized by UV radiation screening compounds or organic UV filters. Though sunscreen falls under the category of PPCPs, new emerging facts are forcing special attention regarding the potential ecological threat. This is because production of personal products is ever still increasing and discharged into the environment in staggering quantities. On the other hand, new organic UV filters have entered the consumer market with severe metabolizing characteristics, turning themselves into potential environmental hazards. In addition, several nanomaterials-based UV absorption sunscreens with a molar absorption coefficient in the UVA range of 320–400 nm and UVB range of 280–320 nm have flooded the new market space, with no control over their production and usage. Therefore, UV radiation-altering products such as skin creams, hair sprays, cosmetics, hair dyes, shampoos, body lotions, etc., composed of organic substances such as ethylhexyl methoxycinnamate (EHMC), octocrylene (OC), butyl methoxydibenzoylmethane (BM-DBM), benzophenone-3 (BP3), etc., are extremely lipophilic and can easily mix with different ecological compartments. Several study reports on such compounds point out their potential endocrine-disruption potentials in humans. Several case studies have detected these

UV radiation altering substances in rivers, lakes, sea water, groundwater, sediments and biota in moderate to very high concentrations.

Studies now focus on quantification and identification of toxicity levels rather than on just qualitative measurement, which has been confirmed by several studies. Therefore, once these substances in the water sources reach significant levels, the quality of wastewater effluents is significantly deteriorate, and the reusable possibilities of treated wastewater come to be narrow.

For many decades now, organic-origin PCPs, industrial chemicals and food additives are in extensive use. Once these EPs are released into the ecosystem, natural degradation and accumulation of organic pollutants depend on numerous progressions such as dilution, hydrolysis, biodegradation, bioaccumulation, partial degradation and sorption, among others. Sorption of EPs into surface and subsurface aquatic bodies depends on both their physicochemical properties and ecological dynamics. The fate of non-ionizable substances is determined by lyophilic or hydrophobic sorption, whereas the charge state of ionizable molecules can influence several key characteristics such as volatility, hydrophilicity, wettability, reactivity and sorption affinity, which take pH into consideration as an influential factor. On the other hand, external influential factors like temperature and atmospheric conditions play an important role in organic EPs attenuation. Similarly, redox potentials of organic substances can also promote or limit biodegradation in an aquatic environment. Therefore, an uneven dissemination of both regulated and non-regulated organic EPs in an aquatic environment is likely to occur at a large scale given the wide variety of dynamic factors involving distribution of potential sources in the specific area, ecological features, hydrogeological topographies of the region and, most importantly, the physico-chemical characteristics of the EP compounds. Upon reaching water bodies, EPs adopt or interact with native species based on their physico-chemical parameters like pH, temperature, electrical conductivity, counter element and their concentrations, redox potential and dissolved oxygen. This may lead to the formation of several metabolites, intermediates and unknown species of EPs in the natural environment.

Today, PCPs including soaps, fragrances, UV lotions, insect repellents, anti-bacterial detergents and some new flame retardants, among others, are produced in millions of tonnes annually. Due to this, PCPs degradation is balanced by a continuous input into the environment. A very small portion of treated and concentrated PCPs are either landfilled or left in catchment areas; however, a large portion of them enter in septic tanks and sewerage. This process causes indirect entry of PCP-class EPs through groundwater or surface water exchange processes. To make matters worse, the combination of extreme rainfall and flooding raw wastewater comprising EPs can cause a sewer overflow, resulting in spreading potential hazardous substances in agriculture lands, urban low-lying areas and several water supply systems. Nevertheless, future monitoring practices and analyses can be optimized by focusing on selected contaminants and physico-chemical and chemical factors that are likely to govern the distribution of the different EPs in the environment. As discussed in this section, these physical, physico-chemical and chemical parameters influence their existence, but ecological and other natural arrangements also accelerate their transformation and transportation in the food chain or the ecological system.

3.6 PHYSICAL METHODS AND THEIR LIMITATIONS IN REMOVING EPS

Physical methods of wastewater treatment consist of filtration techniques screening, grit removal, primary settling, sedimentation, aeration or activated sludge, secondary settling, floating and advanced membrane filtration techniques. This section covers all available physical treatment process and research activity that leads to efficient removal EPs from wastewater of different sources. Physical treatment processes are predominantly categorized as primary treatment techniques in the conventional water treatment regime. However, advanced separation processes like membrane techniques can also be included as physical treatment techniques for EPs-contaminated wastewater. This step plays an important role in reducing the potential large-sized pollutants, blockers and chocking materials, which may cause serious technological threat to further treatment processes. Modern wastewater streams carry a wide variety of solids of various shapes, sizes and densities in their flow. The primary treatment techniques used for removing EPs take the form of suspended solids, odour-causing substances, colour pigments and biological pathogens at various concentrations and pH conditions in industrial and domestic wastewaters. This process of physically removing EPs from wastewater utilizes physical or chemical properties of the contaminants. Therefore, physical treatment techniques target large wastewater targets large tanks, pre-settling basins or catchment tanks, sedimentation tanks, clarifiers, etc. [26].

3.6.1 SCREENING

As a first step, all wastewater treatment plants use a screening section to retain larger particulates in a multi-step treatment process. The screening method uses perforated platform with screens fitted to retain suspended solids found in the influent wastewater, as shown in Figure 3.4. The screening step is mainly used to remove solid materials that could cause damage to other process equipment, cause reduction in efficiency of the whole system, and contaminate waterways. The materials that are removed using screens are called screenings.

Screening panels in the devices are used to remove coarse solids from wastewater. Coarse solids consist of human hairs, plastics, COD/BOD-causing lumps, rags, discarded paper waste and other large objects that often and unaccountably find their way into wastewater reservoirs or catchment tanks. Even though the main purpose of the screening step is to protect pumps and other mechanical equipment, this also helps in preventing clogging and corrosions in valves, appurtenances of wastewater treatment plants. Conventionally, screening was normally used as a first operation step on the influent wastewater; however, advanced screening steps are also adopted to prevent polymeric, microplastic clogging and macromolecular EPS. Wastewater screens are further classified as coarse or fine screening depending on their construction, nature of wastewater subjected and purpose of application. Coarse screen panels generally involve vertical bars spaced approximate 20–60 mm apart and inclined away from the influent wastewater flow. When wastewater passes through these screening bars, suspended

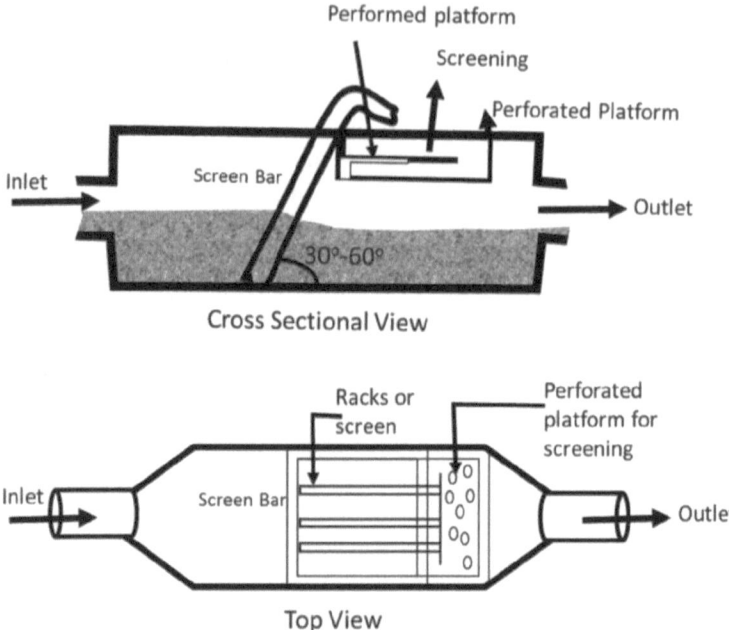

FIGURE 3.4 Schematics of screening techniques used in wastewater treatment, both in top and cross-sectional view.

solids, plastics and macromolecular contaminants or particulate pollutants are retained by the bare which are usually removed by manual raking in small plants or semi-automatic arms, while fully automatic or robotic arms are used to remove retained solid waste from medium and larger-size wastewater treatment plants [27, 28].

On the other hand, fine screening panels having a vertical arrangement of 10–20 mm separation are made up of woven-wire cloth or perforated plates mounted on a rotating disk or drum partially submerged in the flow, or on a travelling belt. Unlike coarse panels, fine screens are mechanically cleaned frequently to remove the retained suspended solid. The amount of suspended solids and other contaminants removed by screening depends on the nature of wastewater, screen-opening size and flow rates of incoming wastewater. Nevertheless, all solid wastes retained at screen panels also accumulate various other organic EPs on their surface which need to be promptly and carefully taken care of. These secondary pollutants generated on the surface of suspended particulates attain a very unpleasant and hazardous status over a period of time, necessitating the prompt discharge of mechanisms to prevent damage to the environment or to humans. Generally, these screened pollutants are currently disposed into sanitary landfills, incinerated via a thermal treatment method or ground into minute particles and returned to the wastewater flow, none of which is considered a safe disposal method [29].

3.6.2 COMMINUTING

Comminutors are another primary wastewater treatment device used in large WWTPs where suspended solids, colloidal pollutants, microplastics and other coarse pollutants in wastewater can be broken or ground. This step facilitates the removal of larger pollutants and prevents the clogging and jamming of pumps during downstream operations. At present, several kinds of comminutors are available in use. Basic comminutors include a screen and cutting teeth or blades in a continuous rotational drum configuration in the vertical plane. Stationary teeth or blades then shred materials in the incoming wastewater that is intercepted by the screen, as shown in Figure 3.5. In general conditions, retained tiny pollutants are shredded repeatedly and returned to the wastewater influents for further treatment processing. A mechanical hammer mill device is most often for the purpose of breaking screened pollutants. Most frequently a shredding device, a comminutor is located across the flow path, and it intercepts the coarse and suspended solids and shreds them to approximately 0.8 mm in size. These tiny particulate solids remain dispersed in the wastewater. In advanced devices, stationary semi-circular screen and rotating or oscillating cutting blades or teeth are commonly used for efficient cutting of larger pollutants. A new class of comminutor devices use advanced barminutor's teeth arranged in a vertical momentary mode in which bars screen with a cutting head moving up and down, shredding the intercepted pollutants going their way [26].

Once the suspended solids and particulate pollutants are condensed into smaller and more uniform sizes, they are returned to wastewater for subsequent processes of manual removal to clear up the trapped solid waste particles which might clog the screens. This device shortens the overall wastewater treatment process in WWTPs with less commotion to the influent; however, persistence of tiny, shredded solids necessitates subsequent agitation that usually takes place in grit chambers. Even with

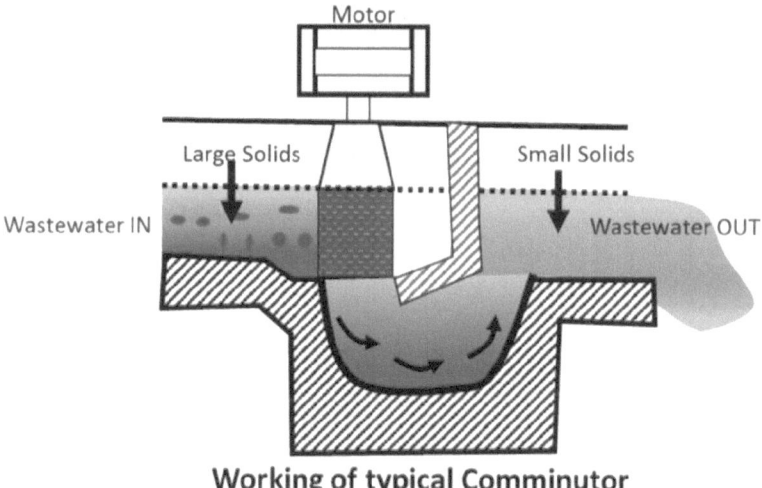

Working of typical Comminutor

FIGURE 3.5 Schematical illustration of a cross-sectional working of a typical comminutor used in wastewater treatment plants.

advanced comminutors and mechanical blades, shredded solid pollutants will remain, posing potential glitches associated with clogging in the subsequent steps of heat exchangers and air diffusers in the aeration pond, and they may also chock the transportation pipelines.

At present, several types of comminutor models are in use in both conventional and advanced WWTPs ,with configuration and designs depending on the size and type of treatment plants. Selection of specific comminutor is made depending upon the types and nature of suspended solids and particulate pollutants present in the wastewater. In most of the WWTPs, a comminutor with a vertical rotating configuration, having a revolving drum on the top driven by a motor with multiple sets of moving teeth inside and shearing bars to shred the incoming suspended solids, are used. In advanced devices, a comminuting mechanism will have an individual circular screen rotating in the counter clockwise or opposite direction to the influent wastewater, and as large solids passes across the blades, they undergo mechanical disintegration. Suspended solids upon reaching the blades will be subjected to the shearing force exerted by both the outer and inner screens. However, the small window between the moving and stationary chopping bars will tear apart the solids reducing its size to micrometre ranges.

Conventional comminutors contain a grit chamber in order to extend the life of the moving parts and to ease wear and tear occurring on the surfaces. A grit removal assembly fitted ahead of the shredder provides a safety mechanism against wear on the cutting head. Typically, grit chambers are designed at or above ground level to facilitate grit handling and smooth management, and additional pumps are fitted in design to lift the sewage to the chamber. In this case, shredding is done ahead of pumping the wastewater. Recently designed advanced WWTPs will comprise both manual bar screens and the comminutors functioning together in parallel and depending on the influent flow rates of wastewater. The major operational challenge is to operate these comminutors with caution, as devices need extensive maintenance against high wear and tear which often require frequent changing of the moving parts. This maintenance activities are frequently adopted step-ahead rock traps prefilters at the upper channel which prevents material from damaging the cutting blades. In severe-category wastewater treatment, head loss occurs in comminutors resulting frequent replacement of parts. However this disadvantage can be overcome by adopting an improved device design in consultation with innovators, designers and manufacturers.

3.6.3 GRIT REMOVAL

Municipal or domestic wastewater is generally characterized by a wide assortment of inorganic solids such as pebbles, sand, silt, egg shells, vegetable residues, oil granules, glass and metal fragments. Before pumping the highly contaminated wastewater into secondary treatment process, removal of these inorganics of larger particulates, heavier organics such as bone chips, seeds, etc., is essential. A combination or mixture of organic and inorganic pollutants are known as grit in wastewaters. Most of the constituents in grit are abrasive in nature, which will cause accelerated wear on pumps and sludge handling equipment when they come in contact for a long period of time. In most domestic wastewater sources, grit deposits absorb greases, subsequently solidifying in areas of low hydraulic shear in pipes, sumps and clarifiers. However, complex grit are generally non-biodegradable in a short time and occupy significant space in sludge digesters, necessitating frequent removal of organic suspended solids. In most of the advanced grit removal devices, a velocity of flow between 0.15 and 0.3 m/s is practically sufficient to remove these complex suspended solids economically.

Grit removal facilities principally consist of an enlarged channel area where reduced flow velocities allow grit to settle out, as shown in Figure 3.6. At least two separate chambers are designed to incorporate in any conventional grit remover in which one facilitates low flow and the other is for the high flow wastewater streams. A period of one minute of retention time is commonly employed to settle grit inside the chamber. As grit accumulates, chambers are frequently cleaned manually in conventional devices (small treatment plants), and mechanically or hydraulically in advanced facilities. Hand clearing in small to medium treatment plants is considered less hygienic; to avoid such health risks, modern devices adopt mechanical cleaning arms or hydraulic cleaning systems. In the mechanical cleaning or hydraulic-cleaning step, the deposited complex solid pollutants are flushed out below fire-streams through pipes in the sidewalls or the bottom of the chamber directed from a centralized monitoring and control system.

A grit chamber is typically mounted before primary sedimentation tanks which is well before wastewater dispatch pumps. There are principally three different types of grit chambers in use for different WWTP applications: (1) the horizontal flow type, (2) the aerated grit chamber and (3) the vortex type. In the horizontal flow type design, wastewater flows through in a horizontal direction at a certain velocity of approximately 0.3 m/s, at which pollutant particles start to settle at the channel before reaching the outlet point. This is typically constructed and used considering removal of the particles that would otherwise be trapped or clogged in a 0.21 mm diameter mesh size. With this configuration, grit settles easily at the bottom of the chamber which can then be removed easily from the system using a conveyor, or manually with buckets or ploughs.

On the other hand, an aerated grit chamber is comprised of an aeration tank, which is intended to create a spiral flow of wastewater within the device cavity as it moves through the chamber. Flow of wastewater at positive flow rates creates velocity at

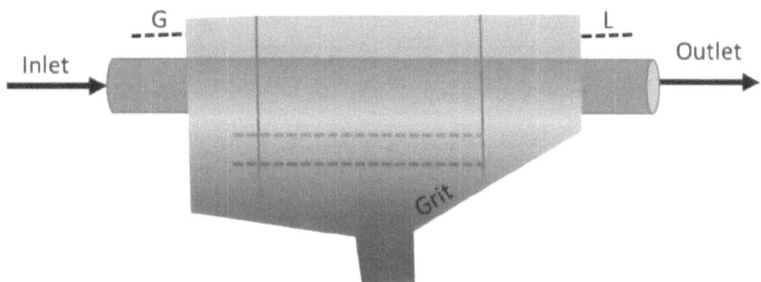

FIGURE 3.6   A cross-sectional view of a grit removal system used for wastewater treatment.

which a certain particle-sized grit will settle at the bottom. A coherently designed device will have the incoming velocity controller which maintains a precise speed in order to avoid the grit exiting out of the chamber with the wastewater to further circumvent the remixing in the chamber. These process adjustments are optimized depending on the size of the device, volume of wastewater and the concentration of the pollutants in combination with accurate fine-tuning of air feeding to the system. Here also, settled particulate pollutants at the bottom will be removed using conveyor buckets, similar to a horizontal flow type system. A vortex type grit chamber operates analogous to the previous version, with one major change in which this type of chamber consists of a cylindrical tank which is designed to create a vortex flow pattern. In order to vertex flow, wastewater is made to enter the device chamber tangentially and, with high-speed flow, a centrifugal force will ensure that the grit is removed out of the system efficiently.

3.6.4 SKIMMING TANKS

The skimmer tanks are used for removing oil, grease and fats from contaminated domestic and industrial wastewater. This tank is designed with a long, trough-shaped structure to narrow down at a steep angle. The wastewater retention period in the tank is a minimum of 3 min for uniform distribution and the availing skimming mechanism. During the retention period, compressed air is blown through the diffusers situated in the floor of the tank to inhibit heavy solids from settling in the bed. Two perpendicular baffle walls are provided in the tank, which divide it into three compartments, as shown in Figure 3.7. A continuous supply of air floats the pollutant matter like oil, fat and grease and remains on the surface of the wastewater in the skimming tank until it is removed manually or mechanically [30, 31].

Meanwhile, it is essential to remove floating pollutant matter in regular intervals from the wastewater tank, or else it may appear in the form of nasty scum on the surface or interfere with the activated sludge process of sewage treatment. The compartment in the device is deliberately designed to be a long trough-shaped assembly separated into two or three sub-lateral sections. These lateral sub compartments fitted with vertical baffle walls having slots for a short distance below the sewage surface

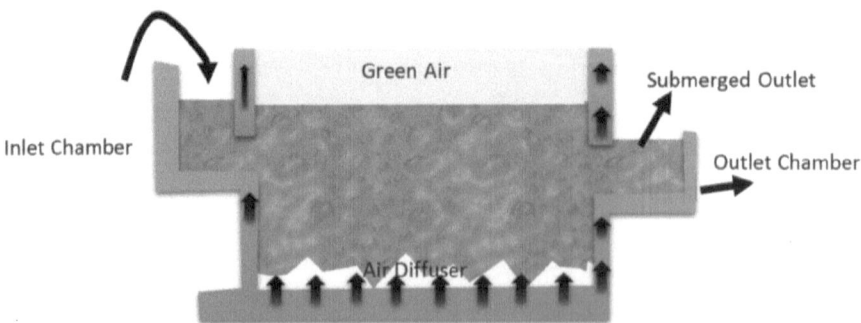

FIGURE 3.7   A cross-sectional view of a working skimmer tank.

permit oil and grease contaminants to escape into stilling compartments. The intensification of floating greasy pollutant matter is brought on the upper surface by a continuous flow of air into the bulk of wastewater from diffusers placed at the bottom. Accumulated greasy lumps, having other inorganic pollutants trapped inside with sufficient capacity, are left for cooling before separation. This is followed by frequent cleaning through detachable covers, which is essential for continuous and reasonable operation of the skimmer. In advanced skimmers, continuous greasy waste collectors or traps are implemented to reduce human intervention and to keep the device working uninterruptedly.

3.6.5 SEDIMENTATION OR CLARIFIER TANK

A sedimentation tank or wastewater clarifiers are used as a primary treatment technique to remove EPs-contaminated wastewater either after or before biological treatment processes. This step is efficient in removing dense and heavier sludge solids by means of settling and separation from the liquid phase wastewater. When used ahead of secondary treatment technique like biological method, clarifiers help in significantly reducing BOD levels. This reduces complex contaminant entering into the aeration pond. However, in the conventional primary treatment step, these types of sedimentation tanks are designed to deal with higher concentrations of wastewater loading. However, conventional clarifiers fail to handle higher rates of wastewater inflow due to the tank design allowing only shorter retention time.

Several factors influence the design and working of sedimentation tanks. Conventional wastewater clarifiers are designed considering the constraint of the surface area of the tank, and depending on wastewater quality, tank size is optimized for wastewater retention time. The sedimentation tank is designed in various configurations and sizes depending on the application and size of the WWTPs. These wastewater clarifiers are designed in (1) rectangular horizontal flow, (2) circular radial flow and (3) vertical flow basis configurations, as shown in Figure 3.8. In each case, the settleable suspended solids, inorganic pollutants in oil aggregates, and other biologically active EPs are removed by gravitational settling under quiescent environments. During this process, the sludge formed as a result of densification of suspended solid accumulates at the bottom of the tank and is detached to the discharging tank under flow either by vacuum suction or by raking. After sludge separation, the clarified liquid is recovered via overflow assembly.

In a rectangular tank configuration, wastewater is fed into the clarifier at one end along with the width of the tank and sludge separated; the overflow is collected at the surfaces either across the other end or at different points along the length of the tank. In all cases, a circulatory conveyor scrapes the floating pollutants through a screen, further settling solids onto a sludge hopper. However, the depth in a clarifier does not contribute to the horizontal flow velocity. The slower flow rates more than enough to prevent settled sludge solids from being scoured from the bottom of the tank. Nevertheless, sludge removal efficiency of clarifiers are mainly calculated based on the hypothesis that all the sludge solids existing in the wastewater have a uniform size, shape, density and specific gravity. However in reality, wastewaters will have varying concentrations of suspended solids with random size and distribution.

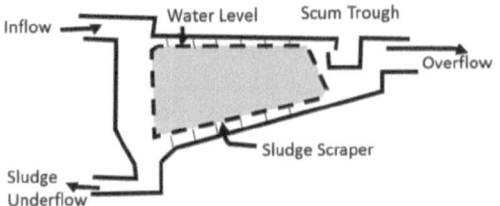

Rectangular Horizontal Flow

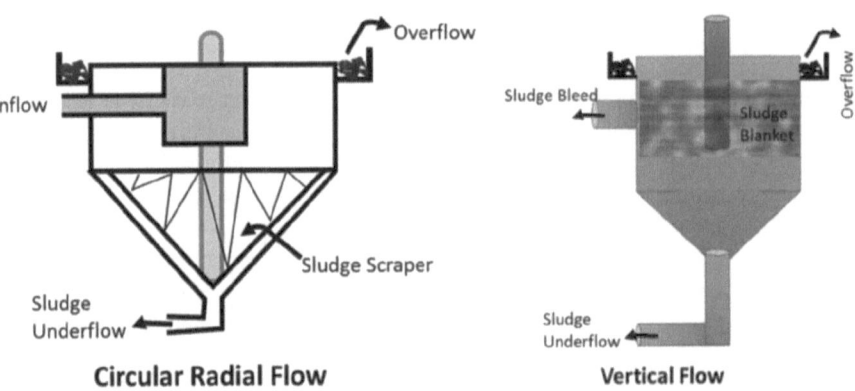

Circular Radial Flow **Vertical Flow**

FIGURE 3.8 Schematic illustrations showing the sedimentation tanks or wastewater clarifiers in three main types of arrangements.

Therefore, in real-time applications, the sludge settling time will be prolonged, and these parameters need to be considered in designing futuristic clarifiers for efficient and robust removal of EPs contaminated sludge from the wastewater.

3.6.6 FLOTATION

The flotation technique is used as alternative to sedimentation or clarifiers as a part of the primary process for industrial wastewater comprising finely divided suspended solids, organic matters of an SVOC nature, PCPs and oily matter. This technique is very effective in removing particulate or suspended contaminants such as fine fibres recovered in paper industry wastewater, suspended metal particulates from the metal finishing industry, proteinases SS recovery from tannery wastewater, retaining oil/greasy granules in a screened effluent, cold-rolling and pharmaceutical industries. This technique significantly retains particles of EPs nature originating from various industries which are very difficult to settle or separate using conventional clarifiers or sedimentation tanks. These types of EPs bear a fine particulate size and take a long time for separation in conventional screening methods. Flotation accelerates the separation of suspended EPs from wastewater through various mechanisms. The aerating techniques are generally used in the flotation tank by passing compressed or pumped air bubbles which influence the suspended matter to float on the surface of wastewater. Once the particulate matter floats on the surface of a floating tank, solid waste is removed manually in conventional devices and mechanically in advanced systems. On the other hand, chemical coagulants such as aluminium and ferric salts or polymer coagulant are frequently used to assist the flotation process. Principally, three different dissolved air flotation systems combining various clarification technologies are in use for medium- to large-scale wastewater treatment. Depending on the type of feed waste streams, the selection of system and processes can be made as follows: (a) dissolved-air flotation system without recycle facility, (b) dissolved-air flotation system with recycling facility and (c) partial aeration of wastewater, as shown in Figure 3.9 [32].

In dissolved air flotation without a recycling facility, air is introduced directly into the wastewater contained in the flotation tank through a revolving impeller or through diffusers. The air bubbles generated using a pumping system in dispersed air flotation systems are generally controlled to approximately 1 mm in diameter. These air bubbles in the flotation tank create turbulence which breaks up fragile floe particles and floats the particulate matter of different compositions. However, dispersed air flotation is not a preferred technique in domestic wastewater treatment containing oil, grease and fine powders. Therefore, dissolved air flotation with a recycling facility is particularly designed to bring EPs-containing wastewater with air bubbles at a different pressure when air is circulated. Air bubbles in the wastewater releasing

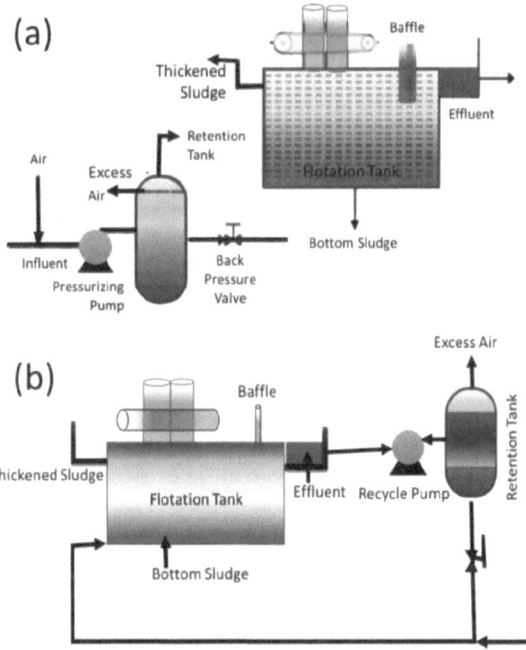

FIGURE 3.9 A cross-sectional illustration of flotation devices with (a) a dissolved-air flotation system without recycle facility and (b) a dissolved-air flotation system with recycling facility.

micron-sized bubbles carrying suspended solids matter covered with oily waste agglomerates to the surface of the flotation tank. In this case, complete circulatory air flow is pressurized and seized in the retention in the tank so that complete air dissolution takes place in wastewater. On the other hand, partial aeration with recycle system is used effectively against the materials that produce fragile floc characteristics. In this process, suspended and partially dissolved EP waste in water sources easily disintegrate due to the intense mixing happening in the pressurized vessel. In this recycling system, air is dissolved and the continuous stream is mixed with the wastewater feed inlet at a point just right before pressure is released. Continuous recycling mode will allow more contact between the air bubbles and suspended solids of different EPs type pollutant thus leads to better efficiency of separation. However, compared to the total aeration described earlier, a partial recycle system requires a larger space build-up because of the combined feed, and bubbling operates at a much lower volume. On the other hand, the rigorous mixing of air in wastewater under pressure often degrades flocculent suspensions or creates oil emulsions which demand chemical treatment for further separation. However, at every stage a portion of the clarified wastewater is recycled for pressurization to prevent intermediate degradations.

3.6.7 Parallel Plate Separator

Parallel plate separators are similar to grit separators, but they include tilted parallel plate assemblies, also known as parallel packs, as shown in Figure 3.10 for oil and grease separation from the wastewater. The parallel plates provide more surface for suspended oil droplets to coalesce into larger globules. Such separators still depend upon the specific gravity between the suspended oil and the water. However, the parallel plates enhance the degree of oil–water separation. The result is that a parallel plate separator requires significantly less space than a conventional grit separator to achieve the same degree of separation [33].

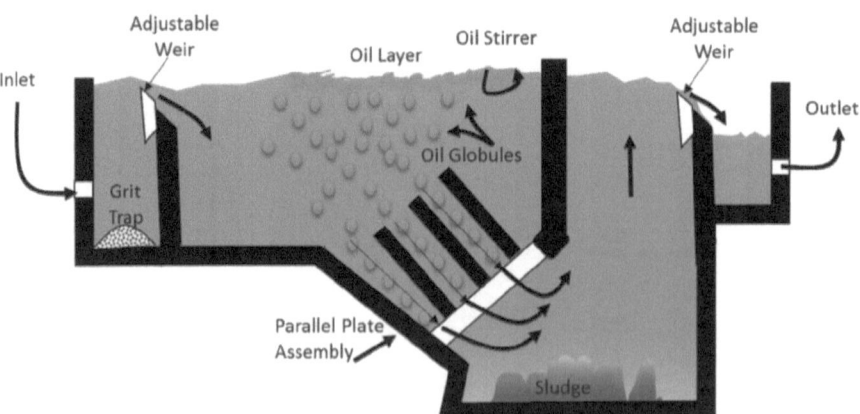

FIGURE 3.10 A cross-sectional view and working of parallel plate separate for sludge and oil/grease separation.

3.7   CHEMICAL METHODS AND THEIR LIMITATIONS IN REMOVING EPS

Several distinct chemical wastewater treatment processes are used for conventional and EPs including chemical neutralization, chemical coagulation, chemisorption, Fenton, photo-Fenton degradation, chemical precipitation, chemical oxidation, advanced oxidation, ion exchange and electrolysis membrane process. On the other hand, ozone disinfection can be used as both chemical and biological microbial treatment applications.

As described in the introductory chapter, odour emanating from contaminated waste sources is being considered under a category of EPs. Typically dynamic features that contribute to taste and odour in wastewater are generally attributed to presence of chemical compounds and their continuously changing forms. Presence of high concentrations of chlorides (> 500 mg/L) causes a salty taste, and presence of hydrogen sulfide gives away the foul odour of rotten eggs, which are being re-evaluated for their potential human risk. Having established a clear relation between odour and the content of the wastewater, while it may not be true for all, research has established that bacterial activity is the leading contributor to the quality of treated water. These bacterial activities render undesired taste and odour to water at unacceptable level. Several studies have confirmed that algae, fungi and protozoa have been responsible for releasing several by-product constituents that are negatively charged colloids into the water. However, when these microbes die in large numbers, it leads to the sudden build-up of undesirable by-products that induces severe odour and further deterioration of water quality. Conventionally, to mitigate the odour and unwanted taste in treated wastewater, activated carbon is added to clarify the water. Activated carbon is typically introduced prior to the chlorination process. Depending on the size and design of the treatment plant, activated carbon or granules are added in the coagulation tank.

3.7.1   Neutralization

A wastewater stream of almost every industry and domestic origin is unfavourably affected by low or very high pH values. These wastewaters generally attain their various pH conditions at their origin and are generally characterized by their raw materials. Characteristic pH of wastewater also adversely influences the deterioration of water quality over a period of time. Absence of a suitable treatment technique aggravates the condition of wastewater, also treatment tech becomes complex and critical when sudden slugs of acids or alkalis are imposed upon the waste stream. Therefore, the purpose of neutralization is fundamentally to adjust the pH value to bring down the severity of wastewater so that further effective wastewater treatment can follow.

Several conventional methods are in use for neutralizing extreme acidity or alkalinity of wastewater, such as limestone treatment, caustic soda treatment, CO_2 treatment, acid treatment and so on. Further, there are random mixing wastes followed conventionally to bring down the net pH effect to near-neutral.

(a) Limestone Treatment: Extreme pH conditions in domestic and industrial wastewater upon release into the ecological environment impose the extreme harsh effects. Extremely pH conditions directly affect aquatic life and potentially harm the vegetation surrounding the aquatic environment. Therefore, it is important to neutralize the wastewater of extremely high or low pH condition, and merely neutral pH wastewater should be discharged into the drain or wastewater treatment plants or environment, in general. A limestone bed is used for acidic wastewater that is made of calcium compounds depending upon the presence and amount of acid. Passing acidic wastewater through beds of limestone is a simple method of neutralizing the acidic wastewater. In another method, wastewater is mixed with lime slurries or dolomitic lime slurries in the proper proportions which helps in reducing the acidity of wastewater. Here, the flow parameter like equalization and chemical neutralization influences the neutralization process. Flow equalization is the step in which the limestone slurry is hydraulic velocity or flow rate is controlled through a wastewater treatment tank. Further, the equalization controls the flow rate, which prevents overflow of solids and organic material out of the treatment process. Chemical neutralization is followed to balance the excess acidity or alkalinity in wastewater.

(b) Caustic Soda Treatment: In this process, acidic wastewater is neutralized using caustic soda. Even though this process is economically intensive, it is an effective method to neutralize the wastewater. Large-volume wastewater is neutralized using a small amount of caustic soda or sodium hydroxide (NaOH) to make the pH neutral. Addition of NaOH will instantly increase the pH of the wastewater. This method will also limit the use of calcium ions in the form of calcium carbonate that are required for precipitation to occur.

(c) Carbon Dioxide Treatment: In this method, carbon dioxide is passed through highly alkaline characteristic of wastewater to bring down the pH to near 7. This is one of the cost-intensive processes, and the excessive carbon production unit can be diverted to treat highly alkaline wastewater.

(d) Sulphuric Acid Treatment: In this simple process, extremely alkaline wastewater is treated by adding sulphuric acid to reduce the pH of the wastewater to near 7 without creating by-products or intermediates.

(e) Utilizing Waste Boiler-Flue Gas: The boiler glue gas, also known as stack gas, contains nearly 12% carbon dioxide and is utilized to treat alkaline wastewater. Industrial wastewater containing EPs like novel synthetic organic materials, such as solvents, paints, pigments, pharmaceuticals, pesticides, surfactants, coking products, etc., are difficult to treat directly. For such wastewaters, a multi-step treatment process needs to be devised to clean, and treatment techniques are often specific to the material being treated. These methods include activated sludge treatment, flocculation, coagulation, advanced oxidation processing, distillation, adsorption, vitrification, incineration, chemical immobilisation and landfill disposal.

However, all chemical methods produce intermediates, by-products and several other secondary pollutants. Some EPs such as surfactants detergents are capable of inducing biological degradation over a period of time. In such cases, an advanced and targeted wastewater treatment needs to be designed for efficient and sustainable treatment of wastewater.

Acids and alkaline-characteristic wastewater are typically neutralized under controlled environments. Every neutralization process frequently produces a precipitate that will necessitate further treatment as a solid residue that may behave like secondary pollutants and toxic. Primarily chemical treated wastewater becomes rich in corrosive by-products which gradually build up as precipitate and can cause severe damage to pipes and, in extreme cases, cause the blockage of disposal pipes.

3.7.2 TREATMENT OF TOXIC MATERIALS

On the other hand, wastewater contaminant EPs such as toxic HMs including zinc, silver, cadmium, mercury and thallium, as well acids, alkalis and non-metallic elements such as arsenic or selenium, are generally resilient to biological processes unless they are found in low or trace concentrations. These metallic EPs can often be precipitated out by changing the pH or by treatment with suitable counter chemicals. However, many of these inorganic EPs are resistant to treatment or mitigation and may require concentration, activation or conversion into solid waste like sludge, followed by landfilling or recycling. On the other hand, dissolved organics are concentrated and incinerated within the wastewater followed by advanced oxidation processes.

3.7.3 COAGULATION AND FLOCCULATION

In principle, coagulation is a process of adding coagulant to disrupt a stabilized charged particle in wastewater and influence the aggregation. On the other hand, coagulation promotes agglomeration and contributes to the sedimentation of largely aggregated particles from wastewater to the bottom of the coagulation bath. In this process, the nature of coagulant, mixing and their characteristics influence the coagulation and flocculation process [34].

Ideally, a lower concentration of suspended solids in a wastewater stream poses resistance to proper mixing and interaction of coagulant. However, in such a case, intensified mixing may yield the separation of charged EPs in the separation bath. On the other hand, high concentration of coagulant may also easily clarify the wastewater. In advanced coagulant tanks, flash mixing is achieved using either mechanical in-line hydraulic mixing or high-speed mixing. However, flocculation is a slow process, and mixing at this stage typically occurs in gently stirred flocculation compartments. Conventionally, flocculation tanks come in the configuration of horizontal reel and in the turbine mixer form. In this design, reel-type flocculation units are combined in parallel, with rectangular shaped sedimentation tanks working together. In an advanced coagulation bath, mechanical stirring or rotating motors are used to achieve constant mixing and flocculation to continue in a controlled clarification of wastewater. However, preventing clogging in the transportation pipeline process requires constant back-flushing, which limits the use of this process to only low volume of wastewater treatment.

3.8 BIOLOGICAL METHODS AND THEIR LIMITATIONS IN REMOVING EPS

Wastewater ozone disinfection is a simple and effective way to kill microorganisms in wastewater before final discharge to the environment. This method is also very effective in comparison to chlorination in disinfecting waste, as ozonization can directly cause lysis of bacterial cell walls and instantly killing them in wastewater. Ozonization also helps in degrading chemically harmful substances, in addition to pathogenic disinfection which significantly reduces unpleasant odour and colour in the treated water. Conventional ozone disinfection efforts were once limited to surface water purification; however, advancement in ozone generation technology has made it a much more cost-effective and efficient pathogenic disinfection method for EPs-contaminated wastewater. Generally, ozonization is a safe method with minimum environmental side effects. However, this process creates toxic disintegrate pollutants in the wastewater, requiring careful discharge strategies. Also, intermediate compounds and by-products formed due to the introduction of ozone to the wastewater are typically characterized by harmful and highly unstable properties. Therefore, for an effective and comprehensive approach, a suitable separation and removal technique needs to be adopted in combination with ozonization.

The secondary wastewater treatment or biological waste treatment process mainly involves removing, stabilizing and rendering harmless characteristics to very fine suspended matter, as well as solids in the wastewater that continue to persist even after the primary treatment, since most of the suspended and dissolved organic matter in wastewater is colloidal in nature. In such cases, the primary treatment processes are largely ineffective in removing dissolved and colloidal organic matters in the wastewater. Therefore, these pollutants remain in the wastewater which has passed the primary stage. Typically, organic matters in the wastewater represent a high chemical and biological oxygen demand before it is considered or rendered suitable for discharge into the environment. In biological treatment, oxygen is constantly supplied under controlled conditions so that most of the BOD is removed in the wastewater treatment plant.

The two most commonly used systems in use for biological waste treatment are the activated sludge system and the biological film system. In the activated sludge system, the wastewater is brought into contact with a diverse group of microorganisms in the form of a flocculent suspension in an aerated tank. In the biological film system, also known as trickling filters, the wastewater is brought into contact with a mixed microbial population in the form of a film of slime attached to the surface of a solid support system. In both cases, the organic matter is metabolized to more stable inorganic forms.

3.8.1 THE ACTIVATED SLUDGE SYSTEM

The essential features of activated sludge process are (1) an aeration stage, solids-liquid separation following aeration, and (2) a sludge recycle system. Wastewater containing residual and sometimes high concentration of organic matter even after primary treatment enters an aeration tank, where organic contaminants are brought into close contact with the sludge of high microorganism content from the secondary clarifier. These microorganisms are at high concentration and in an active state of growth. In such conditions, air is introduced into the tank either in the form of bubbles through diffusers or by surface aerators, as shown in the flow diagram (Section 3) for a typical activated sludge plant given in Figure 3.11. In this process, microorganisms utilize the oxygen in the air and convert the organic matter into stabilized, low-energy compounds such as NO_3, SO_4 and CO_2, resulting in synthesis of new bacterial cells. The organic matter-rich wastewater from the aeration tank in combination with microbial flocculent mass, known as sludge, is separated in a settling tank, sometimes called a secondary settler of a clarifier. Furthermore, in the settling tank, the separated sludge withdraws without contact with the organic matter and becomes activated. This passes through the aeration tank as a seed, and the remainder is wasted. If all the activated sludge is recycled, then the bacterial mass would keep increasing to the stage where the system gets clogged with solids. Therefore, it is necessary to remove additional microorganisms with wasted sludge for smooth operation of the plant [35].

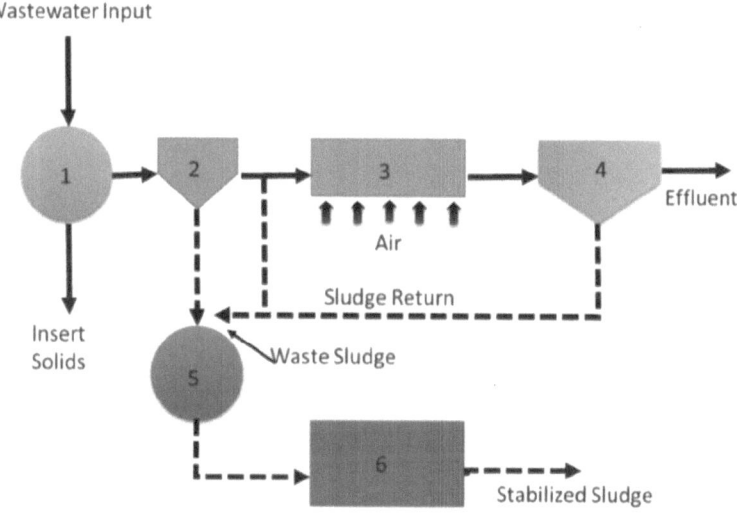

Activated Sludge Plant

FIGURE 3.11 Schematic of an activated sludge treatment plant used for treating organic matter-contaminated wastewater.

3.8.2   BIOLOGICAL FILM SYSTEM OR TRICKLING FILTERS

After activated sludge treatment, the trickling filter process is the most-used method to treat organic contamination in wastewater. Trickling filters, also called percolating filters, offer good adaptability to handle peak shock loads with low to high influents and the ability to function satisfactorily after a short period of time. This method is commonly used wastewater emanating from the dairy industry, paper mills and pharmaceutical outlets. Conventional trickling filters normally consist of a rock bed of 1–3 metres in depth with enough openings between rocks to allow air to circulate easily. In this method, the inflowing wastewater is sprinkled over the bed packing, which is coated with a biological slime, as shown in Figure 3.12. As the liquid trickles over the packing, oxygen and the dissolved organic matter diffuse into the film to be metabolized by the microorganism in this slime layer, resulting in the formation of end products such as NO_3, CO_2, etc. which are diffused out of the film [36].

Inside the sprinkling chambers as the microorganisms eat up the organic matter, the thickness of the slime film increases to a point where it can no longer be supported on the solid media and gets detached from the surface, which is known as sloughing. Once the sloughing yields thick film, the trickling filter removes the detached bacteria film and some suspended matter. However, continuous removal of sludge from the biological wastewater treatment plant is a cost-intensive step in most sewage treatment plants. The concentration of organic pollutants in the primary sludge is about 5%, whereas the activated sludge process yields < 1% and trickling filters yield ~2% removal. This process takes place following sludge treatment and disposal, involving concentration or thickening, digestion, conditioning, dewatering, oxidation and safe disposal.

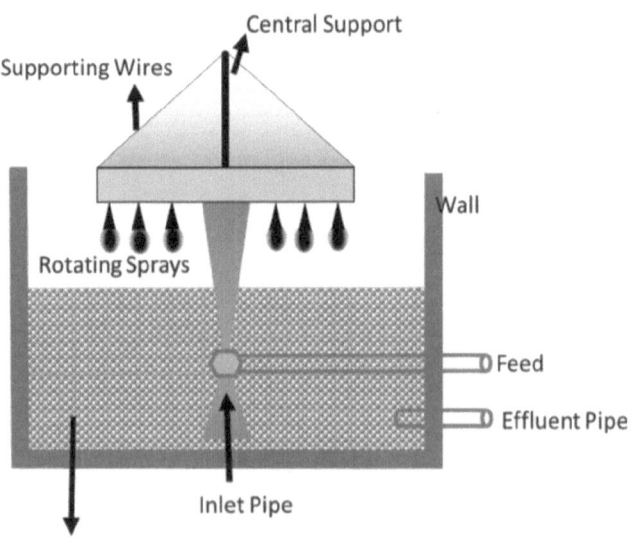

Conventional Trickling Filters

FIGURE 3.12   Conventional trickling filter used for organic matter removal.

3.8.3 Advanced Activated Sludge and Trickling Treatment Process

Biodegradable organic material of plant or animal origin are typically treated using advanced wastewater treatment processes when the influent is extremely diluted with washing water or is highly concentrated with EPs like neat blood or milk. The presence of surfactants, antibiotics and other cleaning agents in domestic wastewater, pesticides in agricultural wastewater can have detrimental impacts on conventional treatment processes.

Therefore, the advanced activated sludge process for domestic and industrial wastewater uses air/oxygen and microorganisms to biologically oxidize organic pollutants, producing a waste sludge or floc containing the oxidized material. This process includes (1) an aeration tank for injection of air/O_2 into the wastewater, and (2) a settling tank, also known as a clarifier or settler, which allows the treated waste sludge to settle, as shown in Figure 3.13. Here also, a minor portion of sludge is recycled to the aeration tank, and the major portion is removed for further treatment and ultimate disposal.

A trickling filter consists of a bed of rocks, gravel, slag, peat moss or plastic media over which wastewater flows downward and contacts a layer or film of microbial slime covering the bed media. Aerobic conditions are maintained by flowing continuous air/oxygen through the bed either by natural convection or mechanically. The trickling filter system consists of: (1) a bed of filter medium upon which is facilitated a layer of microbial slime to form, or promoted and developed; (2) an enclosure or a container which houses the bed of filter medium; (3) a system for distributing the flow of wastewater over the filter medium; and (4) a system for removing and disposing of any sludge from the treated effluent, as shown in Figure 3.14.

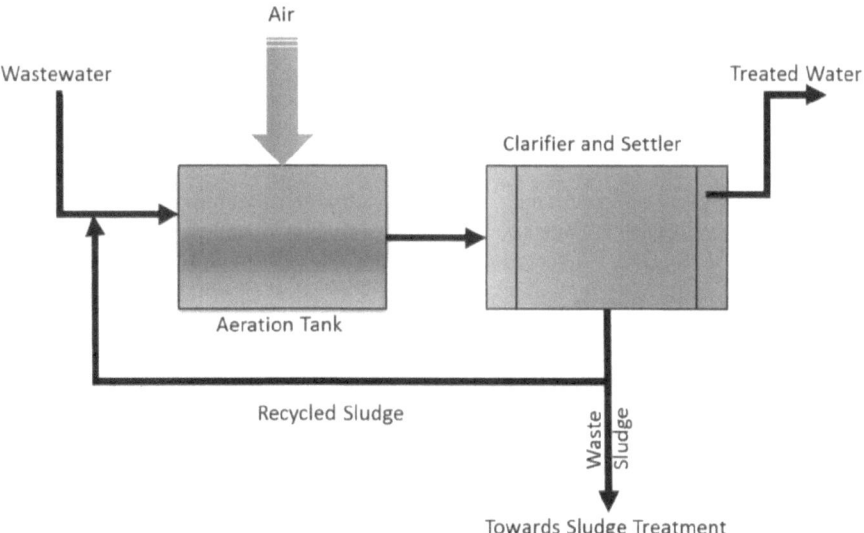

FIGURE 3.13   Schematics of an advanced activated sludge treatment system.

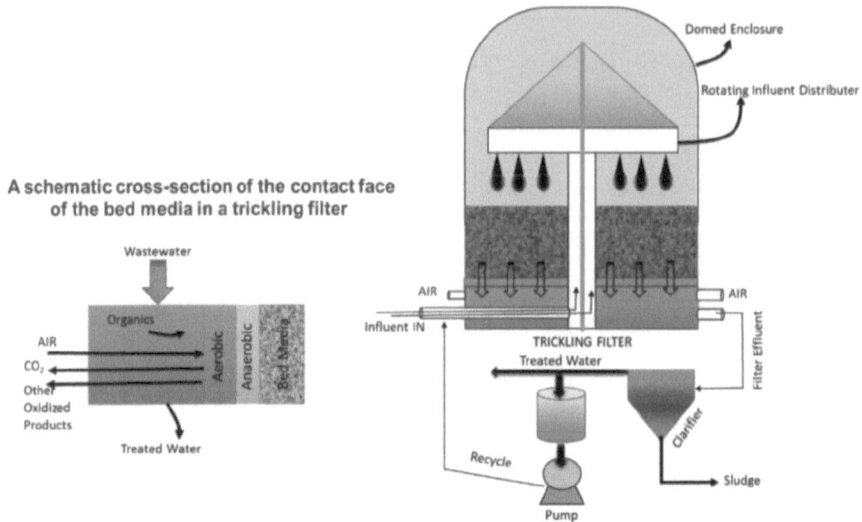

FIGURE 3.14   A cross-sectional working view of the advanced trickling filter process.

The trickling process involves adsorption of organic matter in the wastewater by the microbial slime layer followed by diffusion of air into the slime layer to provide the oxygen required for the biochemical oxidation. Finally, the end products include carbon dioxide gas, water and other products of the oxidation. As the slime layer thickens, it forms a barrier layer for air to penetrate, which results in creating an inner anaerobic layer in absence of air. This also necessitates continuous removal of layer. The treatment of sewage or other wastewater with trickling filters is among the oldest and most efficient treatment technologies. The most advanced trickling filters have an automated spraying, pumping, aeration and sludge removal system, also often called a trickle filter, trickling biofilter, biological filter or biological trickling filter.

3.8.4   Aerated Lagoon and Oxidation Pond

Usually the primary and secondary wastewater treatment techniques produce reasonably good-quality treated wastewater with acceptable standards to discharge into the environment. However, EPs-contaminated wastewater is complex to handle and requires higher water quality standards for direct discharge into water bodies or, in some cases, to some direct reuse. This necessitates advanced wastewater treatment methods such as aerated lagoon, oxidation pond and anaerobic digestions. The aerated lagoon system consists of a large pond that is equipped with machine aerations to maintain an aerobic environment and to prevent settling of the suspend biomass. The population of microorganisms in an aerated lagoon is much lower than that in an actual sludge system because there is no sludge recycle. On the other hand, oxidation ponds, also known as lagoons or stabilization ponds, are large, shallow ponds designed to treat organic pollutant-rich wastewater through the interaction of sunlight, bacteria and algae, as shown in Figure 3.15. These large ponds facilitate

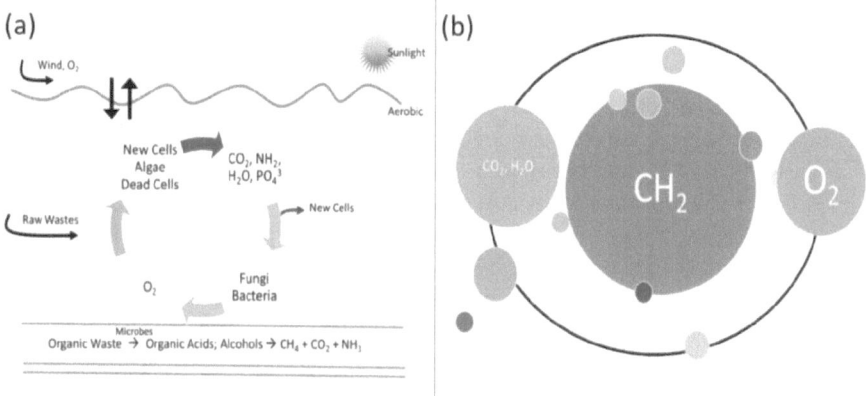

FIGURE 3.15 Advanced secondary wastewater treatment process shown schematically, (a) aerated lagoon and (b) oxidation pond processes.

photosynthesis using organic matter. In these ponds, algae grow using energy from the sun and CO_2, during which inorganic compounds released by bacteria in water making water less toxic.

3.8.5 ANAEROBIC DIGESTIONS

Anaerobic digestion is the process by which organic matter in wastewater is disintegrated to produce biogas and biofertiliser. The anaerobic process takes place in a closed oxygen-free tank called an anaerobic digester. It is one of the advanced wastewater treatments in which various methodologies are followed depending upon the volume and nature of wastewater. These methods are introduced at any stage of the comprehensive treatment process for complete removal of suspended solids, BOD, plant nutrients, dissolved solids and toxic substances.

However, these secondary wastewater treatment processes are mostly used for concentrating or thickening biofilms resulting from the removal of suspended and dissolved organic matter in a microbial processes. Here, most of the organic impurities remained after primary process are concentrated into solid form and are then separated from the bulk liquid as sludge.

3.9 NEED OF ADVANCE MATERIALS AND METHODS TO REMOVE EPS

Among several methods and techniques used for treating domestic and industrial wastewaters, the conventional method fell short in addressing several issues associated with complex pollutants. These primary and secondary methods pose some important disadvantages such as (1) requirements of high energy to operate and many of the primary and secondary treatment techniques requires continuous supply of compressed air/oxygen and turbines, (2) several treatment methods are designed to suit particular waste and volumes annual maintenance however slight deviation demands elevated operation costs, and (3) these technologies need continuous solid waste disposal system at place and large space. However, there are several individual methods and steps among primary and secondary treatment processes which are easy and economically viable. Those techniques are being used in conventional wastewater treatment and are still popular among the research community. However, when the wastewater is contaminated with emerging pollutants, all these processes fall short. This prompted many innovators and researchers to adopt novel advanced separation techniques such as reverse osmosis, ion exchange, gravity and adsorption processes using conventional adsorbents as well as nanomaterial-based cleaners. Among these membrane separation processes and nanomaterials-based adsorption has been termed a promising alternative for future wastewater treatment and management. So far, membrane technology is termed as capital intensive and needs regular maintenance, whereas adsorption methods are known as affordable and easy to use. Many adsorbents are already in use for various applications such as desalination, industrial process applications and removal of water contaminants from surface and ground water due to their ease of operation and efficiencies. Different adsorbents that have been used include use of magnetic nanoparticles, activated carbon, nanotubes and polymer nanocomposites which can remove different contaminants including heavy metals, pharmaceutical and organic molecules like dyes that are very harmful even at low concentrations. Even though adsorption can remove most water pollutants, it has some limitations such as lack of appropriate adsorbents with high adsorption capacity and low use of these adsorbents commercially. Hence, there is a need for more efficient techniques such as membrane technology with low-cost membrane and adaptabilities.

3.10 WATER TREATMENT TECHNOLOGIES STATUS AND SUMMARY

Water scarcity now becomes an important topic in international diplomacy. From village to the United Nations, water scarcity is a widely discussed topic in decision making. Nearly 3 billion people in the world suffer from water scarcity. According to World Health Organization (WHO) sources, a combination of rising global population, economic growth and climate change means that by 2050, five billion (52%) of the world's projected 9.7 billion people will live in areas where freshwater supply is under pressure. Researchers expect about 1 billion more people to be living in areas where water demand exceeds surface-water supply.

Scientists, environmentalists, and biologists worldwide are now alarmed that industrial wastewater production, treatment and disposal can have an impact on the life cycle and the quality pattern and hydrological cycle on the earth, thereby severely affecting the surface and groundwater availability. Industrial wastewater production is believed to rise due to increased population, urbanization and industrial growth, which may lead to the global temperature rising at an increasing pace. Temperature increase affects the hydrological cycle by directly increasing evaporation of available surface water and vegetation transpiration.

The level of water in the ponds and rivers is going down, and in some cases water bodies have dried up completely. Due to increased human and industrial activities, there has been great emphasis on treatment of water and wastewater contaminated with emerging pollutants. Public- and private-funded research activities are going on to reduce the disposal of contaminated wastewater into the environment. This section pays attention only to R&D status on industrial wastewater treatment. Source of water, usage of water, typical use of water and conventional treatment methods still being used and adopted for R&D purpose to produce the high quality of reusable water from industrial sources and typical contaminant characteristics of major polluting industries were discussed in the previous section. Many industries have a need to treat water to obtain very high quality water for demanding purposes. Post water reclamation of high reusable quality, water treatment produces high amounts of organic and mineral sludges from filtration and sedimentation. Ion exchange using natural or synthetic resins removes calcium, magnesium and carbonate ions from water, replacing them with hydrogen and hydroxyl ions. Regeneration of ion exchange columns with strong acids and alkalis produces a wastewater rich in hardness ions which are readily precipitated out, especially when in mixture with other wastewaters.

REFERENCES

1. H. Wang, Response of microorganisms in biofilm to sulfadiazine and ciprofloxacin in drinking water distribution systems. *Chemosphere*, 2019, 218, 197–204.
2. I.B. Gomes, L.C. Simões, and M. Simões, The effects of emerging environmental contaminants on Stenotrophomonas maltophilia isolated from drinking water in planktonic and sessile states. *Science of the Total Environment,* 2018, 643, 1348–1356.
3. I.B. Gomes, D. Madureira, L.C. Simões, and M. Simões, The effects of pharmaceutical and personal care products on the behavior of Burkholderia cepacia isolated from drinking water. *International Biodeterioration and Biodegradation*, 2018, 141, 87–93.
4. I.B. Gomes, et al. Prolonged exposure of *Stenotrophomonas maltophilia* biofilms to trace levels of clofibric acid alters antimicrobial tolerance and virulence. *Chemosphere*, 2019, 235, 327–335.
5. L.C. Pereira, et al. A perspective on the potential risks of emerging contaminants to human and environmental health. *Environmental Science and Pollution Research*, 2015, 22, 13800–13823.
6. J. Wang, et al. PAHs accelerate the propagation of antibiotic resistance genes in coastal water microbial community. *Environmental Pollution,* 2017, **231**, 1145–1152.
7. L. Proia, S. Morin, M. Peipoch, A.M. M. Romani, and S. Sabater, Resistance and recovery of river biofilms receiving short pulses of triclosan and diuron, *Science of the Total Environment*, 2011, 409, 3129–3137.
8. J.R. Thelusmond, E. Kawka, T.J. Strathmann, and A.M. Cupples, Diclofenac, carbamazepine and triclocarban biodegradation in agricultural soils and the microorganisms and metabolic pathways affected. *Science of the Total Environment*, 2018, 640–641, 1393–1410.
9. L. Lv, X. Yu, Q. Xu, and C. Ye, Induction of bacterial antibiotic resistance by mutagenic halogenated nitrogenous disinfection byproducts. *Environmental Pollution,* 2015, 205, 291–298.
10. Huchuan Yan, Cui Lai, Dongbo Wang, Shiyu Liu, Xiaopei Li, Xuerong Zhou, Huan Yi, Bisheng Li, Mingming Zhang, Ling Li, Xigui Liu, Lei Qin, and Yukui Fu, In situ chemical oxidation: peroxide or persulfate coupled with membrane technology for wastewater treatment, *Journal of Materials Chemistry A*, 2021, 9, 11944–11960.

11. S. Zhang, S.W. Gitungo, L. Axe, R.F. Raczko, and J.E. Dyksen, Biologically active filters—an advanced water treatment process for contaminants of emerging concern. *Water Research*, 2017, 7114, 31–41.

12. Y. Yoon, P. Westerhoff, S.A. Snyder, and E.C. Wert, Nanofiltration and ultrafiltration of endocrine disrupting compounds, pharmaceuticals and personal care products. *Journal of Membrane Science*, 2006, 270, 88–100.

13. Francisco G. Calvo-Flores, Joaquin Isac-Garcia, Jose A. Dobado, *Emerging Pollutants: Origin, Structure, and Properties*, ISBN: 978-3-527-33876-4, Hoboken, NJ: John Wiley and Sons, Inc..

14. Violette Geissen, Hans Mol, Erwin Klumpp, Günter Umlauf, Marti Nadal, Martinevan der Ploeg, Sjoerd E.A.T.M. van de Zee, Coen J. Ritsema, Emerging pollutants in the environment: a challenge for water resource management, *International Soil and Water Conservation Research*, 2020, 3(1), 57–65.

15. P. Westerhoff, Y. Yoon, S. Snyder, and E. Wert, Fate of endocrine-disruptor, pharmaceutical, and personal care product chemicals during simulated drinking water treatment processes. *Environmental Science & Technology*, 2005, 39, 6649–6663.

16. Erick R. Bandala, and Markus Berli, Engineered nanomaterials (ENMs) and their role at the nexus of Food, Energy, and Water, *Materials Science for Energy Technologies*, 2019, 2, 29–40.

17. Carlo Giacomo Avio, Stefania Gorbi, and Francesco Regoli, Plastics and microplastics in the oceans: from emerging pollutants to emerged threat, *Marine Environmental Research*, 2017, 128, 2–11.

18. Sudip K. Samanta, Om V. Singh, and Rakesh K. Jain, Polycyclic aromatic hydrocarbons: environmental pollution and bioremediation, *TRENDS in Biotechnology,* 2002, 20(6), 243–248.

19. Jangsun Hwang, Daheui Choi, Seora Han, SeYong Jung, Jonghoon Choi, and Jinkee Hong, Potential toxicity of polystyrene microplastic particles, *Scientific Reports*, 2020, 10, 7391.

20. Y. Peng, L. Gautam, and S.W. Hall, The detection of drugs of abuse and pharmaceuticals in drinking water using solid-phase extraction and liquid chromatography-mass spectrometry. *Chemosphere*, 2019, 223, 438–447.

21. D.D. Snow, et al. Detection, occurrence and fate of emerging contaminants in agricultural environments. *Water Environment Research*, 2017, 89, 897–920.

22. Laura Martín-Pozo, Blancade Alarcón-Gómez, Rocío Rodríguez-Gómez, María Teresa García-Córcolesa, Morsina Çipab, and Alberto Zafra-Gómeza, Analytical methods for the determination of emerging contaminants in sewage sludge samples. A review, *Talanta*, 2019, 192(15), 508–533.

23. Gammarus Pulex, Thomas H. Miller, Gillian L. McEneff, Lucy C. Stott, Stewart F. Owen, Nicolas R. Bury, and Leon P. Barron, Assessing the reliability of uptake and elimination kinetics modelling approaches for estimating bioconcentration factors in the freshwater invertebrate, *Science of the Total Environment*, 2016, 547, 396–404.

24. Yunkun Wang, Ines Zucker, Chanhee Boo, and Menachem Elimelech, Removal of emerging wastewater organic contaminants by polyelectrolyte multilayer nanofiltration membranes with tailored selectivity, *ACS ES&T Engineering*, 2021, 1(3), 404–414.

25. Marta Llamas, Iñaki Vadillo-Pérez, Lucila Candela, Pablo Jiménez-Gavilán, Carmen Corada-Fernández, and Antonio F. Castro-Gámez, Screening and distribution of contaminants of emerging concern and regulated organic pollutants in the heavily modified Guadalhorce River Basin, Southern Spain, *Water*, 2020, 12, 3012.

26. Wastewater treatment | Process, History, Importance, Systems, and Technologies. Encyclopedia Britannica, October 29, 2020. Retrieved 2020-11-04.

27. U.S. Environmental Protection Agency. *Operation of Wastewater Treatment Plants—A Field Study Training Program*, 5th edn, Washington, DC: Prepared by California State University, Sacramento; Office of Water Programs, Vol. 2, 2001.

28. EPA. "Secondary Treatment Regulation: Special considerations." 40 CFR133.103; and "Treatment equivalent to secondary treatment." 40 CFR 133.105, 1984.

29. Nicholas P. Cheremisinoff, *Handbook of Water and Wastewater Treatment Technologies*, Butterworth-Heinemann, Woburn, MA, USA, 17-December-2001 – Technology and Engineering, p. 576.

30. Frank R. Spellman, *Handbook of Water and Wastewater Treatment Plant Operations*, 2nd edn, CRC Press, Boca Raton, Florida 33431, USA, 18-November-2008 – Technology and Engineering, p. 872.

31. Sanjay Patel and Jamaluddin, Treatment of distillery waste water: a review, *International Journal of Theoretical and Applied Sciences*, 2018, 10(1), 117–139.

32. E.A. Deliyanni, G.Z. Kyzas, and K.A. Matis, Various flotation techniques for metal ions removal, *Journal of Molecular Liquids*, 2017, 225, 260–264.

33. J.A. Zeevalkink and J.J. Brunsmann, Oil removal from water in parallel plate gravity-type separators, *Water Research*, 1983, 17(4), 365–373.

34. Chee Yang Teh, Pretty Mori Budiman, Katrina Pui Yee Shak, and Ta Yeong Wu, Recent advancement of coagulation–flocculation and its application in wastewater treatment, *Industrial and Engineering Chemistry Research*, 2016, 55(16), 4363–4389.

35. Duncan Mara and Nigel Horan, *Handbook of Water and Wastewater Microbiology*, ISBN 978-0-12-470100-7, 2003, Elsevier Ltd, Amsterdam, Netherlands.

36. Murray Moo-Young, *Comprehensive Biotechnology*, ISBN 978-0-444-64047-5, 2019, Elsevier B.V, Amsterdam, Netherlands.

4 Membrane Processes for Removal of Emerging Pollutants

4.1 INTRODUCTION

Emerging pollutants (EPs) are the substances or chemicals of synthetic or natural origin that have recently been revealed but for which the environmental or public health risks are yet to be established. Even though there is little to no evidence available on the real threat from EPs due to inadequate available information on their interaction and toxicological impacts on receptors, low levels of contaminants of emerging pollutants in water bodies have raised countless concerns because of EPs' toxicity and endocrine disrupting effects. However, several studies reveal that the presence of EPs in the environment show a potential threat to the ecological system. EPs are expected to cause simple interruption in body functions which may lead to devastating impacts in the course of time. EPs in contaminated water is a growing problem mainly because of the uncontrolled release of new classes of pollutants into the water bodies. This contributes to the worsening of the overall water quality which is blamed for the deaths of several thousand children annually.

As of today, a majority of EPs that are used in our daily lives are dumped into the aquatic environment without regulation. Daily used personal products like surfactants, disinfection products, kitchen waste and antibiotics that are released or excreted by humans and transmitted to water systems end up harming a large portion of aquatic life. Other excreted pharmaceuticals like norfloxacin (NOR) which is a fluoroquinolone antibiotic drug molecule have high chemical stability and complexity. The highly stable chemicals in wastewater pose a challenge to the environment and to conventional wastewater treatment techniques. Most of the conventional techniques and methods have proven inefficient against the removal of a series of EPs in wastewater which allows it to reach the environment instantaneously. This causes unlisted and unmonitored risks both to the environment and human health. On the other hand, over 80% of industrial wastewater contains highly polluting EPs which are discharged into water systems with only partial or no prior treatment. Among these EPs, there are many organic pollutants, in particular plastic pollutants, which are difficult to degrade due to their complex structure, which gives EPs a stability and subsequently renders them non-biodegradable. Another class of EPs such as nanomaterial substances pose when in water a risk to human health and the environment, since they accumulate in water sources causing the loss of aquatic biosystems [1].

DOI: 10.1201/9781003214786-4

Limitations of conventional treatment technologies in EP removal such as oxidation, coagulation/flocculation, sedimentation and biological processes, compelled researchers to develop innovative treatment technologies and methods, such as advanced oxidation processes (AOPs). AOPs include the oxidation of pollutants or contaminants through their reaction with free radicals, such as hydroxyls, thereby resulting in degraded pollutants. There have been several AOP catalysts adopted with varying levels of success. Among them the photocatalysis route is one of the most used treatment methods to degrade organic pollutants. Several photocatalysts have been successfully used, both for the UV-assisted and visible-light-assisted degradation of EPs and general pollutants. This phenomenon takes place as a result of the absorption of a photon in an organic-compound-contaminated wastewater system with energy greater than the band-gap energy of photocatalysts. This leads to the creation of electron hole pairs that react with H_2O, OH and O_2, creating highly oxidizing species that degrade organic EPs via a cascade of redox reactions. Nonetheless, oxidation degradation leaves behind several secondary pollutants. These pollutants have been difficult to trace and identify on a regular basis. In many cases, it has been estimated that degraded smaller organic residues are much more dangerous than their parent pollutants. This again necessitates a secondary treatment step to reduce the environmental and health risks upon their use. On the other hand, membrane separation processes have been successfully employed for conventional wastewater treatment, though the removal efficiencies and overall flux are still limited. However, improved physical methods like the membrane-based separation process have shown greater potential at removing several EPs from industrial wastewater [2].

Membrane separation or a treatment process mainly depends on three basic principles, namely adsorption, sieving and electrostatic phenomena. The adsorption mechanism in the membrane separation process is based on the hydrophobic interactions of the membrane and the solute, such as organic analytes in wastewater. These interactions normally lead to more rejection which leads to a decrease in the pore size of the membrane. The separation of materials through the membrane depends on the pore and molecule sizes. However, this phenomenon in the membrane reduces its life and that of the separation process, drastically prompting higher maintenance costs. Based on these observations, researchers and innovators have designed various membrane processes with different separation mechanisms. These include microfiltration (MF), ultrafiltration (UF), nanofiltration (NF), reverse osmosis (FO) and forward osmosis (RO) [3].

4.2 MEMBRANE SEPARATION SPECTRUM

Therefore, overviews of several conventional wastewater and EP characteristics and methods available currently reveal the need for advanced wastewater treatment techniques. Many conventional wastewater techniques fall short in addressing the EP-contaminated effluents. However, challenges/limitations associated with the use of each technique, including advanced membrane technology, show the several limitations of existing methods. Membrane processes such as MF, NF, UF and RO are currently used for water reuse, brackish water and seawater. These membrane-based separation processes also gain more importance in industrial wastewater treatment. Polymer-based membranes are mostly used-membrane material; however, the hydrophobic nature of most of the commercial membranes, which are made of polymers such as polysulfone, polyamide, polyvinylidienfluoride and polyethersulfone, are relatively hydrophobic which make polymeric membranes vulnerable to fouling. This leads to the blockage of membrane pores and decreases membrane performance, it also increases operation costs by requiring extra cleaning processes. There are factors causing membrane fouling, such as the deposition of inorganic components onto the surface membrane/solute absorption, pore blocking, microorganisms, feed solution constituents and their nature of interaction. These results in either reversible or irreversible membrane fouling. Reversible fouling is formed by the attachment of particles to the membrane surface; irreversible fouling occurs when particles strongly attach to the membrane surface and cannot be removed by physical cleaning. When there is formation of a strong matrix of the fouled layer with the solute during a continuous filtration process this will turn reversible fouling into an irreversible fouling layer.

Conventionally, membrane technologies have been successfully used for desalination, dairy product processing and organic solvent filtration applications. Among these, RO has been efficient at desalinating seawater; NF is gaining importance in organic solvent filtration (OSNF); and UF and MF are largely popular for domestic wastewater treatment. However, recent studies have significantly focussed on the use of these technologies for EP removal from wastewater. This chapter will discuss the importance of the size and surface properties of various EPs and their effect on the removal efficiencies of different membrane processes. This chapter will also evaluate the advantages and limitations of different membrane materials and processes in EP removal [4].

Polymer-based membranes in a series of processes offer several applications due to polymers being used as a precursor which can be turned into a flexible design and module adaptation. The use of a polymer-based precursor makes membrane technology a cost effective, easy operation and a high pre-concentration factor with a high degree of selectivity. Membrane separation processes use several types of membranes depending on the size and nature of the pollutant to be treated, as shown in the infographics diagram of Figure 4.1. The membrane separation broadly describes the process involved, with the size and specific contaminant rejection potentials. The membrane process spectrum is classified based on the pressure required to separate the contaminants from the mixture/feed. Microfiltration membranes have the largest pore size and characteristically reject large particles and various microorganisms at low pressure. Ultrafiltration membranes have comparatively smaller pores than MF membranes and therefore, in addition to large particles and microorganisms, they can reject bacteria and soluble macromolecules such as proteins at slightly high pressure. Nanofiltration is a relatively new class of membranes which falls between dense membranes and a UF pore regime which requires moderately high pressure to operate. Therefore, NF membranes are generally called 'loose' RO membranes. Nanofiltration membranes are porous and generally prepared in the asymmetric format of membrane orders. Thus, NFs are porous membranes in which the pore size

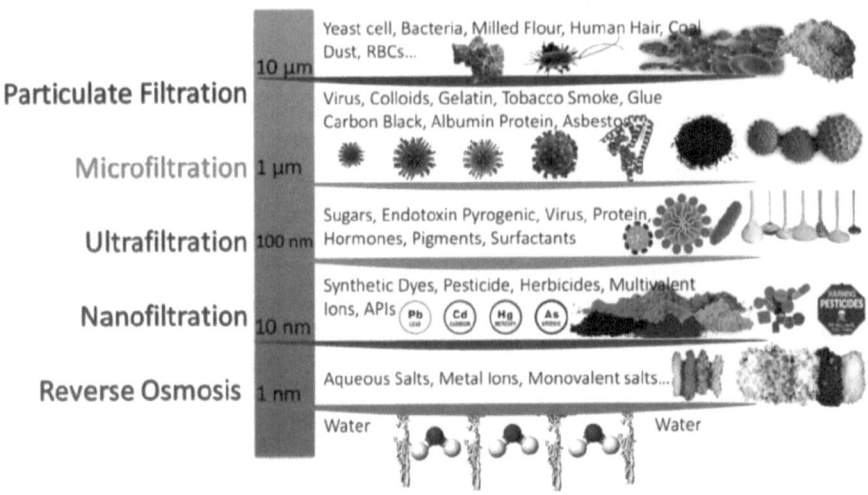

FIGURE 4.1 Infographics showing membrane separation processes with pore size and potential solute or contaminant (EP) separation spectrum.

falls by an order of ten angstroms or less; they exhibit a performance that lies between that of RO and UF membranes. RO membranes are effectively non-porous with a pore size less than 1 nm. The almost dense membrane therefore excludes particles and even many low molar mass species such as monovalent salt ions and even low molecular weight organics [5].

4.2.1 Microfiltration

Microfiltration refers to the large size contaminant removal membranes category where pore size actively ranges between 1 μm and 10 μm. This regime requires the lowest operating pressure to generate permeate fluxes. Separation by MF is largely based on molecular or size exclusion; nonetheless, process conditions can influence the separation efficiency for different contaminants. MF is widely used for several process applications like: dairy processing and food product separation where the de-fatting of whey in the production of whey protein isolates; the separation of casein and serum proteins from skimmed milk; and the clarification of cheese brines is an important step. Conventional and commercially successful MF membranes are prepared using polyvinylidene fluoride (PVDF) polymers, poly(ether sulfone) (PES) and polyacrylonitrile (PAN). The other widely used applications of MF membranes are found in sugars and sweeteners as well as industrial bioprocessing industries. In this case, MF is used for the concentration of saccharification liquors before further refining processes. In the industrial bioprocessing industry, MF is frequently used to clarify bulk fermentation mixtures in order to improve the performance of downstream processing steps. In various applications, MF is also used as a pre-treatment step for other separation processes such as NF and RO.

4.2.2 ULTRAFILTRATION

A UF membrane process regime falls in the pore range from 0.01 μm to 0.1 μm (commercially defined between 10 nm and 50 nm); it is further classified based on its molecular weight cut-off (MWCO) efficiency, which is stated in Daltons (Da) or kilodaltons (kD). The MWCO of UF membranes are characteristically measured as in the range of 1000 Da (1 kD) to 100,000 Da (100 kD). Similar to MF, the separation regime of an ultrafiltration process is largely based on molecular or size exclusion; however, process conditions such as applied pressure required depends on the pore range and distribution. On the other hand, membranes can be selected for specific applications based on the contaminant's size or molecular weight and based on their MWCO. The most frequently used polymers for UF applications are made of PVDF, PES or polysulfone (PS) polymers. The plentiful number of PS-based UF membranes are used as support membranes in RO and NF thin-film composite (TFC) membrane preparation. The widely used application of UF membrane processes includes dairy processing, the bioprocessing industries, electrocoated paint industries to recover pigments and resins, domestic wastewater treatment, industrial wastewater treatment, process water recovery and macromolecular enrichment.

4.2.3 NANOFILTRATION

According to commercial and academic debates, the NF process is a relatively new addition to the membrane separation spectrum. Industrial reference to NF was as a tight or low MWCO ultrafiltration regime. The NF membranes were generally characterized based on the rejection features of ionic solutes like magnesium sulfate ($MgSO_4$) and sodium chloride (NaCl). The range of rejections for ionic salts $MgSO_4$ and NaCl for NF is usually anticipated to be ~90.0–99.5% and ~30–70%, respectively. The separation mechanism of NF is mostly based on the diffusion of dissolved species through the membrane and can be heavily impacted by the pH and chemical charge near or at the membrane surface. The most frequently used NF membranes are thin-film composite membranes made of polymers of polyamide originated from polypiperazine and trimesoyl chloride as a skin/selective layer and a PES or PS UF substrate. Recently, NF has been largely utilized as an alternative to both UF and MF processes, combined for concentrating macromolecules or the demineralization of industrial process intermediates. Over the years, the NF process has progressed as a prominent separation and purification process for applications like desalination, molecular separation, demineralization, low-molecular weight organic separation and wastewater treatment applications. In the sugar and sweetener industry NF is used for eliminating impurities and the enrichment of primary sweeteners and additive substances.

4.2.4 REVERSE OSMOSIS AND FORWARD OSMOSIS

Commercially the most popular regime membranes in a separation process for water treatment, in particular for desalination, is a reverse osmosis process where the separation range is categorized based on the rejection characteristics of a known solute, conventionally sodium chloride (NaCl) which is also popular among commercial membrane characterization for rejection standards. The range of NaCl rejections for RO membranes is expected to be nearer 96.0–99.8%. Like NF, the separation mechanism in RO is largely based on the diffusion of solvent species through the membrane and overpowering osmotic pressure of the process fluid or feed. All the advanced RO membranes are mostly prepared in thin-film composite morphology with selective thin layer and ultrafiltration support. For all commercial purposes, the polyamide thin layer is popularly used on polysulfone ultrafiltration support. RO is widely used for desalination and surface water treatment compared to UF or NF as an efficient means to reduce hardness in drinking water. Improved RO processes with high flux parameters are also gaining wide ranging applications like molecular separation and wastewater treatment. Recent advancements in nanotechnology modified to RO composite membranes have transformed the use of the RO process domestically throughout the spectrum. Nonetheless, demand for household applications like personalized RO plants have also forced innovative ways to use RO modules in miniature form with superior separation efficiencies.

4.3   SIZE AND SURFACE CHARACTERISTICS OF EPS

Most of the pressure driven membrane processes, namely MF, UF and NF, mainly use physical separation following the size exclusion principle to separate contaminants from wastewater. On the other hand, RO FO and gas separation membranes are semi-permeable in nature and the separation mechanism follows with a combination of solution diffusion and molecular sieving. Membrane separation processes are mainly characterized by their ability to separate molecules of different sizes and surface characteristics in various pollutant mixtures. However, surface charge, polarity and size place an important role on the separation of EPs from complex wastewaters, even though applied pressure across membranes places an important role on driving the separation of contaminants. However, the nature of polymer and membrane surface characteristics also contribute greatly to the efficient separation of EPs. Therefore, separation of different contaminants occurs through membranes on the basis of solute size, the nature of the membrane (hydrophobicity/hydrophilicity) and the porosity of the membrane.

Membrane technology also has shown some serious issues during the separation of EPs in wastewater. The diverse physico-chemical properties of EPs and the wide range of variable parameters in combination face series of encounters during their separation and cleaning. Different polymeric precursors and their compositions used for the preparation of membranes, namely homo-polymeric, hybrid, inorganic or supported, blends and composites, undergo several physical and chemical transformations before and after being converted into membranes. These criteria need to be taken into consideration when selecting precursors to avoid undesirable physico-chemical interactions with EPs during their separation. For example, a thin polymer or nanomaterial coating on the large porous surface allows alteration of the membrane surface properties; however, these modifications may induce varying separation characteristics. On the other hand, creating a molecular layer of different surface modifying agents on the substrate induces several functional and chemical characteristics to the membrane. These modification methods overall transform the separation and removal mechanisms of targeted EPs and many times narrow the choice of membrane. In addition to this, the use of nanomaterials or charge polymers for surface modification results in inducing unfavourable charge modulations on the surface exposed to various EPs. Also, surface pore filling attempts may yield a narrowing of the pore channels and overall pore size reduction. Even though pore size alteration routes are followed based on desired end results, this sometimes leads to adverse separation performance. For example, a steric exclusion separation membrane should have pores smaller than the target pollutant; however, reduced pore size may yield a drastic flux decline.

Nonetheless, modifying the membrane surface with a specific polymer, nanomaterial, thin film or organic compound may improve the overall EP separation efficiency. Further, in addition to surface modification, process parameters such as pH, composition, feed concentrations, temperature, water movements or pressure of the feed play major roles in overall separation performance of a membrane. But it is evident from several studies that surface modification precursor, method and feed conditions altogether contribute to the separation mechanism, simultaneously. However, in some cases individual factors can immensely contribute preponderantly over other process parameters in the separation mechanism. For instance, an electrostatic repulsion on the membrane surface directly causes EP rejection, though over a period of time the membrane surface could lose its surface properties during separation due to the counter-ion adsorption on the surface. Further, this phenomenon leads to surface accumulation and diffusion causing severe damage to permeable property.

Even though the ideal performance of a membrane is to achieve near-zero surface accumulation and permeation for pollutants through membranes, the several variable parameters restrict an ideal performance of any membrane. Therefore, every membrane process and membrane in particular experience the phenomenon known as a 'breakthrough curve' which limits its ideal performance in the rejection of EPs to 100%. However, over the years economic and energy consumption have been considered definitive parameters in developing an optimally performing membrane and process. Here, researchers have paid special attention to tuning the surface and pore characteristics of membranes, tuning RO to the NF range, making it more economical and viable for the removal of a range of EPs. This has greatly influenced the energy consumption of the overall process in removing EPs during water and wastewater treatment applications.

Therefore, surface charge, morphology and pore characteristics greatly contribute to the membrane process and its efficient applications. Interestingly, membrane processes like UF and MF may not experience the influence of surface charge and pore filling effects upon surface modification. This makes them less vulnerable to surface fouling due to charged pollutants. Nonetheless, the nanofiltration pore regime has greatly contributed in replacing RO in rejecting a large number of EPs from water and wastewater, making it economically viable and an alternative process. These considerations also required tuning the optimal performance membranes with suitable surface characteristics that are beneficial for full scale process application with minimal energy consumption. Generally, most of the commercially successful applications consider a membrane as core for the overall performance of the process. Membrane separation characteristics greatly contribute to the overall process economics. Nonetheless, at bottom, membrane synthesis, the membrane's material cost and environmental assessment are the key considerations for efficient output. In the general case, RO with the smallest pore regime gives higher performance in EP rejection (~100%) in optimal conditions, compared to NF, though NF offers more economical and environmental benefits compared to the RO process.

Therefore, NF membranes are increasingly utilized for a wide variety of applications from chemical industry process applications, petrochemical fractionation, biological contaminant treatment and the desalination industries. This is mainly due to NF technology overcoming the operational difficulties, problems and issues that are associated with conventional RO technology for similar applications. Moreover, NF comes in between RO and UF in its pore size regime, making it suitable for tuning a range of MWCO ranges. Like the other two processes, NF uses moderately applied pressure to operate; however, it can be tuned to separate at the molecular level colloidal size contaminants. NF can separate EPs of dissolved components with an MWCO of about 200 to 2000 Da and a molecular size of about 1 to 10 nm. For the same reason, NF has already gained importance as a pre-treatment technology at present assisting further processes like FO, RO and membrane distillation (MD). Now, most of the estimated EPs fall in the range of 100–5000 Da which makes NF a suitable choice for the future.

The main influencing factors in membrane separation and selectivity parameters that influence EP treatment can be categorized as: (1) the charge of the molecule; (2) the hydrophobicity of the separated molecule; (3) the size of the separated molecule; (4) the pore size, its distribution and dimension in the membrane's selective layer; and (5) the affinity of the molecule to the membrane, which indicates the closeness of the EP molecule that is able to diffuse through the membrane or is able to penetrate its pores. In every case, the solvent is the permeating or product species, because of its small molecule size, in the most employed processes such as NF, RO or FO. The separation mechanism in RO and UF take place due to solution diffusion and sieving effects, respectively. However, NF takes into account solution diffusion as well as the sieving effect such as Donnan interaction, dielectric exclusion and electro-migration, simultaneously, which makes it a complex, but efficient mechanism to separate both charged as well as uncharged organic solutes.

In addition to these membrane surfaces and morphological characteristics, experimental conditions could strongly influence the affinity of the target EP molecule to the membrane. In such a case, the separation mechanism can be explained based on the flux, selectivity and permeation data which give specific trends and cumulative effects of the electrostatic repulsion, steric hindrance and the solubility and diffusivity of the molecule in the membrane. The general phenomenon commonly observed between solute and membrane is the hydrophilic–hydrophobic forces that also greatly contribute to the separation mechanism [6].

In membrane science and technology, the performance of a membrane is evaluated based on its permeability and solute selectivity. In general, polymer-based membranes with both high permeability and selectivity are desirable. For commercial benefits, a high permeability membrane helps to reduce the use of membrane precursors and overall membrane technology plants thereby drastically decrease the capital cost of membrane units. On the other hand, higher selectivity results in a higher purity product permeate. Therefore, overall performance of the membrane is measured using substantial datasets. However, a general trend in the reported literature indicates that high selectivity membranes tend to be low in permeability. Therefore, several academic and industrial studies adopt a general trade-off relation that is recognized for two important parameters, permeability and selectivity. Accordingly, polymer-based membranes that are more permeable are generally less selective and vice versa. Materials with the best performance would be in the upper-right-hand corner of Figure 4.2(a). Materials with high permeability or selectivity combinations beyond and to the right of the line drawn in the middle (known as the 'upper bound' in this figure) are extremely rare. Therefore, ultimately permeability and selectivity are fundamental parameters which govern transportation and are finally used to evaluate membrane performance in contaminant separation. Conventional polymer-based membranes frequently suffer from this trade-off, while the performance of membranes composed of porous membrane structures both of a symmetric and asymmetric nature can occasionally exceed this limit.

On the other hand, except in some high permeability membranes, resistance to permeate flow in different sections of the polymer materials is considered as one of the important process operation parameters for overall performance evaluation. The well-established resistance model is used for estimating the single layer performance of an asymmetric composite membrane with distinct layers of thickness. The total resistance (R_t) can be expressed as:

$$R_t = R_s + R_i + R_p \tag{4.1}$$

where R_s, R_i and R_p are the resistances of the selective layer (top), intermediate internal layer (intermediate) and porous support layer, respectively, as shown in Figure 4.2(b). The intermediate (gutter) layer comprises material introduced between the thin top selective layer and the mechanically strong porous support to prevent intrusion of the selective layer into the porous support. The resistance arising from the porous support is minimal or negligible (R_p) and is typically ignored. Hence, the total resistance can be written as the sum of the thickness over permeability of the selective top layer and intermediate layer:

$$R_t = R_s + R_i = \frac{l_s}{P_s} + \frac{l_i}{P_i} \tag{4.2}$$

where l and P are the thickness and permeability of the corresponding layers, respectively. As such, it can also be used to calculate the permeance of each layer.

On the other hand, permeate transport mechanism in different membranes is a complex phenomenon. Different types (porous/non-porous) of membranes follow different transport mechanisms under applied pressure and process conditions. Many RO and FO processes use non-porous or dense membranes and the transport of components through the dense membrane will be a solution–diffusion mechanism. In microporous membranes such as MF and UF having a relatively large porous structure, the transport mechanism involves the pore size of membranes to diffuse or reject the solutes. On the other hand, an NF membrane regime demonstrates a behaviour between dense and porous membranes. In addition to diffusion (Knudsen diffusion), size exclusion and electrostatic interaction are the two fundamental phenomena that govern the solute rejection and permeation in NF membranes. On the other hand, in porous membranes molecular diffusion follows different mechanisms, depending on the relative size differences between the solute molecule and the pore mouth. The mechanism of pore properties on the inner pore walls on selective layer surfaces or the solute condensability is shown in Figure 4.2(d). In such a pore structure, Knudsen diffusion is the main transport mechanism in which the mean free path of the solute molecules is of the same order as the pore size. Inside the pore, collisions between solute molecules and the pore walls are more frequent than those between the molecules themselves. This mechanism is often substantial at low to moderate pressure ranges for membranes in NF and UF regimes with a mesoporous pore size of 2–50 nm. As the adsorption capacity of the pore wall is considered the same for all the solute species in this case, differences in diffusion will arise from the diffusion coefficient (D):

$$D = \frac{d_p}{3} \sqrt{\frac{8RT}{\pi M_i}} \tag{4.3}$$

where d_p is the mean pore size of the pores, R is the gas constant, T is the operating temperature and M_i is the molar mass of the gas. Therefore, the larger the molar mass the smaller the Knudsen diffusion rate. Another separation mechanism in porous membrane separation is molecular sieving, as shown in Figure 4.2(e), in which the smaller dynamic diameter solute penetrates through the pore channel while the larger solute size is rejected, due to which high selectivity for a larger solute occurs. However, in surface diffusion which is also known as a hopping or jumping mechanism there is an exceptional mechanism in porous membranes which occurs when the solute molecule has strong interactions with the adsorption sites on the pore walls of a pore as shown in Figure 4.2(f). In another mechanism, the solute molecules or gas molecules from a gas mixture in a membrane pore occurs via adsorption through successive jumping to adjacent sites under driving forces such as temperature, pressure or chemical potential gradients. In this mechanism, the condensable mass diffuses through the capillary condensation as shown in Figure 4.2(g).

4.4   MF AND UF FOR REMOVAL OF MICROPOLLUTANTS

MF membranes have larger pore sizes than UF membranes, which typically reject materials in a size range of 100–1000 nm, typically micron size contaminants, both inorganic (aggregated colloids, red blood cells (RBCs), etc.) as well as of a biological nature such as bacteria and pathogens relatively less able to operate with moderate to high fluxes. On the other hand, a UF membrane pore regime typically between 10 nm and 100 nm rejects large colloidal particles, macromolecules, biopolymers and

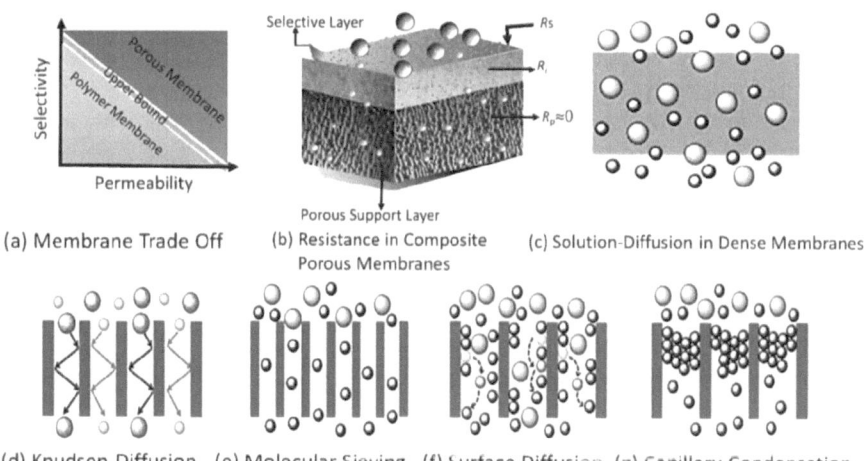

(a) Membrane Trade Off

(b) Resistance in Composite Porous Membranes

(c) Solution-Diffusion in Dense Membranes

(d) Knudsen-Diffusion (e) Molecular Sieving (f) Surface Diffusion (g) Capillary Condensation

FIGURE 4.2 (a) Schematic showing membrane trade-off which is a direct indication of the quality (selectivity) and quantitative (permeability) separation of solute/solvents, (b) the schematic cross-section of an asymmetric composite membrane with the corresponding layer resistance, (c) diffusion mechanisms in dense/non-porous membranes via solution diffusion and diffusion in porous membranes, (d) Knudsen diffusion, (e) molecular sieving, (f) surface diffusion and (g) capillary condensation.

viruses whose sizes generally range from 100 nm to 200 nm and where the process is based on the size exclusion principle. Both MF and UF have been widely used as pre-treatment steps for many other processes like NF, RO and FO processes. However, as a standalone process both MF and UF have been used for the successful treatment of domestic wastewater, industrial wastewater, recovery of surfactants in industrial cleaning, dairy production fractionation, food processing, protein separation, and so on. The UF membranes are fabricated using a variety of natural, synthetic polymers as well as inorganic precursors such as cellulose derivatives and typical synthetic polymers such as PAN, Poly(sulfone amide) (PSA), PES, PVDF and inorganic materials such as TiO_2, Al_2O_3 and ZrO.

4.4.1 MICROFILTRATION (MF)

The large-pore-size-range membrane process, MF, has been conventionally used to remove or concentrate EPs of colloidal particles, dye-contaminated wastewater, organic matter and other high molecular weight soluble personal care products from waste streams. MF has been generally successful as a part of a hybrid process in treating complex EPs in domestic wastewater and industrial EPs. For instance, domestic wastewater comprising hormones such as 17β-estradiol (E2), estrone (E1) and 17αethynylestradiol (EE2) as well as the industrial chemical bisphenol A (BPA) that are classified as endocrine disrupters are efficiently removed in a series of hybrid processes. EPs of an endocrine disruption nature, even at the trace level, need to be removed for a safer environment for which several attempts have been made using

MF with reasonable success. However, MF membranes prepared from PES, CA, nitrocellulose, polyester, regenerated cellulose and polyamide-66 (PA) in cross-flow configuration with pore size ~200 nm have successfully removed estrogens even in concentrations as low as one microgram per litre from domestic wastewater sources. One of the limitations of microfiltration membranes is the surface adsorption of organic solutes on the membrane which leads to accumulation and subsequent fouling. The surface adsorption of some of the organic or biological EPs in both pre-treatment as well as in a standalone MF at higher concentrations severely affects the membrane performance. The MF membrane process is particularly susceptible to biological and organic contaminant scaling, causing membrane fouling that significantly disturbs the flux of membranes. In many cases, most EPs in wastewater show a steady interaction with the membrane material like the polyamide membrane via hydrogen bonding, thus showing their efficient removal from wastewater feed streams. Due to this reason, cellulose-based, relatively hydrophilic membranes are used to remove severely contaminated pathogenic EPs in wastewater.

Several regulatory authorities worldwide constantly monitor the health and environmental risks associated with micro-EPs, in particular biological contaminants like progesterone. Conventionally, standalone membrane processes are ineffective at removing hormones like EPs, therefore several researchers use an adsorption assisted membrane process for such EP removal from wastewater. Studies also reveal that the presence of biological contaminants and EPs on an adsorbent also causes clogging in the long term, creating high-pressure drops and slow mass transfer. These hybrid processes also arouse less interest as the post-recovery of an adsorbent and the separation of EPs is a complex and cost intensive process. Further, researchers developed a zeolite imidazolate metal-organic framework (ZIF-8) with nanoparticles that incorporated nanocomposite MF membranes in a spiral-wound configuration to induce EP separation efficiencies in a poly(tetrafluoroethylene) (PTFE) matrix. Nanocomposite MF (PTFE/ZIF-8) membranes greatly improve anti-fouling and anti-scaling properties, making it an effective medium for separating biological EPs from wastewater. Here, the ZIF-8 nanoparticle-induced MF membrane exhibits a high product flux with a rejection rate of as much as 95% for EPs like hormones and bacteria, even at low operating pressures. High rejection EPs from wastewater also accounted for the weak chemical interaction between hormones and the membrane surface via H-bonding. These separation efficiencies were also enhanced using a high surface area of ZIF-8 nanoparticles in a membrane matrix. Therefore the use of nanocomposite membranes and their preparation approaches have proven beneficial for a high rate of rejection of micro-EPs, such as hormones and bacteria from wastewater streams.

On the other hand, active pharmaceutical ingredients (APIs) such as diclofenac and ibuprofen in the concentration ranges of 0.14–1.48 µg/L and 160–169 µg/L can be easily degraded using UV-active TiO_2 photocatalysts. However, the photocatalyst produces several secondary degraded pollutants which need further processes to separate from the contaminated wastewater in order to obtain clean potable water. In such cases, a hybrid MF can not only be useful for removing or effectively extracting pharmaceuticals but also can be used to recycle the photocatalyst TiO_2. For this purpose, several strategies have been followed to prepare photocatalytically active membranes, such as an in situ TiO_2 synthesis process to develop composite membranes with PVDF polymers using titanium tetra-isopropoxide (TTIP) and the post-blending of nanomaterials in polymers before casting them into MF membranes. Likewise, photoactive TiO_2-induced composite MF membranes were used to remove the degraded methylene blue, diclofenac and ibuprofen in MF processes. Further, the high concentration pharmaceutical EPs, like carbamazepine, diclofenac, atenolol, azithromycin erythromycin and pesticides (~200 ng/L), were removed from wastewater in a treatment plant to the extent of ~98% using an MF-assisted RO process. These approaches demonstrate the performance in wastewater treatment plants on the medium to large scale on-site effluents of an urban wastewater facility. In addition to this, the multi-step approach can effectively remove micro-EPs and with a careful pore tuning approach even nanopollutants such as pesticides can also be eliminated.

The major advantage of using an MF process is that it can be widely used for the removal of domestic wastewater contaminants of varying physical, chemical and physico-chemical properties. Domestic wastewater is considered a major source of EPs originating from pharmaceutical and personal care products (PPCPs), pathogens, hormones, microplastics, oil contaminated wastewater and human discards. All these EPs result in inducing wastewater of a severity making it highly complex to treat. It is evident from several studies and reports that a new class of EPs originated from PPCPs and which have been identified in conventional wastewater treatment plants. Conventionally, domestic and industrial wastewaters are characterized by a high concentration of EPs including sewage, surface water, sludge, sediments, soil, wild animals and a series of PPCPs like surfactants. These complex wastewaters are generally treated in series of primary, secondary and advanced methods to achieve a desirable discharge quality. Among them are screening, trickling, skimming, parallel plate separators, comminutors, and aerobic and anaerobic treatment. However, these conventional methods are time consuming and lengthy and need to follow a series of monitoring steps to achieve good quality recovery. However, advanced treatment processes like chemical adsorption, advanced oxidation and neutralization use harsh and toxic chemicals to separate EPs of domestic origin. These processes also generate a large number of secondary pollutants, necessitating a further treatment step. In many cases, secondary pollutants pose complex problems including detection and characterization. Therefore, direct removal methods of EPs are highly desired. In this direction, composite MF membranes having single-walled and multi-walled carbon nanotubes (SWCNT and MWCNT) onto PVDF membrane surface have shown great potential. The presence of carbaneous fillers in PVDF significantly enhanced the removal rate of PPCPs, thus demonstrating the potential of MF technology for water polluted with surfactants, PPCPs, pathogens and colloidal contaminants. This approach also helped in removing APIs like triclosan (TCS), acetaminophen (AAP) and ibuprofen (IBU) in varying concentrations. It is also noticed that in some cases modulation in wastewater pH could lead to the effective separation of EPs against the membrane matrix. Therefore, researchers have also altered the pH of wastewater as a pre-treatment step between 4 and 10 to achieve a PPCP and API removal rate up to 70%. Interestingly, in a carbaneous nanofiller incorporated membrane, API and PPCP removal was enhanced to a near neutral pH, which may be due to the reduced electrostatic repulsion. In another prediction, the removal of charged EPs like PPCPs and APIs may be due to complementary API/CNT and PPCP/CNT interactions.

4.4.2 ULTRAFILTRATION (UF)

Unlike the microfiltration process which has large pore size ranges, UF has a solute rejection regime of above 2 kDa molecular weight, which can cover the removal system of a large number of micro-Eps, making it a technologically wide range process. Considering small pore sizes which can accommodate low MWCO limits stretching to near and below 2 kDa, many of the EPs in the category of the macro-molecular range can be separated from wastewater using a UF process. However, UF may not be sufficiently helpful to produce high quality potable drinking water from wastewater sources, as the EPs below 2 kDa still can pass through pores. Using UF, the removal of macro-EP low molecular weight organics, heavy metals, and a large number of APIs, pesticides and ionic species from inorganic categories cannot be achieved effectively. This regime may not achieve ultra-pure quality water for reuse purposes from industrial wastewater for the same reason. In such situations, the UF of a low MWCO or tight UF regime can be adopted to achieve a desirable rejection pattern for micro-EPs from wastewater. Recent studies name these tight UF membranes as loose nanofiltration membranes (LNF) which can accommodate the rejection of pollutants with 200 Da. However, in many cases a high MWCO UF is used in combination with commercial NF membranes for the removal of a series of EPs from wastewater streams. This approach is particularly popular for the recovery of primary and secondary EP prior and post-secondary treatment processes. Such studies included the removal of a series of EPs present in the domestic wastewater treatment plants where secondary effluents containing various pharmaceuticals and pesticides are handled effectively in hybrid UF–NF processes. In these hybrid processes, EP separation is achieved mainly by adsorption (electrostatic interaction) of EPs on the membrane surfaces, since adsorption is the principle mechanism for micro-EP retention by UF membranes, while size exclusion and electrostatic repulsion at varying pH values are dominant for NF membranes. Even though a UF–NF combination is a cost intensive approach, studies reveal that both hybrid processes are highly effective in treating secondary EPs from a series of wastewater streams.

Another set of studies recommended that the use of UF as a pre-treatment step greatly improves the efficiencies of the RO process in treating EPs from various sources. The RO process requires a high operating pressure to separate EPs from effluents which make them vulnerable to bio-fouling and subsequently leading to concentration polarization. To avoid severe fouling several physico-chemical approaches are conventionally used with an RO process such as combined coagulation-disk filtration/UF/RO (CC-RO) and activated carbon/UF/RO (AC-RO). Using this route, large-scale water reclamation was achieved comprising a combination of CC-UF membranes to remove EPs such as atenolol, carbamazepine, caffeine and sulfamethoxazole. However, this method was time consuming and tedious and ineffective at removing EPs. However, a combination of CC-RO achieved a desirable water quality in treated wastewaters. Interestingly the negatively charged EPs such as APIs and some PPCPs were efficiently retained with the use of low MWCO UF membranes. On the other hand, activated carbon in both powder and granular form as a pre-treatment as well as a post-UF adsorption step for the retention of EPs was relatively effective. In this direction, powder-activated carbon-UF or granular-activated carbon-UF was highly effective in retaining low molecular weight EPs, which otherwise would be difficult to remove using a UF process alone. This approach also substantially reduced the fouling of UF membrane surfaces. Use of activated carbon was used as a combined treatment technique via adsorption and/or coagulation in a step-wise configuration to retain EPs like acetaminophen, metoprolol, caffeine, antipyrine, sulfamethoxazole, flumequine, ketorolac, atrazine, isoproturon, 2-hydroxyphenyl and diclofenac. This process is also considered cost effective as the activated carbon was in a low dose sufficient to remove the EPs from wastewater.

A surface charge on membranes and a charge on solutes plays an important role in the rejection pattern of micro-EPs. Knowing this important factor, researchers have altered the surface charge of UF membranes using several surfactant modifiers for the cost-effective removal of charged EPs from wastewaters. In this pursuit, different micelles, such as sodium dodecyl sulphate (SDS), Triton X-100 (TX-100), Tween 20 (TW-20), cetylpyridinium chloride (CPC) and cetyltrimethylammonium bromide (CTAB) have been grafted on to the surface of a UF membrane to improve the separation of APIs like acetaminophen, metoprolol, caffeine, antipyrine, sulfamethoxazole, flumequine, ketorolac, atrazine, isoproturon, 2-hydroxybiphenyl and diclofenac. It was noticed that cationic micelles like CPC and CTAB modified UF membranes effectively remove negatively charged and hydrophilic APIs from wastewaters.

Nevertheless, the combination of a UF–NF system has been successfully applied to remove APIs like amoxicillin, naproxen, metoprolol and phenacetin from secondary effluents. In both UF and NF, a permeate flow significantly depended on the membrane's nature and morphology, the applied pressure and the feed operating temperature. Interestingly, the retention coefficients for different APIs in UF membranes varied and were higher for naproxen than metoprolol with the lowest being phenacetin. Further, the retention properties in commercially successful tight PVDF UV membranes showed a trend for the highest for amoxicillin and the lowest for phenacetin which was due to the electrostatic repulsion associated with the membranes. However, as expected, surface fouling in the UF membrane was pronounced which drastically reduces the flux decline while treating the secondary effluent.

A relative assessment of the existing combination of treatment techniques with a UF process reveals the limitations of the standalone process. The hybrid UF-RO process significantly contributes to removing a series of APIs to the extent of over 95%, making them an effective combination for handling the EPs of PPCP and pharmaceutical origin. However, as discussed earlier, the use of a carbon filter step significantly adds to the performance of a UF process in retaining conventional contaminants. Therefore, the use of UF as pre-treatment to RO was found to be useful, particularly in drinking water treatment plants for the treatment of an effluent containing oxy-chlorinated compounds and their by-products and intermediates. Interestingly, similar to an advanced oxidation process, electrochemical oxidation (ECO) processes were efficiently used for retaining versatile compounds from wastewater treatment plants. However, the efficiency of electrochemical oxidation greatly depends upon the conductivity of the wastewater to achieve an efficient and cost-effective treatment. The relatively high conductivity or ionic species contaminated wastewater can be effectively treated using an ECO process. However, the presence of high molecular weight organic compounds hinders the ECO process by interrupting its ionic mobility which necessitates the use of UF for the removal of such substances. However, ECO approaches have the advantage of coupling electricity-driven reactions with the in situ generation of oxidants, which makes this method an attractive treatment alternative. So, ECO in combination with membrane processes like UF was able to eliminate the interrupting organic EPs for further ionic species elimination. Similarly, an ECO method was used to successfully eliminate APIs in a point-of-use hybrid UF-RO from wastewater sources. With this approach, EPs like atenolol, bezafibrate, caffeine and diclofenac using UF obtained a moderated retention of up to 20%; however, this approach helped in enhancing the effectiveness of the ECO process.

More interestingly, the physico-chemical and mechanical properties of UF membranes are found to play an important role in fouling control and the efficient washing of EPs from the membrane surface. The impact of surface stress on membrane UF fouling was studied under controlled conditions while removing series of important organic EPs such as APIs and endocrine disrupting compounds (EDCs). The rejection patterns in the UF process shows EP retention on the surface of membranes was dependent on a specific water matrix in which there was improved retention with higher concentrations of organic matter. A simulated mixture of macromolecule solutions was used to artificially create a secondary pollutant layer on the membrane surface to assess the effect of fouling on EP removal. A meticulously designed study showed that macromolecular layer/scale formation did not act as the secondary selective barrier for retaining the hydrophobic micro-EPs from the mixture. However, under higher shear stress conditions, a lower fouling retention was improved for water tainted with higher concentrations of organic matter and biopolymers. Also, the interaction between hydrophobic, neutral compound, organic micro-PEPs and biopolymers in a solution was responsible for the enhanced retention of EPs in a series of wastewater streams.

However, UF membranes frequently fail to achieve the required product standards, and hence, various hybrid technologies have been adopted to design efficient treatment technology for EP-contaminated wastewater treatment. The design of a UF-advanced oxidation process (AOP) was one such alternative hybrid system used for the removal of EPs 1-H-benzotriazole, chlorophene and nortriptyline, dissolved in different streams. In this hybrid process, the UF process was efficient in removing major portions of EPs and subsequent steps disinfected traces of permeated molecules, making the product water reusable. In this hybrid set-up, conventional chlorination and ozonation contributed greatly to reduce EPs from the concentrated (retentate) stream of UF rejects or retentates, making this combination a potentially viable option for setting up as a low to moderate scale wastewater treatment plant (WWTPs). The performance of the UF process was also improved using an ultrasonication assisted AOP system in the form of a hybrid UF-AOP. This ultrasonication assisted AOP process was particularly effective in removing priority organic EPs from wastewater and solid or sludge wastes. Conventional ultrasonication water treatment systems are known to be effective for treating both biodegradable and non-biodegradable stubborn organic compounds. However, this method has found limited large scale application in WWTPs. However, in recent times, ultrasonication in combination with a UF process is gaining importance in treating highly water soluble but toxic organic substances like 4-chlorophenol (4-CP) from wastewater. This method leads to the Sono–Fenton interaction with the pollutant which degrades 4-CP in a short period of time which was then removed using a UF process, making it one of the economically viable techniques for organic EP treatment. Interestingly, ultrasound also helps in drastically reducing and controlling membrane fouling, which reduces the regular cleaning steps. Several hybrid technologies have been adopted involving an ultrasound facility in the UF process to control the membrane fouling during EP-contaminated wastewater treatment. The constant introduction of ultra-low (frequency of 35 kHz) doses of the ultrasonic waves were sufficient to continuously clean UF membrane surfaces during the removal of active pharmaceutical EPs such as diclofenac, carbamazepine and amoxicillin from wastewater. Standard comparison studies also validated the usefulness of ultrasound, which showed the standalone UF process was ineffective for API removal purposes.

Active pharmaceutical ingredients or substances are vital to restoring health, preventing diseases, treating infections and maintaining energetic productivity in humans and animals. Biomedical and pharmaceutical science and technology have advanced to the extent where APIs are designed with the utmost efficacy. This ensures the complete utilization of active agents until they perform their intended action in humans or animals, resulting in their partial metabolism in the human body. However, in reality, a high concentration of pharmaceutical molecules, partially metabolized, conjugate forms of APIs enter WWTPs. On the other hand, the use of APIs has dramatically increased in the last few decades which is causing widespread occurrences of these excess APIs in the environment, which are now listed as EPs of future concern. In recent times, several advanced treatment techniques like AOPs have raised the removal efficiencies of APIs, PPCPs and other organic contaminants to a great extent. However, these chemical degradation methods pose a serious disadvantage as they produce a lot of by-products and degraded residues which are considered as secondary EPs. These secondary EPs are of great concern as the toxicity of the effluent and the producing of toxic transformation products have not been listed or monitored regularly. It is evident from several studies that the standalone ultrafiltration process is ineffective for EP removal due to its relatively high molecular weight cut off and the inevitable disadvantage of fouling by EPs on the membrane surface when used for wastewater treatment. This necessitates the use of auxiliary remedies to achieve effective EP removal. Therefore, UF has been effectively used for EP removal in combination with activated carbon processes, AOPs and ultrasound irradiation, which considerably improves EP treatment from different wastewater sources. Also, the development of a low-cost, effective and tight UF or nanofiltration membrane is gaining importance. A nanofiltration segment is considered to be an effective means to treat large segments of organic and charged EPs from wastewater streams.

4.5 NF AND RO FOR TREATMENT OF EPS

Among several exiting membrane processes, nanofiltration with a membrane pore size ranging from 1 nm to 10 nm in which the rejection of contaminants, or solutes or pollutants greater than 2 nm, can be easily achieved. Most of the toxic EPs listed today by the WHO have been considered to fall under the NF rejection regime, which includes most organics, biomacromolecules, APIs, PPCPs, SVOCs, heavy metals and a variety of metallic salts beyond divalent salts. Though the performance of NF falls between the RO and UF process regimes, it is considered remarkably effective in treating charged, neutral as well as low molecular weight EPs. The RO membranes are non-porous, prepared from dense polymers and voids, have free volume channels originated from the self-assembling of polymers and are considered as pores which fall in the size range of ~0.1–1.0 nm. Due to its practically non-permeable nature, an RO membrane rejects the majority of contaminants including ultra-low molecular weight organics, minerals and metal ions. The most common applications of RO are in seawater and surface water desalination to produce potable water. Polymer-based RO membranes are frequently used to produce high water flux, high solute rejection, an engineering design, stability under applied pressure and stress, chemical resistance and operating temperature. However, in recent times, ceramic or zeolite-based nanocomposite polymer membranes have emerged as high-performance RO and NF membranes that are successfully commercialized in wastewater treatment. Also, these RO and NF membranes are successfully used in tubular, flat sheet and spiral wound configurations in large scale purified water production from both domestic and industrial wastewaters.

Currently, RO technology has extended its footprint to household applications to supply high quality drinking water to a large population from surface treatment. RO and NF technologies have seen many innovations for desalination application to address several issues pertaining to high energy requirements, membrane fouling and concentration polarization. Usually, the product recovery and solute separation in conventional RO processes greatly depends on the mass transportation mechanism. In RO seawater desalination, the product recovery is restricted to ~30% at an operating pressure of around 65 bars. However, when NF is used as a pre-treatment to an RO process it significantly improves the product recovery rate to >35%, making it more viable. This is due to the separation of larger solute molecules which are eliminated at the NF pre-treatment state which significantly improves the RO mass transportation efficiency. This step of the hybrid process adoption drastically reduces the applied pressure dependency below 50 bars, making it economically viable compared to standalone. Besides desalination, the NF process is also used widely for the removal of organics and arsenic, the molecular separation of organics from salt, organic solvent filtration and the recovery of high value organometallic catalysts from the reaction mixtures used for catalytic applications that need to be recovered and reused.

Nevertheless, the NF process is considered as a futuristic option for a series of EP treatment. NF has the potential to separate EPs of an ionic nature from long chain molecules like polymers. In this direction, NF is gaining increased importance in inorganic salts from organic pollutants in rich wastewaters. For this purpose, NF uses pore characteristics and their charges to separate one type of contaminant from another via electrostatic repulsion. On the other hand, wastewater contaminated by pharmaceutical and personal care product residues can also be effectively treated using NF and the surface charge and tunable pore characteristics as a means to reject EPs. Some of the EPs already found in pharmaceutical ingredients, drugs and antiviral molecules (namely N-acetyl-d-neuraminic acid, clindamycin, sodium cefuroxime, cephalosporin-C, cephalexin etc.) are effectively removed from pharmaceutical as well as domestic wastewater using advanced NF processes. In another important segment, the NF process is effective in removing dye molecules and semi-volatile organic compounds found in textiles and distillery wastewater, respectively. Usually, the charge is assessed using the streaming potential technique.

The high-performance membranes are evaluated using established theoretical models which are available, according to which the surface charge and pore characteristics play prominent roles. The presently used NF membranes are typically negatively charged or neutral. On the other hand, the ionic transport and selectivity of NF membranes largely depend upon three effects: (a) charge repulsion, (b) steric/hydrodynamic effects and (c) dielectric effects as shown in Figure 4.3. The charge repulsion effect is caused by the charged nature of the membrane and electrolytes, while the steric effect is caused by the relative size of the ions to the membrane pores. On the other hand, the dielectric effect is instigated by the differences in the dielectric constant between the bulk and membrane pores. However, several other studies also predict that additional phenomena like ion-membrane affinity, specific adsorption and hydration may also contribute greatly to NF membrane separation performance. The dielectric exclusion effect occurs due to the variations in polarization charges, which results in the altered dielectric constant between the bulk and the nanocavity

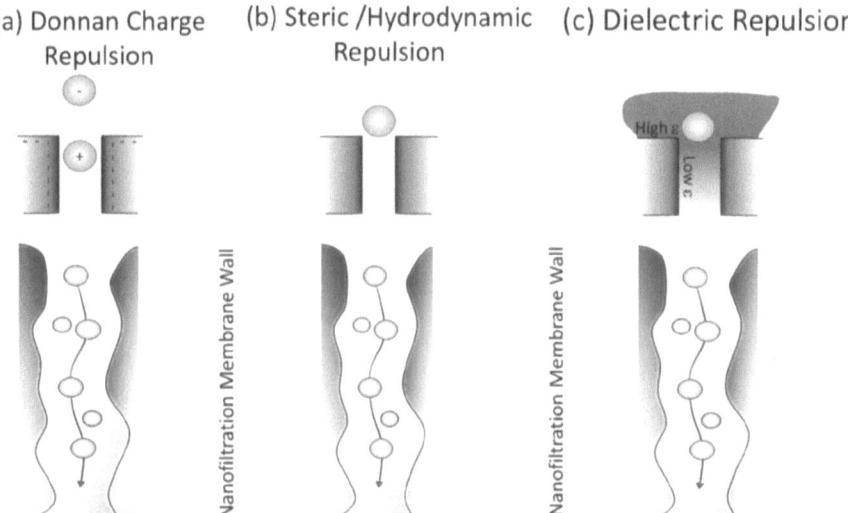

FIGURE 4.3 Schematics of NF membrane solute rejection mechanism predicted using the Donnan–Steric Pore Model with the Dielectric Exclusion (DSPM-DE) model.

(nanopore volume). This difference in the dielectric constant existing between the aqueous solution in the pores and the membrane material is assumed to be prominent in determining the rejection mechanism in NF membranes [7].

On the other hand, the transport properties of NF membranes are predicted using an improved transport model, which includes the dielectric exclusion in the form of both the Born dielectric effect and image force contribution. This prediction clearly shows that dielectric exclusion cannot be ignored in the analysis of filtration performance and the properties of the NF membranes. Thus, charge density on the surface NF membranes is generally measured using zeta measurement tools in a streaming potential pool of electrolytes. Therefore, the average charge of the selective skin layer that determines the selectivity towards ions of different charges can be quite low, consequently an average charge is used to further usage, which behaves like a dielectric medium. During membrane characterization, experimental techniques like nuclear magnetic resonance (NMR) and Fourier transform infrared spectra (FTIR) are used for measuring the functional groups on a membrane surface which shows weak signals, and which may occur due to the presence of residual and unreacted monomeric species during interfacial polymerization. The charge density on an NF membrane surface has direct correlation from ionic radii. The theoretical prediction using the Stokes–Einstein, Born and Pauling equations illustrate a profound effect in predicting charge density and the influence of ionic radii at the surface of the NF membrane. The Donnan steric pore model (DSPM) has been well established for theoretically investigating the effect of surface charge and radii on the rejection pattern of various solutes like sodium chloride, sodium sulphate, magnesium chloride and magnesium sulphate. According to the DSPM, differences in predictions are observed when the NF membrane is positively or negatively charged [8, 9].

Based on these theoretical models, the EP rejection pattern of different NF membranes can be listed as follows: (a) non-ionic hydrophilic EPs like paracetamol, caffeine and methylparaben in NF membranes takes place as a result of physical sieving; (b) the rejection of hydrophobic non-ionic Eps such as carbamazepine and estrone takes place by the influence of the initial adsorption of the molecules on the membrane, subsequently as the NF membrane surface gets saturated with the solute concentration, the rejection decreases; and (c) positively charged EPs like propranolol, metroprolol and negatively charged ones like ibuprofen, naproxen and diclofenac are rejected through electrostatic interactions between the molecule and the membrane surface, and with a sieving mechanism.

4.5.1 REVERSE OSMOSIS (RO)

RO technology has emerged as a leader among many modern technologies in producing instant potable water from various contaminated sources. RO technology in recent times has transformed into an innovative, cost effective and most reliable means for desalination application. Today RO technology reach has extended from industries to household application due to its compactness, improved flux at low operating pressure and reduced fouling propensity. Conventionally, RO has been widely used for industrial scale process water recovery and wastewater treatment. However, recent interest in EP removal from wastewater pushed water treatment industries to adopt advanced RO and NF technologies for large scale applications. Initial effort was only focused on studying the feasibility of RO in eliminating certain drugs and metabolites from secondary treated wastes in a pilot plant. Conventional commercial membranes were ineffective in removing EPs from wastewater. Recognizing this drawback, further innovations in RO membrane development focused on improving retention efficiencies for series of EPs such as PPCPs and EDCs. Novel membranes were developed to retain series of EPs via simultaneous adsorption and size exclusion in the RO and NF membranes. In some cases, the altered pH in feed solutions leads to the improvement in charged EP rejection through electrostatic repulsion which is due to the solution pH being higher than solute pK_a values.

Initially, simple approaches like activated carbon filter assisted RO hybrid systems were effectively used to remove micro-EPs from contaminated waters. The activated carbon filter–RO hybrid process effectively removed highly water soluble, hydrophilic and low molecular weight EPs such as nitrosodimethyl amine, dioxane and 2-methylisoborneol. Interestingly, the removal of low molecular weight organic EPs through activated carbon filters was robust; on the other hand, the hybrid NF/RO process was effective in removing antibiotics as high as over 98%. This led to innovations in tight NF or low MWCO NF membranes, which showed a promising future in replacing high applied pressure RO membranes.

Bisphenol-A (BPA) is another possible toxic EP that has been widely removed by RO membrane technology. The low operating pressure RO has been effective in removing BPA from contaminated water at the critical range of pressure of 4 to 5 bars, making it one of the economical alternatives. Interestingly, the pH of the feed plays an important role in the removal efficiencies of BPA, where the BPA rejection rate was lowest when the pH ranged between 8 and 10; however, when BPA was ionized the interaction between the ions results in a higher rejection rate. Disadvantages in BPA removal using the RO process were encountered when the higher concentration in the feed showed significant surface accumulation leading to concentration polarization and BPA permeation in permeate product water. Similarly, cyclophosphamide (CP), an anti-cancer drug frequently used in chemotherapy, was also found in increasing high concentrations in biomedical waste. The presence of CP in water causes adverse effects to living organisms at higher concentrations. Several membrane-based separation processes like NF, RO and MBR have been tested with limited success (less than 40%) for their CP removal efficiencies in feed water. However, the RO process was found to be effective in CP, retaining up to 90%.

Similarly, the removal of micro-PEPs like APIs and EDCs have been successfully removed using the RO process from wastewater recycling plants. In addition to this, high BOD and COD in combination with micro-EPs like salicylic acid (~50 µg/L) and BPA (~25 µg/L) pose serious threats to the environment. Therefore, the MF pretreatment assisted RO process was effectively used to remove such APIs and EDCs from contaminated wastewater. Here, MF acted as a primary treatment step where the concentration of all EPs drastically reduced and a subsequent secondary treatment step eliminated these EPs to the extent of being >97%. This led to the production of high-quality recycled water production with a permissible level of EPs (<0.1 µg/L) in permeates. In another approach to improve membrane performance, AOP was combined with the RO process to remove APIs from organic matter and inorganic species contaminated wastewater. In this hybrid process, RO was used as a first step to concentrate APIs; subsequently concentrated retentates were subjected to AOP where degradation was initiated by a hydroxyl ($\cdot$OH) radical. In this study, to track the degraded APIs, an in situ degradation process was monitored using excitation-emission matrix spectroscopy. Similarly, a new class of EPs known as β-blockers such as metoprolol and propranolol, which are classified as potentially toxic to aquatic organisms, were removed from wastewater using an RO–AOP hybrid system. Here, EPs were degraded using an ozonation process for the mitigation of β-blockers. Again, in this approach, the concentrated retentates of β-blockers were produced using an RO process and were further degraded using an ozonation step. Further, advanced chemical methods like chlorination was also effectively used in combination with an RO process in removing EPs like β-blockers. In this comparative study, both chlorinated and non-chlorinated WWTP discharges showed improved ozone stability, giving a decrease in $\cdot$OH radical exposure, proving the effectiveness of RO for the removal of β-blockers.

Most of the pressure-driven membrane processes fail to address highly contaminated sewage or municipal wastewater. Modern day municipal wastewater is characterized by EDCs like N-nitrosodimethylamine (NDMA) and antibiotics (ciprofloxacin). Also, high organic content in sewage wastewater induces higher COD and BOD making it tedious to handle in membrane systems. However, with previous understanding, researchers have adopted advanced oxidation processes to reduce organics from contaminated streams. Further, such AOP treated sewage water was efficiently clarified using the RO process. The adoption of AOP–RO was effectively designed to take out >99% of EDCs in a pilot-scale experiment. Remarkably, an H_2O_2 dose in the AOP process played a crucial role in the degradation of EDCs such as NDMA. Inspired by these studies, several researchers and industries developed suitable commercial scale spiral wound RO membrane modules to remove (>90%) antibiotics like ciprofloxacin and EDCs like NDMA from wastewater.

4.6 ELECTRODIALYSIS FOR REMOVAL OF CHARGED EPS

Electrodialysis (ED) is an ion-exchange membrane-assisted effective process widely used to remove undesirable ions from wastewater streams. The ED process uses a pair of ion-exchange membranes for a separation process in which electrical potential is used as the driving force for the separation. ED is conventionally used to treat relatively low concentration brackish water desalination, juice concentration, electroplate industry process water recovery and precious metal recovery for value addition applications. Ionic transportation in ED takes place based on the transport of the dissolved ionic salts through a stack of cationic and anionic exchange membranes using an optimized applied electric potential. The applied potential drives the ionic species out of the concentrated chamber diluting feed stream. As the potential difference across ion-exchange membranes causes migration, it can be used for the concentration of charged EPs in aqueous solutions. Ionic EP separation using ED follows a simple process which requires minimum energy intake. ED semi-permeable membranes are specially designed to be cation- or anion-selective, which essentially means that either positive ions or negative ions will flow through [10].

The ED membranes which are cation-selective are also known as polyelectrolytes with a negatively charged substance, which rejects negatively charged ions and permits positively charged ions to flow through. Similarly, anion-selectivity with a positively charged polymeric matrix rejects positively charged ions and allows only negatively charged ions to flow across. ED membrane stacks are prepared by placing multiple membranes in a row, which alternately allow positively or negatively charged ions to flow through; the ions can be removed from the wastewater. In an ED stack, cation- and anion-exchange membranes are positioned alternately between the cathode and anode which are placed at the extreme ends. When a potential difference is applied between both electrodes, cations move toward the cathode and anions toward the anode. There are several advantages in using ED for desalination purposes which attract special attention to the process. In a simple operation, the cations under the influence of an applied potential migrate through cation-exchange membranes (CEMs), which have negative stationary groups, and are rejected by anion-exchange membranes (AEMs) and vice versa. This drive produces a rise in the concentration of ions in adjacent compartments across membranes, known as the concentrate, and a decrease in the adjoining compartment, known as the diluate. However, some of the external influencing factors such as temperature may not have any impact on the rate of separation of salts from the mixed media and therefore ED can be operated continuously without interruption. Conventionally, ED has been used for the production of high-quality table salt, the concentration of organic acids and heavy metal extraction from polluted soil samples, sugar demineralization, blood treatment and wine stabilization.

Among the many advanced water treatment techniques, membrane-assisted separation techniques like MF, UF, NF and RO have been most widely used for treating a variety of industrial effluents. Many of these processes use applied pressure as the driving force to separate contaminants from wastewater streams. Even though pressure driven membrane processes are known to offer a high level of purification, these processes are cost intensive as relatively high energy is consumed for the separation purpose. Another major disadvantage is limited product recovery which necessitates adopting a different version of the process. In addition to this, efforts are being made across the spectrum of separation and purification technologies to ensure that maximum water is recovered from the waste streams, targeting zero liquid and zero solid waste disposal into the environment. Strategies for obtaining maximum feed concentration also serve the purpose of extracting value-added products from wastewater sources.

Now, the scope of the ED process is expanded to handle emerging pollutants from various wastewater streams. For this purpose, ED has been adopted as both a standalone and hybrid step process in combination with other membrane processes. In one such application ED and MF processes have been designed to treat paper industry wastewater. In this hybrid process, an MF step was designed to remove large constituents like residual pulp, suspended particles and biological contaminants. On the other hand, an ED step was applied to extract ionic species used for pulp digestion purposes. After initial success in handling paper industry wastewater, an ED stack was designed to treat bio-chemistry constituents and charged active ingredients from pharmaceutical industries. Unlike other processes, ED was effective in recovering a large number of industrially important chemicals used as process precursors or other valuable products as well as a process effectively retaining toxic components. Some of the industrial effluents, such as from the leather industry, metal finishing, pulp and paper processing, have a complex composition with a number of pollutants and/or valuable and recoverable components. These include heavy metal ions, acids, skin proteins, organic matter, precious metals and reusable electrolytic components. Likewise, treated discharges from municipal or domestic farming sources contain recoverable and reusable nutrients and a high quality of water. In conventional pressure driven membrane processes, desalination plants reject high quantities of brines, which is also considered as secondary waste, and during this process large quantities of natural resources are rejected with elevated contamination into the environment without adequate use. On the other hand, ED can serve as a highly efficient wastewater treatment system due to its ability to separate charged particles along with the recovery of several products and possible electrical energy generation.

Even though the performance of different types of membranes like polymeric, ceramic, blend and composites in the range of MF, UF, NF and RO have been relatively successful in purifying pulp and paper, and sugar industry and pharmaceutical wastewater, charged pollutants have always been of concern due to their fouling susceptibility. In the absence of suitable standalone processes, hybrid processes have also been tested with varying levels of success. On the other hand, electrodialysis is effective in treating low concentration pollutants and recovering value added minerals in a single unit operation.

In spite of the fact that ED has been in use for over a century, intense research in recent times is reinventing the effectiveness of the ED process for EP removal applications. Most industrial wastewater contains different compositions; however, it often contains dissolved ions. For such ionic species, ED has proven to be effective at varying concentrations. Therefore, ED treatments for industrial effluent of an EP nature can be classified into (a) heavy metal (HM) ion removal, (b) recovery and/or regeneration of the acid/base, (c) salt conversion, (d) desalination, (e) protein recovery and (f) charged organic or API removal.

The separation of HM ions, which are a dangerous class of EPs, is characterized by toxicity, carcinogenicity, non-biodegradability and persistence in the environment and in living beings. Among the different advanced treatment processes used for HM contamination, electrodialytic approaches have been tested and shown to be promising for industrial effluents in several processes like metal finishing and the leather industry. Interestingly, unlike other processes, ED can concentrate HM brines to facilitate the recovery and recycling of ions easily. Also, ED can recover water and metal ions from spent baths or the rinse waters of plating processes. However, ED also depends on several factors for efficient mass transportation and resource recovery, such as IEM properties, complex formations, pH effect and ion competition. Therefore, IEMs are designed to keep several influential process factors in consideration. The ED process is particularly advantageous in concentrating the EPs of HM such as Ni, Cu, Zn, Cr, Fe and Pb from aqueous wastewater mixtures [11].

On the other hand, real industrial wastewaters frequently contain mixtures of metal ions in various concentrations or in impurities. ED processes can be an effective method to recover water from low to moderately polluted wastewater. Therefore, a suitable combination of CEM and AEM can easily concentrate ions and separate different charged organic species from each other. In this way, the concentration of Cu and Zn was achieved from cyanide-free electroplating baths with EDTA by adjusting the pH of the feed. Remarkably, ED also facilitated HM and EDTA simultaneously in the over-limiting regime with 0.0006 M $CuSO_4$, 0.0014 M $ZnSO_4$, 0.0015 M EDTA and 0.03 M NaOH. The Cu, Zn and EDTA were concentrated (~3.45 against ~2.94 fold with an under-limiting operation) likely due to (a) water dissociation generating protons which reacted with complexes and insoluble species, and (b) electro-convective mass transfer enhancement. A major advantage with the ED process is both CEM and AEM possess anti-fouling and anti-scaling properties. These combined properties make ED a high performing process for real electroplating effluents containing Ni^{2+} and Cu^{2+} even at concentrates >25 mg/L. ED was also used as part of the tertiary treatment step to reclaim electroplating industrial wastewater with a mixture of heavy metal ions at low concentration (~1 mg/L). In this approach, MF and UF steps were designed to remove organics and suspended solids, then ED was used for concentrating metal ions. The ED step was particularly effective in removing >97% of Cr^{3+}, Cu^{2+} and Zn^{2+}; on the other hand, counter-ions like SO_2^{-4} and Cl^- were concentrated at over 95 and 85%, respectively, from an initial concentration of 1000 mg/L. ED was also effective in concentrating feed brine which helps in the chemical precipitation of a real Cr(VI), Cu^{2+}, Zn^{2+} and Cd^{2+} from electroplating wastewater with minor concentrations of feed mixtures. The effects of feed pH on concentration was used to alter the toxicity from Cr(VI) to Cr^{3+} making it less toxic which was then chemically precipitated along with Na_2S and $FeCl_2$, using an NaOH dosage (pH = 9) [12].

Many industrial manufacturing units produce large quantities of acidic/alkaline wastewater during product development. These industrial wastewaters are extremely corrosive and their neutralization is essential before disposal into the environment. Conventionally, chemical neutralization is followed to treat alkaline/acidic effluents which elevate their salinity. Therefore, a physical method of extraction gives economic and environmental benefits for contaminant recycling and reuse. In this perspective, ED processes can be useful for treating diverse industrial wastewaters such as acidic wastewaters with HM ions.

Industrial and domestic wastewaters are generally characterized by high organic matter, including organic acids. Industrial wastewater from tanneries, sugar manufacturing and paper and distillery processing units also possess a high content of organic matter along with high salt or ionic compounds. In recent times, ED has been successfully used for acid/base recovery and organic matter separation, in spite of such solutions being challenging and carrying potential EPs that cause organic fouling. However, recent innovation has adopted an ultrasound assisted mechanism to avoid scaling or fouling on CEM at the base side which was initiated by Ca^{2+} and Mg^{2+}, along with minor fouling on the other surface. Also, glyphosate recovery and HCl/NaOH production were obtained by ED of alkaline glyphosate with a feed concentration of 12.8 g/L, and NaCl with ~175 g/L by neutralizing a liquor from the pesticide industry. Using ED, the glyphosate recovery achieved was as high as >98%, and with the help of an NaOH solution with nearly 1.45 M with a recovered compound purity was >96%. This makes ED an economical and sustainable option to operate. In another attempt, industrially important and low to medium concentrations of monomers like aniline (1000–3000 ppm) and salt (0.1 M NaCl) leached into wastewater are recovered using the ED process. Initially, wastewater with aniline was converted to a base form which was then completely transported to the base chamber at η up to 80% and an E_{spec} of ~3 kW h/kg. Following this, NaCl desalination was achieved at over 94% at η = 90% and $E_{spec} \approx 1$ kW h/kg. These possibilities pave the way to using conventional ED for recovering industrially important substances like aniline and H_2SO_4 during wastewater treatment which adds great value for treatment and recovery processes from fine chemical industrial effluents.

Electrodialysis process-assisted desalination can produce high quality reusable water from treated municipal wastewater which can be utilized for many applications. Furthermore, ED and ED-assisted hybrid methods are economical for recovering nutrients, organic acids, HMs, volatile fatty acids (VFAs) as well as water from treated wastewater and similar related domestic and agricultural effluents. Conventionally, wastewater from industrial and municipal WWTPs, secondary or even tertiary effluents are typically not reusable in irrigation, aquifer recharging or industrial processes. However, ED-based hybrid methods have been proven economically beneficial for desalinating drainage wastewaters for agricultural reuse [13].

4.7 FORWARD OSMOSIS FOR CONCENTRATION AND REMOVAL OF EPS

RO technology has seen many innovations of desalination application to address several issues pertaining to high energy requirements, membrane fouling and concentration polarization. Further, rapid industrialization has led to the production of enormous quantities of severely polluted effluents including bio-wastes and series of EPs causing severe environmental threats. In many cases, discharged toxic untreated effluent waste of various origins and types, along with EPs, which enter into clean water bodies is a serious cause of concern. Large amounts of fresh water are consumed in industrial manufacturing processes such as tanning or leather production, textile dying processes, sugar manufacturing, mining, automobiles and electroplating, of which >90% of used water is discharged into the environment as wastes carrying unused and unreacted raw materials and by-products in them. However, the absence of appropriate wastewater management technologies has led to a deteriorating of the existing problems pertaining to the discharge of highly polluted effluents and the production of reusable water [14].

This has driven the scientific community to develop osmotically driven FO in which the osmotic gradient across the semi-membrane plays an important role in mass transport and EP separation. Compared to all other pressure driven and potentially driven membrane processes, FO has shown most promise due to its simple operational mechanism. Therefore, FO is more suitable and energy efficient for treating EP contaminated feed with a high organic and bio-fouling tendency which it may not be reasonable to do with RO. However, standalone FO has some niche applications, such as sugar cane juice concentration, biomacromolecule enrichment, fertilizer dilution and fruit juice concentration. During its initial phases, FO has been regarded as a pre-treatment step to reduce contamination levels in feed solutions that are subsequently fed into either RO, NF or membrane distillation processes. However, in the absence of suitable draw solute/solutions, FO proved complex and there was a need to design or adopt a suitable diluted draw solution recovery step.

According to a recent study, approximately 80% of global industrial and municipal effluent is discharged without appropriate treatment. Although the discharge percentage may vary from industry to industry along with the level of contamination, it is estimated that >90% goes into the environment untreated/partially treated. To minimize the amount of waste water flowing into the environment, various physico-chemical and biological treatment technologies have been designed and adopted to treat extremely contaminated wastewater before being discharged into the environment. Very recent studies have characterized these wastewaters for moderate to high concentrations of series of EPs. In this direction, a series of advanced physico-chemical treatment processes and membrane-based technologies have been tried and tested with varying levels of success. These membrane-based separation processes are considered to be the most sustainable among other water treatment technologies. Further, the evaluation of standalone RO, NF, UF, MF and ED have also led to the understanding of their process-related limitations. However, forward osmosis has received considerable attention as it offers outstanding potential in water reclamation and reuse prospects from severely polluted wastewater streams. A major advantage of FO is its potential at concentrating feed contaminants near 90%, subsequently reclaiming 90% of reusable water from complex EP contaminated wastewater. This also provides an important opportunity in value added product recovery from concentrated feed. However, a series of FO studies and innovation points towards the need for dedicated FO membrane manufacturing, though process development and operational optimization abilities demonstrate the large-scale potential of FO and FO-based technologies for EP treatment considering several sustainability criteria [15].

In this direction, forward osmosis has shown several advantages over more popular membrane-based separation processes for wastewater treatment with a series of EPs as well as resource recovery. The method utilizes engineered osmotic gradients artificially generated by the use of a high concentration draw solute/solution across the membrane. In contrast to other membrane processes, FO membranes are less susceptible to organic and bio-fouling. Therefore, FO can be utilized for a variety of industrial and domestic pollutants including red category wastewater such as tannery/leather and distillery industry discharges. FO operates at a zero hydrostatic or applied pressure but provides sustainable reusable water recovery in the presence of a high osmotic pressure draw solute/solution. Until recent times, the FO method had not achieved commercial success except for seawater desalination, perhaps because of a lack of suitable FO membranes. However, recent progress in FO membranes and compatible draw solutions has led to the wider applications of FO in wastewater treatment, the recovery of valuable metals and the food industry to concentrate bio-macromolecules, remove toxic metals, and so on. Therefore, modern problems like emerging pollutants require careful consideration of both membrane and draw solute/solution to devise effective FO processes for EP removal and recovery. Nonetheless, the essential constituents required for capable FO processes include: (a) membranes that are less inclined to internal concentration, (b) efficient draw solutes or solutions and (c) a mechanism for putting effective draw solution recovery in place.

Quite interestingly, natural water channels called aquaporins formed by the self-assembly of a larger family of major intrinsic proteins that result in ordered pores in the membrane of biological cells have been successfully used as FO membranes. Aquaporins are the naturally occurring protein channels used as semi-permeable water pathways in FO separation processes. Aquaporins infused in a selective skin layer (<200 nm) of thin-film composite (TFC) membranes on an ultrafiltration porous polysulfone support was meticulously designed to act as FO semipermeable channels. These FO membranes made of aquaporins have been tested for >90% rejection of urea with a water flux of 10 L/m^2 h against a draw solution of 2M NaCl concentration. Further, several studies have been conducted utilizing FO for the separation of organic EPs from wastewater sources.

In this direction, initially hybrid FO processes were designed to separate EPs from wastewater for EP removal. For this purpose, a hybrid FO–RO system both at the laboratory and pilot scales was designed to remove more than 30 EPs of different kinds, including non-ionic, hydrophobic and negatively and positively charged species from wastewater. A hybrid FO–RO proved to be effective in rejecting non-ionic compounds. The FO–RO hybrid process also showed the interesting trend of increased removal efficiency for EPs with increasing molecular weight. Interestingly, the RO showed a superior rejection rate for BPA than FO; however, FO has shown an enhanced rejection rate >99% for oxybenzone, methylparaben, amitriptyline and triclosan pesticides compared to RO. This important observation led to the widespread use of FO for the separation of different EPs from various industrial and wastewater resources. An FO-based separation process is particularly effective in removing non-ionic EPs of both a hydrophilic and hydrophobic nature. Predominantly, over 20 hydrophilic and hydrophobic micro-EPs from secondary wastewater effluent (SWWE) were effectively removed along with desalination using a hybrid FO–RO process. Nevertheless, in standalone FO mode, retention of neutral and non-ionic hydrophilic EPs varied between 50 and 85%, whereas for hydrophobic neutrals EPs were retained in the range of 40 to 88%. Interestingly, the retention of ionic EPs ranged between 90 and 97%. These varying retention properties of FO membranes suggest a typical interactive tendency of FO membranes with EPs.

Further studies utilized hybrid technologies with various cellulose acetate (CA)-based commercial FO membranes in pressure retarded osmosis (PRO) mode in combination with an RO process. On the other hand, the NF process is also known to be effective in treating contaminates such as the commercial cellulose triacetate (CTA) membrane with similar MWCO. However, CA membranes showed a higher water flux with better EP retention in PRO mode. In earlier tested cases the RO mode is effective in removing large numbers of electro-active EPs from wastewater through an electrostatic interaction. However, in FO and PRO modes, the retention of active EPs was governed by the electrostatic interactions between the membrane matrix and the solute, while the rejection of neutral compounds was controlled by the size exclusion in which retention was higher for PEPs with a high MW. In all the cases, retention of neutral EPs was higher in FO compared to RO processes.

Among several surfactants, disinfected substances and personal care products, haloacetic acids (HAAs) are the well-recognized disinfection by-products (DBPs) present with the highest concentrations which are found in chlorinated or chloraminated sewage treatment plant effluents. Another disinfection by-product is trichloroacetic acid (TCAA) whose concentration can be as high as ~500 µg/L in chlorinated wastewater effluents. These EPs were extensively characterized using ultra-performance liquid chromatography-electrospray ionization in tandem with mass spectrometry (UPLC-MS/MS). Further, an FO process was utilized to remove HAAs from feed wastewater with the help of a high-performance draw solute. Interestingly, the retention ratio for each HAA improved with an increased draw solute concentration for the active layer facing the feed solution (ALFS) orientation. The retention of all the HAAs were more than 95% for an ALFS orientation, but retention reduced to below 90% when FO is operated with an active layer facing the draw solution (AL-DS) using a 1 M NaCl draw solution.

On the other hand, the processing of industrial effluents containing organic EPs such as phenols, aniline and nitrobenzene is a rather challenging task as these easily overcome treatment barriers. Many of the existing technologies prove to be ineffective against organic EPs. However, some activated carbon assisted ones have shown limited success in retaining organic EPs from wastewater. A membrane-based separation process also has limited success at removing organic micro-EPs. Nonetheless, RO was proven to be ideal for the removal of organic micro-PEPs. However, at a high operating pressure RO is costly and the membranes used were susceptible to organic fouling, limiting their large-scale implementation. But, among all these efforts FO is proven to be effective in retaining most of the organic micro-EPs through both lab-scale FO membranes under both FO and RO modes and commercially available RO membranes to remove organic PEPs. The lab-scale FO membrane separation set-up with a thin-film polyamide layer on a porous UF membrane support provided EP rejections in the range of 72–90%. However, FO in pressure retarded osmosis (PRO) mode was relatively less efficient. Still, the effects of pH and membrane orientation on the permeate flux was predominant in rejecting EP-like carbamazepine and sulfamethoxazole. Interestingly, the retention of neutral carbamazepine was pH independent of both the FO and PRO modes, but the percentage rejection was relatively lower in PRO mode than in FO. This is due to a possible steric barrier which was governing the separation patterns for neutral carbamazepine in FO.

Therefore, to further improve the process efficiency in retaining series of EPs from wastewater, direct contact membrane distillation (DCMD) and FO hybrid systems. The hybrid DCMD-FO showed improved retention efficiency for the removal of EPs like estrone and 17β-estradiol from contaminated wastewater. In the first step, DCMD showed more than 99% for the hormone, near 99.9% for urea and 99% for ammonia retention with a constant flux. On the other hand, FO removed estrone and estradiol alike, but hormone rejection was dependent on the initial feed concentrations. The wastewater produced from the processing of olive production and other activities in olive mills are characterized by a high concentration of COD (220 g/L). In addition to this, the presence of EPs like antibacterial phenolic compounds in olive oil mill wastewaters make them complex to handle during treatment processes. In such cases, FO has been proven to be an effective means for phenolic compounds from olive mill wastewater before being discharged into the environment. Similarly, FO was effectively used to treat phenolic EP-contaminated olive oil wastewater using 3.7 M $MgCl_2$ as a draw solution which was effective in concentrating organic EPs including biophenols by 98%. In another approach, an MBR-based pre-treatment was adopted to feed clarified wastewater to FO which helped in improving product flux by 30% with simultaneous EP concentration up to 92%. This facilitated the recovery of reusable water from wastewater feed by 95% with the help of cellulose triacetate (CTA) membranes.

Pharmaceutical and personal care products come in a variety of chemical functionalities. Handling PPCPs using a conventional treatment technique is ineffective in many cases. However, an FO–coagulation/flocculation (CF) hybrid process devised to treat textile dye wastewater showed encouraging results when adopting PPCP treatment as well. FO was used as a primary step for the spontaneous recovery of water from wastewater via FO-CF, which exhibited exceptional advantages in attaining a high-water flux and a recovery rate for API treatment with an initial water flux of 36.0 L/m^2 h and when used for textile wastewater treatment, dye retention was more than 99.9% against 2 M NaCl as the draw solution. The challenging task in treating the trace amounts of APIs from the wastewater comes from the handling of the FO concentrate. In such a situation, the API concentrate from the FO operation was subjected to electrochemical oxidation (ECO) to degrade these pharmaceutical EPs. In this approach, a synergistic effect of FO-ECO was observed and adopted for further treatment of series of EPs such as micro-organics, APIs and PPCPs. In particular, a hybrid method was highly effective in treating EPs like antibiotics in the primary step of FO (98%) and the residual concentration of antibiotics was further degraded and removed using an ECO (99%) step [16].

Therefore, FO in which the osmotic gradient across the membrane necessitates the need to develop an efficient draw solute or solution to enhance mass transport and EP separation. However, so far this has only been of limited possibility, as FO is more suitable and energy efficient for treating EP-contaminated feed compared to a commercially successful technology like RO. Yet, the standalone FO too has some issues to be addressed like a constraint for a dedicated FO membrane. On the other hand, FO offers niche applications, such as fertilizer dilution and fruit juice concentration and can use waste natural resources as a draw solute or solutions. This makes FO advantageous over pressure driven processes in handling EP-contaminated wastewater. Thus, FO has proven to be an efficient alternative to produce or reclaim high quality reusable water from EP-contaminated wastewater. On the other hand, standalone FO and FO-based hybrid technologies have been effective in retaining or removing a series of EPs from various wastewater streams.

4.8 MD AND HYBRID-MD PROCESS FOR REMOVAL OF EPS

Freshwater scarcity and the excessive discharge of contaminated wastewater to the environment is regarded as a serious challenge over past decades. Increased water consumption has left serious and dangerous pollutants in aquatic systems. Therefore, several membrane-based separations have been tried and tested to address contaminants in waste sources. Based on the type of industry, a variety of emerging pollutants such as a series of organic compounds, and inorganic substances such as arsenic, fluoride, cadmium, chromium, mercury, manganese and lead, have been treated with varying levels of competence. However, several pressure-driven separation processes have shown limited success in addressing emerging pollutants. To address this issue, several simple and effective treatment technologies like membrane distillation have been adopted by several researchers. These alternative and effective methods have proven successful over reverse osmosis, coagulation-flocculation, disinfection, granular filtration, gravity separation, air stripping, aeration ion exchange, adsorption and membrane filtration in addressing emerging pollutants [17].

Among several tested separation techniques, the membrane process has become highly popular due to its potential benefits associated with the technology such as quick separation and ease of operations. Pressure driven processes are particularly advantageous in permeating compounds of a non-toxic nature to pass through the membrane based on a driving force such as a concentration gradient, pressure gradient, temperature gradient or electrical gradient. This phenomenon occurs by membrane module characterization such as pore size, pore shape and membrane surface characteristics such as porosity, charge/hydrophobicity and membrane configuration (e.g. geometry), and dimensions, which define the permeation characteristics of a particular membrane and module configuration. In other words, membrane separation processes used in wastewater treatment are considered to be isothermal and non-isothermal processes. The first category includes concentration-driven membrane processes such as pervaporation and membrane extraction, pressure-driven membrane processes MF, UF, NF and RO, and electrically driven membrane processes including ED and electrophoresis. The latter category uses a thermally driven membrane process known as membrane distillation (MD). Interestingly, MD has been developed in various forms and configurations to treat wastewater with an extremely high level of salinity, hazardous contaminants and emerging pollutants in various wastewater sources.

In this situation, extensive research has led to the greater exploration of MD technology for various wastewater treatment and resource recoveries. A detailed overview of the current and prospective role of membrane engineering reveals the importance of achieving the objectives of a process which increases strategic approaches to improving the efficiency and sustainability of novel membrane processes, including MD. This section discusses the evolution of MD as an emerging pollutant removal method for various water and wastewater. MD has shown several advantages over pressure driven processes in utilizing energy input in which MD adopts low-grade or renewable heat sources such as waste heat from industrial processes to make it an economically viable alternative [18].

Further, MD consumes renewable sources and works with low trans-membrane heat loss and increases the proportion of heat recovered from the permeate stream. Simultaneously, the MD process can be operated using low-grade heat and/or waste including solar energy, geothermal energy, tidal, wind and nuclear energy, or low-temperature industrial streams. It should be pointed out that the MD process is also driven by the vapour pressure difference between the permeable hydrophobic membrane pores. Therefore, volatile vapour molecules are permitted to pass through the membrane while non-volatile compounds are retained on the retentate stream. The saturated and permeated volatile vapours are then collected or condensed by various procedures. Lastly, a completely pure product of water that is theoretically 100% free from solid, harmful substances and non-volatile contaminants and concentrated

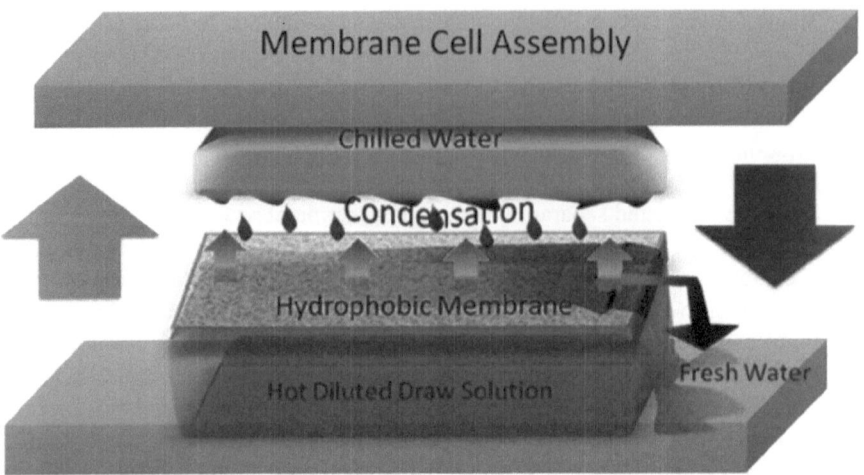

FIGURE 4.4 Schematic working and membrane arrangement in a membrane distillation process.

pollutants are separated. Figure 4.4 illustrates a schematic diagram of the MD process. In this illustration, it is shown that the volatile vapour molecules in the hot feed which are vaporized at the liquid/vapour interface are able to permeate through the membrane pores. On the other hand, the liquid feed is prevented from transporting through the membrane pores.

On the other hand, a new class of emerging pollutants, namely micro-pollutants, occurs in point sources like wastewater treatment plant effluents, surface water, sewage spillage and non-point sources like domestic wastewater, agricultural runoffs and stormwater. In all cases, a major portion of the micro-pollutants enter wastewater treatment plants after being discharged into surface waters. Even though recent wastewater discharge norms have restricted the direct release of EPs like micropollutant removal to the environment, there are no legal discharge limits for micro-pollutants in the environment. Also, there are no conventional wastewater treatment plants subject to any standard emission guidelines for micro-pollutants. At present small portions of micro-pollutants are subjected to treatment process depending upon their physicochemical properties such as hydrophobicity, sorption affinity, biodegradability and volatility using various conventional wastewater treatment processes such as coagulation, precipitation, chlorination and adsorption. The several limitations associated with the conventional treatment processes have necessitated action to explore alternative technologies for the removal of micro-pollutants. In such a situation, several potential advanced treatment methods have been explored so far for removing emerging micro-pollutants from water via various techniques such as adsorption, pressure-driven membrane filtration processes, advanced chemical techniques like advanced oxidation processes (AOPs) using heterogeneous photocatalysts, electrochemical oxidation, wet peroxide oxidation, catalytic ozonation and Fenton and Fenton-like reactions through sulphate radicals and hydride processes.

Even though advanced treatment processes have shown promising outcome as alternative methods for some micro-pollutant removal, these methods have also exhibited a series of issues bearing high operating costs, low rejection factors, high osmotic pressure at high concentrations in filtration, fouling and the formation of secondary pollutants or toxic by-products which limit their application. Nevertheless, advances in membrane technology and their adoption in various forms have led to alternative and promising membrane configuration such as the membrane distillation process. Thermally driven MD was initially adopted for desalination applications; however, later it was adopted for application where emerging pollutants can easily be treated with utmost efficiency. MD with the help of a hydrophobic membrane separates emerging pollutants from vaporized feed stock driven by applied heat. The vapour pressure gradient of the hot feed and the cold permeate across the non-wettable porous hydrophobic membrane is the driving force, which is a main advantage compared to pressure-driven membrane processes. At present, there are several configurations such as (1) direct contact membrane distillation (DCMD), (2) sweeping gas membrane distillation (SGMD), (3) vacuum membrane distillation (VMD), and (4) air gap membrane distillation (AGMD) all of which are in commercial use in various capacities to address industrial needs. Among these, DCMD has been widely used for pollutant treatment in different sizes and capacities due to its simplicity in membrane module design which does not uses external condensers and provides increased stability as well as high permeate fluxes. The DCMD process is also a non-isothermal process determined by the vapour pressure gradient existing between both porous hydrophobic membrane sides. Here, permeated water is colder than the feed solution when maintained in direct contact with the permeate side of the membrane. In this configuration, both the feed and permeates are distributed tangentially to the membrane surfaces. However, the transmembrane temperature difference brings the vital water vapour pressure difference across the porous hydrophobic membrane. Due to this vapour pressure difference, water vapours diffuse through the hydrophobic membrane pores from the feed side of the membrane and condense at a lower temperature on the permeate side. Using MD, many researchers have achieved complete removal of all non-volatile emerging micro-pollutants which facilitates mass transfer only in the form of vapours. Nevertheless, recently reported studies have shown the greater potential of MD processes in eliminating micro-pollutants from different contaminated waterbodies. Micro-pollutants were recently classified as compounds or elements which are present in water and wastewater sources at very low concentrations from a few ng/L to µg/L concentrations. These micro-pollutants have been mainly characterized as compounds or degraded by-products of disinfectants, pharmaceuticals (APIs), personal care products, steroids and hormones like synthetic compounds, industrial chemicals, pesticides, heavy metals, radioactive elements and endocrine disrupting chemicals [19].

On the other hand, municipal wastewater treatment plants are generally designed to remove various organic, inorganic, nanomaterial, biological and carbonaceous substances and discarded nutrients. Based on the method and capacity of a treatment technique, micro-pollutant removal efficiency varies significantly. Also, based on the contaminant level and the nature of other contaminants and their properties such as volatility, hydrophobicity and biodegradability, treatment methods and operational conditions are defined. In this direction, a membrane bioreactor (MBR) process can effectively remove some micro-pollutants either via biological conversion or through sorption to filter media or an adsorbent. However, these strategies are ineffective at the removal of certain compounds such as diclofenac, alkylphenols, carbamazepine, atrazine or less hydrophobic compounds. Such limitations associated with many technologies including several advanced methods led to the exploration of alternative technologies such as MD. The MD process has already proven its ability in entirely removing all non-volatile compounds from wastewater sources, which include (1) the complete elimination of all non-volatile compounds like inorganic salts and pathogens since in MD mass transfer occurs in the gaseous phase, (2) MD requires comparatively low temperatures (feed <70°C), (3) MD also uses lower operating pressures than conventional processes, (4) MD is indifferent to feed concentration, (5) an MD membrane can work with a relatively broad range of conditions, and (6) the MD process is less susceptible to fouling. These advantages make MD suitable for treating a variety of emerging pollutants.

Emerging pollutants have also been identified in various industrial wastewaters, namely textile, tannery, distillery and pharmaceutical sources. These industries consume large amounts of potable quality water for their various processes. In the textile industry, the dyeing process leads to the most severe water pollution in textile wastewater in which many of the discharged water organic pollutants have been classified as EPs. These dye substances are normally mixed in a dyeing process at different ranges of pH to generate intensive colours with the help of different salinities. It is projected that ~21–377 L of fresh water is consumed for 1 kg of textile product, making it one of the most water intensive processes, among which >90% of water gets contaminated with various levels of dye concentration. These wastewaters are partially treated using techniques like adsorption, coagulation-flocculation and AOP such as ozone or chlorination. However, MD has been effective for removing organic EP molecules like dyes. Dyeing industry or textile industry wastewater is of special concern because the water used will have a high temperature (70–100 °C) condition. Therefore, organic EPs like dye contamination in wastewater is quite favourable for the application of MD. In such conditions, MD can exploit the free energy retained in the wastewater discharged by the textile or dying industry and can be a sustainable alternative as it uses significantly lower energy externally, which makes MD a feasible alternative. Dyes in wastewater are categorized as non-volatile EP compounds, therefore it is necessary to cease completely from using techniques like the MD process.

Generally, dye contaminated wastewater also contains a mixture of dyeing agents like polymers (alginates) and a salt mixture to facilitate a uniform and stable dyeing process. These mixtures along with dyes in textile wastewater necessitates complete removal rather than recovery. Therefore, a highly concentrated dyebath having mixtures of EPs needs to be further treated for zero liquid and zero solid discharge to make them safe for the environment. On the other hand, among all these, DCMD has been effectively applied for the removal of textile wastewater. In most cases, DCMD has been applied effectively to remove different dyes such as acid, reactive, basic and direct dyes in various concentrations from 1 to 1000 ppm. Some researchers have also used VMD for the removal of various organic dyes such as methylene blue from contaminated wastewater using polypropylene (PP) membranes. Even though PP could remove dye contamination with reasonable efficiency, membranes experienced severe fouling and swelling during their use. However, several studies confirmed that MD processes are better than the pressure-driven membrane separation processes. Some of the limitations with MD systems still need to be documented for a better understanding of process feasibilities for large scale operation. This requires detailed studies on factors affecting MD performance such as the pH conditions of EPs, diverse ions, salts, surfactants, module geometry and system configurations.

On the other hand, the presence of EPs like pharmaceutical and personal care products in sewage wastewater and drinking water require immediate attention, even though there exist several conventional and advanced techniques such as filtration, flotation, coagulation, AOPs, Fenton reactions and pressure driven membrane processes with limited success. Among all these, heterogeneous photocatalysis and AOPs in combination with MD have shown reasonable removal efficiency against pharmaceutical EP such as ibuprofen. In this approach, the photodegradation of ibuprofen sodium salt was achieved in a new hybrid system in combination with photolysis and DCMD. The photodegradation of EPs was achieved in a hybrid photoreactor with a germicidal UV lamp (UV254) in a hybrid system. In this hybrid process, the influence of the initial concentration of EP-contaminated feed/wastewater with various model compounds were tested for effective removal. Further, model compounds were replaced with traces of an organic compound (TrOC) in feed in a hybrid MD process.

Hybrid MD has been shown to be highly efficient at removing a total of 29 TrOCs which also includes 11 pharmaceutically active (APIs) substances typically characterized in municipal and/or domestic wastewaters. Except triclosan every EP in this EP mixture was efficiently removed at the MD step. However, MD was effective at the removal of complex and poisonous EPs like pesticides, toxic metals like arsenite (As(III)) and arsenate (As(V)) in DCMD mode using relatively hydrophobic PVDF membranes. The DCMD with a PVDF membrane resulted in a remarkable permeate flux of 20.90 kg/m^2/h with as high as >99% removal efficiency for TrOCs, including pesticides and toxic metals. Further, another set of hybrid forward osmosis (FO)-MD hybrid systems were tried and tested for the removal of metallic EPs like arsenic (As(III)) removal in which the initial As(III) concentrations of 30–3000 µg/L were decreased directly using FO and further with MD below 10 µg/L in the final permeate without the necessity of involving the pre-oxidation of As(III) to As(V). Further, the MD hybrid system was extended to remove (70–99%) a series of pesticide compounds such as dichlorvos, parathion-methyl, clofibric acid, atrazine and phorate from a model groundwater solution using a DCMD system. This hybrid system also enabled water recovery from the model solution up to 75%. Even though this approach resulted in high water recovery and a high percentage of EP removal, the process suffers from several issues which identify optimum boundary conditions for reducing energy consumption, enhancing flux rates, a high rate of rejection (capable of reaching >99%) and reducing fouling proficiency in membranes in all MD configurations, namely DCMD, VMD and AGMD [20].

In another set of studies, EPs like heavy metals and metalloids have been removed using MD-based techniques from contaminated sources. In most MD studies targeting heavy metals, the MD process in different configurations produced a high-quality permeate with the complete removal of heavy metals. The MD process having PTFE and PVDF membranes of different pore sizes and their distribution (0.22 and 0.45 μm) were used for the complete removal of heavy metals. The modified DCMD process was effective in retaining all heavy metals from metal-plating wastewater. In this study, researchers removed (>95%) highly toxic Hg, Cd and Pb from metal-plating industrial wastewater using a hybrid process of FO and MD. The FO-based hybrid process was effective in achieving high rejection with a stable water flux. In this approach, a generally observed disadvantage of lowering osmotic pressure during water transportation from feed to draw solution noticed in the standalone FO process was overcome using the hybrid process of MD.

Using a standalone MD process researchers have effectively removed and recovered endangered metals like chromium (VI) from contaminated wastewater sources. An MD process with a hydrophobic PTFE membrane combined with polyethylene-terephthalate (PET) demonstrated improved MD performance with greater resistance to emerging pollutants in wastewater. Membranes were also noticed to be stable against the high concentration of solutes in feed against the fouling process. Nonetheless, it is noticed that among all MD-based processes, DCMD has been proven suitable for heavy metal removal and recovery from contaminated sources due to its ability to use alternative energy sources.

Radioactive elements are known to be a new entrant to the emerging pollutants list. Radioactive-element-contaminated liquid waste from nuclear power plants and stations is one of the latent sources of thermal energy. Even though the characteristics, concentrations and potential risk to the environment of the radioactive liquid wastes are yet to be realized, broader pollutant characteristics like radioactivity, composition and salt content are presently considered to categorize radioactive waste into different categories of EPs. These fundamental understandings are helping researchers to devise proper treatment techniques. In one such attempt, a spiral-wound MD module having a hydrophobic PTFE membrane was effectively used to remove different elements present in radioactive liquid waste entirely in a single step. For this purpose, MD has been designed to completely remove radioactive waste constituents from liquid waste and which is better than an RO process. The hydrophobic PTFE membranes in MD have shown reasonably good stability during the treatment of the liquid radioactive waste. One of the major advantages to using MD in treating radioactive waste is that the process does not create waste by-products or secondary pollutants unlike other pressure-driven membrane processes such as RO and NF. The presence of emerging isotopes in radioactive wastewater faces a diffusion effect in MD which increases the separation of H_2O/HDO and $16H_2O/18H_2O$ and results in enrichment. Recent studies also report that the MD process also has been adopted to desalinate nuclear waste in a simple one-step system. Depending on the choice of membranes, and their stability and life, applications of MD processes are designed to treat and manage radioactive waste. In general, hydrophobic polymers like PTFE, PVDF and PP have been successfully adopted to act as an MD membrane for a series of radioactive waste treatment. However, a major disadvantage of MD membranes is that they do not sustain highly radioactive, severe concentrations and acidic radioactive liquid wastes.

Therefore, the selection of the membrane and the mode of MD process become very significant in treating the specific characteristics of radioactive liquid waste. Nevertheless, the characteristics of liquid radioactive waste also influence the choice of membrane and MD process. The MD process also suffers from an economic disadvantage in which high energy requirements and membrane wetting properties also play an important role in devising suitable constituents for the MD process. However, thermal energy consumption depends on the type of membrane, membrane module and MD configuration. However, the energy consumption for the MD process depends on various other factors: a case study reveals that the removal of 273 micropollutants in an MD process using spiral wound MD modules needs thermal energy of 600 kW h/m^3. However, nuclear or radioactive waste treatment processes considerably reduce the energy consumption as the MD process utilizes the high heat energy available for use from radioactive liquid waste. However, recent advancements in nanocomposite membranes in hybrid RO-MD have shown greater advantages in converting the conventional MD process into an energy efficiency process for nuclear and radioactive liquid wastewater treatment.

Nonetheless, scale-up activities need to be extensively studied to make MD-based processes commercially feasible for large scale applications. Most of the literature suggests a large number of lab-scale reports waiting to scale-up their efforts. In this direction, several researchers have devoted time to developing large scale MD set-up activities. On the other hand, more research is also needed for a better understanding of the parameters and process factors affecting MD performance, such as choice of membrane, different ions, surfactants or washing solutions, salt, pH values, module geometry and system configuration. Thus, efforts need to be invested in evaluating pilot-scale and large-scale MD-plant performance in long-term tests under real conditions. Besides, the performance of MD-based processes is limited by process-related issues such as the lack of commercially accessible membranes, membrane module design, low feed flow rate, time dependent flux reduction, adaptable plants, pore wetting, high energy and cost requirements. The replacement of existing water treatment plants with MD processes is still an achievable target once the capital and operating costs of the process are optimized for a specific set of emerging pollutants. In addition to this, sustainable MD operations also need to consider suitable membrane cleaning solutions and procedures to prevent chocking and interruptions while operating. These factors will provide MD-based processes with a sustainable and effective alternative to existing cost-intensive water treatment technologies for emerging pollutants in different wastewater sources [21].

4.9   HYBRID-MEMBRANE PROCESSES FOR REMOVAL OF EPS

Currently membrane technology has been recognized as one of the best options available for water and wastewater treatment of industrial importance. Membrane-based separation processes offer simple, single step operations, technologically an easy scale-up, and uses no added chemicals to operate which makes them possibly attractive as a more rational utilization for modern water related issues like emerging pollutants. Commercially, membrane processes are available in various modes and configurations for a wide spectrum of industrial applications. The pressure-driven

MF, UF, NF and RO are the most popular standalone processes and are well accepted unit operations. On the other hand, potential-driven electrodialysis is also another successful standalone process used for several contaminant concentrations and treatments. In all cases, standalone membrane operations have shown their potential in mineral fractionations, molecular separations, nutrition concentrations, clarifications, and so on in the liquid phase. Standalone membrane processes have already represented an efficient means of separation for desalination, water treatment, process water recovery, value addition and so on; however, the design of hybrid systems in recent times is gaining more importance for various reasons. These hybrid processes may be the combination of different membrane processes with other conventional techniques or integrated with different membrane operations. Based on a wide data analysis, these hybrid processes are considered to be the way forward and a more widespread and rational application for the treatment of emerging pollutants. The synergic combination of hybrid processes has achieved several advantages in many effluent and emerging pollutant treatments, where they can be eventually combined with other conventional unit operations for economical and sustainable operations [22, 23].

Interestingly, these hybrid processes have been developed to address the urgent need for efficient technologies via innovation to avoid the already strained good water supply, the eminent water scarcity and address ever-increasing wastewater discharges. Nonetheless, for decades, distillation-based machineries and technologies have dominated the industrial scale water treatment approaches which are now being gradually replaced with membrane processes for economic and environmental reasons. Membrane-based separation processes provide straightforward solutions to produce clean drinking and reusable water free from emerging pollutants.

Though several of these membrane-based technologies, including some hybrid configurations, have emerged as potential water treatment technologies, only a few are successful for large-scale application. The pharmaceutically active compounds are vital to prevent diseases and restore health. However, the excessive use of medication is causing serious issues in human health and also leading to the discharge of excessive drug molecules into the environment through human discards. Technologies need to be designed to address these new classes of emerging pollutants with utmost efficacy. The complete removal of APIs and PPCPs from the environment or aquatic ecosystems will ensure safer life support. Similarly, dyes, colloidal products used in paints, personal products, surfactants, polymer residues, microplastics and nanomaterial discards also necessitate immediate attention for suitable treatment techniques.

Research efforts on exploring MF, UF, NF, RO, FO and hybrid-membrane processes have concentrated on the removal of dyes, colloidal particles, organic matter and other high molecular weight series of emerging pollutants from waste streams. MF has been commonly employed in combination with other methods and techniques for the removal of complex EPs from various domestic and industrial origins. In this direction, domestic wastewater comprises hormones such as 17β-estradiol (E2), estrone (E1) and 17α-ethynylestradiol (EE2) as well as the widely used monomer bisphenol A (BPA) which is a known endocrine disrupter. These compounds present even in trace quantities can cause extreme damage to the human endocrine system. To solve this problem, a crossflow MF system has been used with a series of membranes prepared from nitrocellulose, polyether sulfone (PES), regenerated cellulose, cellulose acetate (CA), polyester and polyamide (PA). Remarkably, PA membranes with a pore size of 0.2 μm has shown high removal efficiency for the treatment of wastewater containing EPs such as 0.2 μM estrogens around a sorption capacity of 81 L m^{-2} (0.44 μg cm^{-2}) for E1, 150 L m^{-2} (0.82 μg cm^{-2}) for E2, 208 Lm^{-2} (1.23 μg cm^{-2}) for EE2 and 69 L m^{-2} (0.32 μg cm^{-2}) for BPA. These adsorption phenomena on membrane surfaces were characterized by the interaction between polymers and the emerging pollutants. The MF membranes have shown surface adsorption of some of these emerging pollutants at higher concentrations and severely affected the membrane performance, causing membrane deposition, which led to membrane scaling and fouling. This is also due to the presence of organic matter (EPs) in the feed wastewater that has significantly affected the reduction of the flux of the PA membrane. In any case, all the PEPs showed a consistent interaction with the PA membrane via H-bonding, thus resulting in their efficient removal from the feed streams.

This study highlighted the importance of PA membranes compared to other polymeric membranes, in particular cellulose-based membranes which were used for the treatment of domestic wastewater with pharmaceutical EPs. However, a standalone MF process was ineffective to complete the removal of micro-EPs which requires the use of activated carbon to separate them from domestic wastewater sources. However, this approach also showed several shortcomings like clogging which resulted in high-pressure drops, slow mass transfer and also required an appropriate method to recover the adsorbed materials. In another study, zeolite imidazolate metal-organic framework (ZIF-8) nanoparticles incorporated into a poly(tetrafluoroethylene) (PTFE) double layer MF membrane in spiral-wound configuration provided shorter bed height and higher pore ranges for a convective flow. In this configuration, the ZIF-8-induced MF membrane showed a high flux with an improved rejection rate of 95% for EPs like hormones at low operating pressures. The high rejection of hormones in this case accounted for the chemical interaction between an EP (a hormone) and high surface area ZIF-8 nanoparticles via H-bonding. Thus, this approach proves to be efficient in removing micro-EPs like hormones from domestic wastewater.

Further, active pharmaceutical ingredients (APIs) such as diclofenac and ibuprofen present in low concentration ranges (0.14–1.48 μg/L and 160–169 μg/L) and can be degraded using a TiO$_2$ photocatalyst to assist the UV-degradation process. However, the photocatalyst requires post-degradation separation to make the wastewater free from secondary pollutants. For this purpose, a hybrid MF can was used to clarify the polluted water with EPs, secondary degraded products and a photocatalyst, which helps in the simultaneous recycling of the photocatalyst TiO$_2$. Through this approach an in situ TiO$_2$ synthesis process of nanocomposite membrane preparation was followed using PVDF as a base polymer and titanium tetra-isopropoxide (TTIP) to TiO$_2$ nanoparticle in a PVDF MF matrix. Thus, photoactive TiO$_2$ in nanocomposite PVDF membranes are further used for the simultaneous degradation and removal of methylene blue, diclofenac and ibuprofen from wastewaters. However, wastewater having high concentrations (162–240 ng/L) of EPs like diclofenac, carbamazepine, erythromycin, atenolol, azithromycin and pesticides were effectively (~98%) removed from contaminated wastewater using hybrid MF-RO processes.

With this hybrid configuration, the 12 most found APIs were removed effectively. However, trace amounts of pesticides were permeated in the RO process with a reduced concentration in nano-gram levels which are under permissible levels.

Further, PPCPs were effectively treated using composite UF membranes induced with single-walled and multi-walled carbon nanotubes (SWCNT and MWCNT) in a PVDF MF membrane matrix and surface. This approach significantly enhanced the removal rate of PPCPs using a composite MF membrane in which the removal of acetaminophen (AAP), triclosan (TCS) and ibuprofen (IBU) were achieved from 10 to 95%. Interestingly, the variation in aromatic rings and the specific surface area of the fillers influenced the removal of EPs. Process parameters like pH variation (from 4 to 10) also significantly altered the PPCP removal rate, up to 70%, from wastewater sources. However, the higher PPCP removal was achieved at a neutral pH due to reduced electrostatic repulsion. The higher removal rate of PVDF-CNT membranes for EPs is also accounted for by favourable PPCP–CNT interactions.

UF, unlike the MF process, follows a lower pore size regime and a large range of a solute rejection spectrum above 2 kDa molecular weight. Therefore, a UF process covers a large range of macro-EPs like PPCPs and pesticides in municipal and domestic secondary effluents. In standalone UF the limited separation of EPs was achieved due to the surface adsorption of contaminants, since adsorption is the main mechanism for micro-EPs in UF membranes, unlike predominant size exclusion and electrostatic repulsion mechanism observed at high pH values in NF membranes. Some of these inferior performances of UF forced researchers to adopt hybrid UF-NF and UF-RO processes to efficiently remove secondary pollutants like degraded micro-EPs. Studies also reveal the potential clogging and chocking of low pore size NF and RO membranes when used as a hybrid process in combination, which necessitated the use of additional coagulation and disk filtration steps to control the fouling. In one such attempt, while removing a series of EPs large-scale water reclamation was achieved using coagulation and disk filtration (CC-DF) along with UF/RO membranes. In this approach, CC-DF/UF-RO process EPs like atenolol, caffeine, carbamazepine and sulfamethoxazole were effectively removed. However, recent studies have found that standalone RO is significantly effective in removing most of these EPs in a single step operation. Interestingly, the negatively charged EPs use small pore size membrane systems like NF and RO efficiently compared to neutral pollutants. One of the major advantages is that membranes can be easily washed to remove UF, NF and RO membranes by desorbing EPs from the surface of NF and RO membranes. The separation of micro-EPs by the hybrid CC-DF/UF method was independent of the molecular weight of Eps; however, the complete removal of micro-EPs depended on the RO process.

In another approach, micellar-enhanced UF was designed as a cost-effective alternative to separating the EPs from various contaminated sources. In this pursuit, different micelles like sodium dodecyl sulphate (SDS), Tween 20 (TW-20), Triton X-100 (TX-100), cetylpyridinium chloride (CPC) and cetyltrimethylammonium bromide (CTAB) improve the ultrafiltration process for separating PPCPs like caffeine, acetaminophen, antipyrine, metoprolol, flumequine, sulfamethoxazole, ketorolac, atrazine, 2-hydroxybiphenyl, diclofenac and isoproturon. In this approach, cationic CPC and CTAB were effective in removing negatively charged and hydrophilic

PPCPs. However, among all the micelles, CPC showed the maximum of ~95% rejection for most of the EPs in water sources. Further, the pre-treated wastewaters were subjected to a UF membrane process for efficient EP removal. In one such approach, activated carbon was devised as a pre-treatment and UF as a post-adsorption step to retain EP-adsorbed activated carbon. Another study also reported the use of granular activated carbon in a step-wise UF treatment plant to remove 11 emerging pollutants like metoprolol, acetaminophen, caffeine, antipyrine, ketorolac, sulfamethoxazole, atrazine, flumequine, 2-hydroxyphenyl, isoproturon and diclofenac from an effluent treatment plant. In this study it was proved that a low dose of activated carbon was sufficient to remove a series of low concentration EPs from wastewater; a UF step was adopted to achieve complete removal of EP-adsorbed activated carbon.

In another approach, a UF-NF hybrid system was used for separating four APIs like naproxen, amoxicillin, metoprolol and phenacetin from the secondary effluents. However, the permeate flux and rejection pattern of EPs in both UF and NF depends on membrane morphology, applied pressure and the operating temperature. Interestingly, the retention coefficients for APIs like naproxen was higher in UF membranes than metoprolol and was recorded as lowest for phenacetin. The highest removal of amoxicillin and the lowest for phenacetin was accounted for by the tight pore size structures and electrostatic repulsion from NF membranes. Also, the NF process showed a high rejection efficiency of >80% for all APIs in wastewater except for phenacetin. In all cases, moderate to high flux decline was observed in both UF and RO membranes due to membrane fouling. However, a combination of UF-RO has been tested successfully and efficiently (>94%) for removing 29 APIs compared to other conventional and advanced treatment processes. Conventional pollutants were removed more efficiently in the first step than in the membrane step but the hybrid process was proven to have only limited success for EPs.

Electrochemical oxidation (ECO) techniques have been widely used for conventional wastewater treatment. The ECO-based techniques have been efficiently applied for treating versatile charged pollutants in wastewater. However, ECO requires relatively high conductivity wastewater characteristics to achieve an efficient and cost-effective treatment. ECO techniques have the advantage of coupling electron-driven processes or reactions which can generate in situ oxidants, which makes these processes an attractive alternative for EP treatment. Overall, the ECO method was efficient in eliminating the EPs from the RO concentrate streams in a hybrid system. Therefore, researchers used an ECO route to remove APIs in wastewater using a hybrid UF-RO system. Several prominent API-EPs like bezafibrate, caffeine, atenolol and diclofenac were removed by up to 20% in a UF step; however, the ECO of an RO concentrate with diamond electrodes was able to drastically reduce the total concentration of EPs from 149 µg/L down to 10 µg/L below the permissible level [24].

Nevertheless, hybrid processes involving membranes as part of the system still suffer from fouling stress, even though protocols have been developed to remove organic EPs (APIs and EDs) using a UF treatment plant with minimum fouling stress on membrane. Interestingly, a controlled study showed that EP scale formation on membranes did not pose a secondary selective barrier to facilitate the deposition of macromolecular contaminants and hydrophobic micro-EPs, though a higher concentration of biopolymers and organic matter did result in inducing resistance to permeate flow in the long run. However, UF often fails to achieve the required standards of quality permeates and, hence, various hybrid technologies and protocols have been designed in combination with AOPs. AOP has been successfully used to treat a series of EPs like chlorophene, 1-H-benzotriazole and nortriptyline found in different wastewater streams. In this hybrid system, the pre-treatment of wastewater carried out using UF and further permeates as well as concentrates involved ozone or chlorine treated in an AOP approach separately to achieve complete EP removal. Even though a UF step was successful at the partial elimination of all the EP traces except 1-H-benzotriazole, an AOP (chlorination and ozonation) step was efficient in removing all EPs below permissible levels in wastewater effluents.

Recently, ultrasound-treated AOPs have been used for removing several organic EPs from wastewater and solid waste. In this direction, hybrid ultrasonication-UF was applied for the removal of 4-chlorophenol (4-CP), which by the use of the homogeneous Sono–Fenton technique degrades 4-CP in less than 1 hour and up to ~45% efficiency. The ultrasound techniques have been also proven to be effective to control membrane fouling in ultrasonic-UF hybrid processes. The ultrasonic aided UF process with varying frequencies cleans the UF membrane surfaces during EP removal. Interestingly, a lower frequency of 35 kHz ultrasonic dose was enough to reduce the fouling in UF membranes. This hybrid process is effective in removing APIs such as carbamazepine, diclofenac and amoxicillin from wastewater treatment plants by up to ~99.5% [25].

Therefore, several hybrid processes involving membrane techniques and conventional and advanced treatment techniques have shown their usefulness in treating both conventional pollutants and emerging micro-pollutants. Though ultrasonication effectively works for both biodegradable and non-biodegradable and refractory organic compounds, more data and experimental designs need to be adopted at various stages and levels to examine the feasibility.

4.10   LIMITATIONS OF MEMBRANE-BASED SEPARATION PROCESSES IN TREATING EPS

Membrane-based separation processes have shown great flexibility in being adopting for water and wastewater treatment applications. They have been successfully used in the separation, concentration and purification of a wide variety of pollutants, process materials and wastewater across a wide range of industries. High pore size range processes like microfiltration and ultrafiltration can operate as highly efficient sieves which are capable of fractionating and separating particle species according to size. One of the major advantages in membrane processes is that there are no phase changes involved, no added chemicals for the separation process and both feed and product streams remain in the liquid form before and after treatment. A membrane separation process does not require added or applied temperature to separate the contaminants, which results in low energy consumption. In addition to this, all membrane processes are relatively simple to scale up [25].

However, major limitations of membrane processes come from their membrane and operation related issues. All processes generally use polymer-based membranes as the separation medium and are prone to membrane fouling effects which lead to a decrease in the permeate flux and a lower rejection of emerging pollutants. The disadvantages of membrane-based separation processes also come from the operational and maintenance aspects. Constant membrane cleaning and washing is an expensive process in the long run to keep membrane processes sustainable. In the absence of suitable membrane cleaning protocols, a drastic flux decline is noticed. This is due to the fact that the high operating pressure severely damages the pore structures which block and exert resistance to the flow of the permeate. Also, high flow rates used or applied in a cross-flow feed operation can damage shear sensitive materials, resulting in membrane surface damage. Most importantly, membrane modules and operational equipment demand high capital costs and high operational expenses. One more major disadvantage in the preparation of membranes is that the manufacturing of the membrane process is not precisely controlled; at present the commercially followed, phase inversion membrane preparation technique results in wide pore size distribution which is the main drawback. The wide range of pore distribution in membranes may result in poor separation performance.

On the other side, high concentrations of emerging pollutant contaminated wastewater also may pose a serious threat to the performance of membrane modules. The turbid and highly concentrated particulate wastewater cannot be used in membrane filtration. Another major issue associated with membranes is organic and bio-fouling which may lead to bacterial abundance on polymeric membranes and numerous microorganisms. On the other hand, MD processes have shown various advantages over pressure-driven conventional membrane separation systems. Among the several advantages, MD works with low operating temperatures, cost-effective operations, and can use waste heat and renewable energy sources with high permeate purity and minimum membrane fouling. Therefore, these remarkable features make MD an attractive technique for emerging pollutant treatment of wastewater.

However, commissioning MD in different plant sizes and configurations for industrial purposes is limited by some important challenges like membrane wetting in which the risk of total or partial wetting is a serious technological drawback found in most commercial membranes. Also, there is a scarcity of suitable membranes commercially available on a large scale and the limited choices in the market make the MD process technologically risky to adopt. Therefore, before considering MD as an alternative option for emerging pollutant treatment in various feed stocks or wastewater samples, the selecting of appropriate membranes and identification of sustainable energy sources to operate continuously are the two main factors that must be taken into account. Nonetheless, hydrophobicity in MD membranes is the fundamental requirement which takes care of the mass transportation across the barrier compartment. Therefore, MD membranes must be prepared with a synthetic polymer of a hydrophobic nature and low surface energies. With such an approach a few important factors can be modulated on the membrane surface to have a low resistance to mass transmission and low-thermal conductivity to prevent heat loss across the membrane.

In addition, the MD membrane should have high permeability, good thermal stability to tolerate high temperatures and good chemical resistance. To satisfy these features, the membrane selective surface layer is required to be as thin as possible so that the vapours are permeable through the membrane in quick time. Interestingly another main characteristic, the liquid entry pressure (LEP) in the MD membrane, can be enhanced by applying a polymer with high hydrophobicity and a high number of small pores. Nevertheless, the MD process has been extensively used in water and treatment at the lab scale to remove emerging pollutants from the wastewater produced from numerous industries, although advanced research in the membrane-based separation processes have shown greater potential due to the significant availability of funding sources and larger efforts from research community. These cumulative efforts are yielding efficient membranes, economical membrane modules and adaptable plant designs for various emerging pollutant treatments in the near future. Therefore, a membrane-based separation process will show a sharp technological edge for industrial applications.

REFERENCES

1. Jerker Fick, Hanna So¨Derstro¨, Richard H. Lindberg, Chau Phan, Mats Tysklin, and D.G. Joakim Larsson, Pharmaceuticals and Personal Care Products in the Environment: contamination of surface, ground, and drinking water from pharmaceutical production, *Environmental Toxicology and Chemistry*, 2009, 28(12), 2522–2527.
2. J. Rodriguez-Chueca, M. P. Ormad, R. Mosteo, J. Sarasa, and J. L. Ovelleiro, Conventional and advanced oxidation processes used in disinfection of treated urban wastewater, *Water Environment Research,* 2015, 87, 281–288.
3. Lavanya Madhura, Suvardhan Kanchi, Myalowenkosi I. Sabela, Shalini Singh, Krishna Bisetty, and Inamuddin, Membrane technology for water purification, *Environmental Chemistry Letters*, 2018, 16, 343–365.
4. Fan Fei, Hai Anh Le Phuong, Christopher F. Blanford, and Gyorgy Szekely, Tailoring the performance of organic solvent nanofiltration membranes with biophenol coatings, *ACS Applied Polymer Materials*, 2019, 1(3), 452–460.
5. Elorm Obotey Ezugbe, and Sudesh Rathilal, Membrane technologies in wastewater treatment: a review, *Membranes*, 2020, 10, 89.
6. Angelo Basile, Alfredo Cassano, and K. Navin, *Advances in Membrane Technologies for Water Treatment, Materials, Processes and Applications*, ISBN: 978-1-78242-121-4, 2015, Elsevier Ltd.
7. Omar Labban, Chang Liu, Tzyy Haur Chong, and John H. Lienhard, Relating transport modeling to nanofiltration membrane fabrication: navigating the permeability-selectivity trade-off in desalination pretreatment, *Journal of Membrane Science*, 2018, 554, 26–38.
8. A. A. Hussain, S. K. Nataraj, M. E. E. Abashar, I. S. Al-Mutaz, and T. M. Aminabhavi, Prediction of physical properties of nanofiltration membranes using experiment and theoretical models, *Journal of Membrane Science*, 2008, 310(1–2), 321–336.
9. Nataraj S. Kotrappanavar, A. A. Hussainb, M. E. E. Abashar, Ibrahim S. Al-Mutaz, and Tejraj M. Aminabhavi, Prediction of physical properties of nanofiltration membranes for neutral and charged solutes, *Desalination*, 2011, 280(1–3), 174–182.
10. R. S. Keri, K. M. Hosamani, H. R. S. Reddy, S. K. Nataraj, and T. M. Aminabhavi, Application of the electrodialytic pilot plant for fluoride removal, *Journal of Water Chemistry and Technology*, 2011, 33(5), 293–300.
11. Tetra Tech, Inc. Fairfax, Virginia, Emerging Technologies for Wastewater Treatment and In-Plant Wet Weather Management Prepared for: Office of Wastewater Management U.S. Environmental Protection Agency Washington, D.C. EPA 832-R-12-011.

12. Kevin L. Gering, and John F. Scamehorn, Use of electrodialysis to remove heavy metals from water, *Separation Science and Technology*, 1988, 23(14–15), 2006.

13. Luigi Gurreri, Alessandro Tamburini, Andrea Cipollina, and Giorgio Micale, Electrodialysis applications in wastewater treatment for environmental protection and resources recovery: a systematic review on progress and perspectives, *Membranes*, 2020, 10, 146.

14. Bryan D. Coday, Bethany G. M. Yaffe, Pei Xu, and Tzahi Y. Cath, Rejection of trace organic compounds by forward osmosis membranes: a literature review, *Environmental Science & Technology*, 2014, 48(7), 3612–3624.

15. Jiale Xu, Thien N. Tran, Haiqing Lin, and Ning Dai, Removal of disinfection byproducts in forward osmosis for wastewater recycling, *Journal of Membrane Science*, 2018, 564, 352–360.

16. Jianlong Wang, and Xiaojing Liu, Forward osmosis technology for water treatment: recent advances and future perspectives, *Journal of Cleaner Production*, 2021, 280, (20), 124354.

17. S. F. Ahmed, M. Mofijur, Samiha Nuzhat, Anika Tasnim Chowdhury, Nazifa Rafa, AbrarInayat, T. M. I. Mahlia, Hwai Chyuan Ong, Wen Yi Chia, Pau Loke Show, Recent developments in physical, biological, chemical, and hybrid treatment techniques for removing emerging contaminants from wastewater, *Journal of Hazardous Materials*, 2021, 416, 125912.

18. Enrico Drioli, Membrane Distillation, First Edition, 2017, ISBN 978-3-03842-460-4 (Pbk) ISBN 978-3-03842-461-1 (PDF), MDPI AG St. Alban-Anlage 66 Basel, Switzerland.

19. B. B. Ashoor, S. Mansour, A. Giwa, V. Dufour, and S. W. Hasan, Principles and applications of direct contact membrane distillation (DCMD): a comprehensive review, *Desalination*, 2016, 398, 222–246.

20. D. Woldemariam, A. Kullab, U. Fortkamp, J. Magner, H. Royen, and A. Martin, Membrane distillation pilot plant trials with pharmaceutical residues and energy demand analysis, *Chemical Engineering Journal*, 2016, 306, 471–483.

21. A. Alkhudhiri, N. Darwish, and N. Hilal, Membrane distillation: a comprehensive review, *Desalination*, 2012, 287, 2–18.

22. Nurafiqah Rosman, W.N.W. Salleh, A. F. Ismail, and Z. Harun, Hybrid membrane filtration-advanced oxidation processes for removal of pharmaceutical residue, *Journal of Colloid and Interface Science*, 2018, 532, 236–260.

23. C. Y. Wu, S. S. Chen, D. Z. Zhang, and J. Kobayashi, Hg removal and the effects of coexisting metals in forward osmosis and membrane distillation, *Water Science and Technology*, 2017, 75(11), 2622–2630.

24. Daiana Seibert, Camila F. Zorzo, Fernando H. Borba, Renata M. de Souza, Heloise B. Quesada, Rosângela Bergamasco, Aline T. Baptista, and Jonas J. Inticher, Occurrence, statutory guideline values and removal of contaminants of emerging concern by Electrochemical Advanced Oxidation Processes: a review, *Science of The Total Environment*, 2020, 748, 141527.

25. Naresh N. Mahamuni, and Yusuf G. Adewuyi, Advanced oxidation processes (AOPs) involving ultrasound for waste water treatment: a review with emphasis on cost estimation, *Ultrasonics Sonochemistry*, 2010, 17(6), 990–1003.

5 Polymer/Resin-Based Techniques for the Removal of Emerging Pollutants

5.1 INTRODUCTION

An ion-exchange polymer (IEP) or ion-exchange resin (IER) is a polymer or resin that is a crosslinked macromolecule that acts as an ion-exchange intermediate or medium for ionic species in water. Ion-exchange resins or polymers have lately been attracting great attention due to their potential at treating conventional as well as emerging pollutants (EPs). Ion-exchange materials provide a unique property of reversible and interchangeable ions between a solid and a liquid such as contaminated water and wastewaters. Ion-exchange resins have been conventionally used to remove harmful contaminants or hardness from liquids or water in general, replacing ion-exchangeable groups in resins with beneficial, preferred or specific ions. In the past, several polymer-based ion-exchange resins have been developed with ion-exchangeable crosslinking agents. A literature survey suggests that most viable ion-exchange resins were made of polystyrene-based macromolecules. These conventional ion-exchange resins were initially used for the separation and purification of common ionic water pollutants such as calcium, magnesium and potassium. This is generally termed as water softening and ground and surface hard water purification. Ion-exchange resins were first discovered by Robert Gans, a German Chemist in the year 1905 using synthetic aluminosilicate materials, today well known as zeolites. These zeolitic ion-exchange materials were first used for water softening applications [1–3].

The aluminosilicate-based synthetic zeolite ion-exchange material was shortly replaced by a naturally resourced material like greensand. However, greensand soon proved to be less effective in exchanging ionic contaminants compared to synthetic materials. Nevertheless, greensand was much more stable during its application in water purification compared to zeolitic materials at an industrial scale. The ion-exchange capacity (IEC) is the parameter generally looked at in a material to select it for the use of calculating the performance of an ion-exchange material. IEC is defined as the number of exchangeable ions a unit quantity of resin will remove from a solution or contaminated water [4, 5].

DOI: 10.1201/9781003214786-5

Sulfonated coal was prepared as a cation-exchange intermediate commercially known as carbonaceous-zeolite, used for hydrogen cycle operations in the reduction of alkalinity and hardness in contaminated sources. Immediately after this, a series of innovations discovered anion-exchange resin as a result of condensation reaction production between polyamines and formaldehyde. These novel anion-exchange resins were applied as a hydrogen cycle cation resin in an effort to demineralize and remove all dissolved salts from wastewater or contaminated water sources. Yet, these anion-exchange resins were unstable and could not remove weakly ionized acids in the form of silicic and carbonic acid from water sources.

By the mid-1940s, polymer-based ion-exchange resins were developed following the copolymerization technique in which styrene long chains were crosslinked with divinylbenzene (DVB). Polymer-based resins were largely proven to be stable and having higher exchange capacities than inorganic resins. The polystyrene-divinyl-benzene-based (PS-DVB) anion-exchanger was efficient in removing anions, including silicic and carbonic acids from conventional contaminated sources. This innovative approach was able to achieve the comprehensive demineralization of water for all possible ionic species. Even though several advanced ion-exchange resins entered the market, PS-DVB resins still govern the large commercial space in the majority of ion-exchange applications. For very long periods, the basic resin constituents remained the same, but the resins have been improved in many forms to meet the requirements of specific needs and applications. In the meantime, resin constituents and preparation parameters were optimized to induce a longer life span in the resulting resin. These macroreticular or macroporous morphologies have been induced in polymers during crosslinking and structure tuning to make them stronger and higher performing [6].

Standard macroporous resins provide a free flow of permeates when packed in bed form which also acts like a permeable membrane medium. This macroporous structure enables resin to obtain the required chemical and physical parameters to suit a particular application. Conventional resin attains a gel-like physical nature when the polymer is crosslinked with a suitable agent having an exchangeable active site. However, to treat severely contaminated water or wastewater, resin needs to possess mechanical and chemical resistance to a range of harsh conditions. Conventional polymer resins may undergo structural deviation and mechanical deformation under harsh conditions. Therefore, stronger mechanical and chemical properties in resins can be induced by tuning the physical and chemical properties required in the resin structure to sustain high exchange capacities and mechanical strength.

The PS-DVB-based macroreticular resins feature distinct pores within a highly crosslinked polymer network matrix. The highly crosslinked network structure possesses a rigid and higher physical strength than loosely crosslinked gels. Higher crosslinking also provides greater resistance against thermal degradation and oxidizing agents. Most importantly macroreticular anion resins also possess a much needed anti-fouling property against organic and biological contaminants due to their porous structure.

In recent times, the scope has been expanded to accommodate a series of polymeric systems to convert them into ion-exchange media such as acrylic polymers in addition to conventionally popular PS-DVB systems. In addition to ion-exchange capacities, acrylic polymers are also known for their anti-fouling properties when used for organic contaminated water treatment applications. The main feature of stable backbone ionizable functional groups/moieties plays an important role in the ion-exchange property and capacity. These functional moieties or groups comprise both negatively charged anion elements and positively charged cation elements in which one of the ionic species is mobile or exchangeable. For example, the ionic group in sulfonic-PS-DVB has ionic sites consisting of immobile anionic (SO_3^-) radicals and mobile calcium (Ca^{2+}) and sodium cations (Na^+) as shown in Figure 5.1. When this PS-DVB is added to raw water, ions diffuse into the resin bed and facilitates exchange for the mobile portion of the functional group; ions are displaced from the resin and diffuse back into the water solution when the pH is changed.

The IEC of the resins/polymers/membranes can be measured using the classical titration method. In this process, the pre-treated resins/polymers/membranes in their fully protonated form will be immersed in 50 mL of an NaCl (2.0 M) solution for 2 h to exchange H$^+$ in the resin/membrane for Na$^+$ ions. This reservoir solution was titrated with NaOH (0.025 M) to a phenolphthalein end-point. Following titration, the resins/polymers/membranes are then protonated in 0.1 M HCl for 1 h, rinsed with deionized water and dried under vacuum at 80 °C to obtain a constant dry weight (W_{dry}). The IEC (mmol g^{-1}) was calculated from W_{dry}, the volume (V_{NaOH}) and concentration (C_{NaOH}) of NaOH as:

$$\text{Ion Exchange Capacity}\left(\text{IEC}\right) = \frac{V_{NaOH}\,C_{NaOH}}{W_{dry}} \tag{5.1}$$

where V_{NaOH} is the volume of NaOH used in the titration, W_{dry} is the dry weight of the resin/polymer/membrane in g, and C_{NaOH} is the strength of NaOH used for the determination of the IEC.

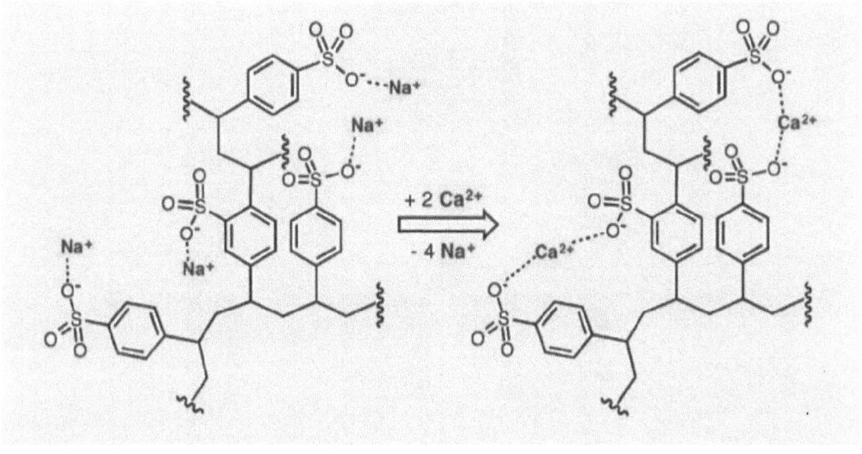

FIGURE 5.1 Sulfonic polystyrene-divinylbenzene (PS-DVB) showing immobile sulfonyl group and mobile cations.

Ion-exchange resins prepared using different polymer backbones have been used for a number of applications in water treatment, metal extraction, heavy metal removal, and fruit and beverage industry processing to improving taste and remove hardness by improving the flavour and taste of surface and ground water through the removal of undesirable components. Other long running conventional applications include the removal of trace metals, decoloration, bad taste and smell.

5.2 TYPES OF ION EXCHANGE RESINS

Ion-exchange is the process through which ions in a solution or contaminated water are transformed into a solid ion-exchange resin which releases ions of a different kind but of the same polarity as its active exchangeable site. This means that the ions in solutions are replaced by different ions originally present in the solid. Ion-exchange resins are mainly classified into two types: cation exchange resin which exchanges positive ions and anion-exchange resin that exchanges negative ions. Further, these industrially important water treatment resins are sub-divided into four basic categories: strong acid cation (SAC), weak acid cation (WAC), strong base anion (SBA) and weak base anion (WBA) as schematically shown in Figure 5.2. In each case, an ionizable active functional group is attached to the resin or polymer backbone to determine the IEC [7].

SAC resins are used to neutralize strong bases and convert contaminated water to neutral salts into their equivalent acids. On the other hand, strong base acid resins are used to neutralize strong acids and convert contaminated sources of neutral salts into their consistent bases. All these variant resins are designed in different concentrations and granular sizes in softening and hardness removal by full demineralization applications. WAC and WBA resins are used to neutralize strong bases and acids containing water sources, respectively. All these four resins are conventionally used for partial or full demineralization.

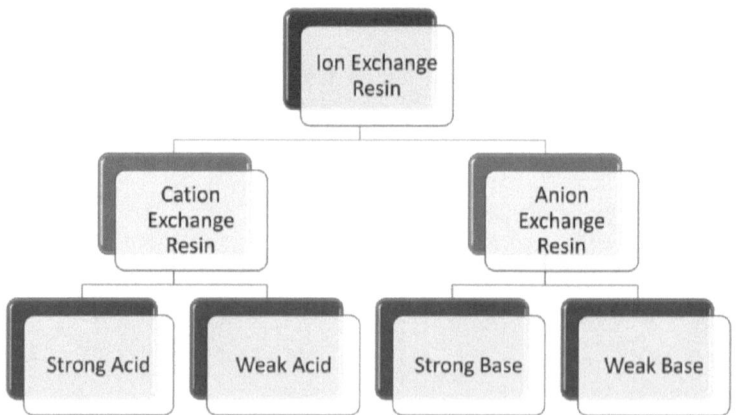

FIGURE 5.2 Classification of resins consists of two main types: cation-exchange and anion-exchange.

SAC resins derive their functionality from sulfonic acid (sulfonyl) groups (HSO_3^-). When sulfonic acid functionalized polymeric substrate or resin is added to contaminated water, these resins completely remove almost all cations from raw or contaminated water, replacing them with H^+ ions as shown in Figure 5.3(a). These SAC resin exchange reactions are reversible, when a change with pH resins can be regenerated. Further, when the capacity of a resin gets exhausted, the ion-exchange resin can be regenerated with the help of an excess of mineral acid. Interestingly, strong acid cation exchangers function very well at all pH ranges making them a robust class of resins. Therefore, SAC resins find their applications in various types of contaminated waters and process systems to recover ionic species. For example, SAC resins are used in the sodium cycle where sodium is the mobile ion for softening, and in the H-cycle for de-cationization. On the other hand, WAC exchange resins derive their exchangeable functional groups from a carboxylic group (–COOH). When these WAC resins are used in the hydrogen form, resins remove cations that are associated with alkalinity, producing carbonic acid as shown in Figure 5.3(b) [8].

Like all ion-exchange resins, WAC is also reversible in nature and allows the return of the exhausted resin to the regenerated form. One of the limitations with WAC resins is that they limit their cation removal efficiency in most water supplies. But WACs can be easily regenerated in comparison with SAC resins with a minimum amount of acid for the process. These reduce the discard of resin media which also helps in minimizing secondary pollutants drastically. WAC frequently in combination with SAC resins are used mainly for the softening and dealkalization of high-hardness ground water, high-alkalinity borewell waters and Na-cycle polishing systems. Also, these two resins are used for the full demineralization of severely contaminated water and wastewater in a cost-effective manner. Unlike other ion-exchange resins, strong base acid resins derive their exchangeable functionality from quaternary ammonium functional groups. This ammonium functionality is also categorized into two types: Type I and Type II. The Type I sites have three methyl groups as shown in Figure 5.3(c).

In a Type II ion-exchange resin one of the methyl groups is substituted with an ethanol group. However, the Type I SBA resin has a greater stability than the Type II resin, due to which these resins can be used to remove the weakly ionized acids. Interestingly, Type II SBA resins provide a better regeneration efficiency and a greater IEC for the same amount of chemical used to regenerate and recycle resins. In addition to this, SBA in the hydroxide form can remove all commonly encountered anions in water and wastewater as shown in Figure 5.3(d). These cation exchange resins can be reversible in which the regeneration takes place with the help of a strong alkali such as caustic soda (NaOH) to reverse the resin to the hydroxide form. Weak base ion-exchange resin functionality mainly derives from primary ($R-NH_2$), secondary (R-NHR') or tertiary ($R-NR'_2$) amine groups. These weak base acid resins are conventionally used to readily remove nitric, hydrochloric and sulfuric acids as represented in the reaction mechanism shown in Figure 5.3(e).

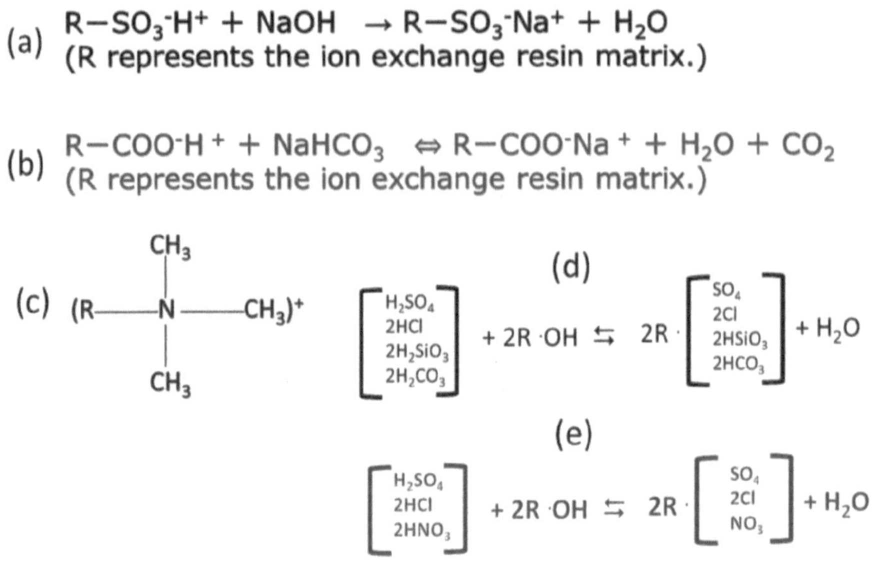

(a) $R-SO_3^-H^+ + NaOH \rightarrow R-SO_3^-Na^+ + H_2O$
(R represents the ion exchange resin matrix.)

(b) $R-COO^-H^+ + NaHCO_3 \Leftrightarrow R-COO^-Na^+ + H_2O + CO_2$
(R represents the ion exchange resin matrix.)

(c)

$$(R-N(CH_3)_3)^+ \quad (CH_3, CH_3)$$

(d)

$$\begin{bmatrix} H_2SO_4 \\ 2HCl \\ 2H_2SiO_3 \\ 2H_2CO_3 \end{bmatrix} + 2R \cdot OH \leftrightarrows 2R \cdot \begin{bmatrix} SO_4 \\ 2Cl \\ 2HSiO_3 \\ 2HCO_3 \end{bmatrix} + H_2O$$

(e)

$$\begin{bmatrix} H_2SO_4 \\ 2HCl \\ 2HNO_3 \end{bmatrix} + 2R \cdot OH \leftrightarrows 2R \cdot \begin{bmatrix} SO_4 \\ 2Cl \\ NO_3 \end{bmatrix} + H_2O$$

FIGURE 5.3 Functionality and ion-exchange mechanism in various type of resins: (a) strong acid cation, (b) weak acid cation, (c) Type-I quarter ammonium, (d) strong base acid in hydroxide form and (e) weak base resin functionality.

5.3 USE OF POLYMERS AND RESINS IN REMOVING EPS

5.3.1 INORGANIC ION-EXCHANGE SYSTEMS

An ion-exchange phenomenon in minerals or clays is dependent on the composition and crystalline structure of the mineral or clay and most importantly on the chemical composition of any contaminated solution or water that comes in contact with the mineral or clay. This section will briefly discuss the chemical structures of clay and zeolitic minerals for their ion-exchange reactions and capacity. An ion-exchange mechanism in inorganic minerals is a reversible chemical reaction that occurs between ions held near a mineral surface by unbalanced electrically charged ionic species inside the mineral or clay framework and ions in a solution in contact with the mineral. Conventionally, the excess charge originated on the mineral or clay surface is negative, which attracts cations from the solution to neutralize the charge via the ion-exchange mechanism. The chemical interactions in ion-exchange resins follow the law of mass action; however, the ionic interaction or reactions are restricted by the number of ionic sites available for exchange on the mineral or clay and by the strength of the bonding of the transferable cations to the mineral surface [9, 10].

On the other hand, zeolitic clay like sodium zeolite is widely used as a softening agent for water treatment applications. Hard water is generally characterized by a high concentration of calcium or magnesium salts. More commonly, calcium sulphate and magnesium sulphate cause severe scaling formation upon long term storage in containers. The zeolite softening agents capture calcium and magnesium when contaminated water passes through a resin bed having a strong acid cation resin. This zeolitic ion-exchange resin in the sodium form prevents calcium and magnesium scaling, effectively, at lower operational and material costs. Nowadays, water softening is widely adopted that uses ion-exchange resins as a pre-treatment technique to obtain hardness-free water. This pre-treatment step is adopted as a complementary step in reverse osmosis systems to produce soft water for various household applications.

The working of zeolitic ion-exchange clay in water softening is as follows. The removal of hardness-causing calcium and magnesium salts from contaminated water by a sodium-zeolite softening process is illustrated in Figure 5.4.

When contaminated water is passed through a zeolite softener, ion exchange takes place to yield softened water free from calcium and magnesium ions. However, the ion exchange in the process leaves small amounts of ionic traces, known as a leakage water source. The extent of hardness leakage depends on the level of hardness and the amount of sodium in the inflowing feed water as well as the amount of salt used for the zeolite regeneration. Repeated inflows of water in the softener produce a low and nearly persistent level of hardness until the IEC of the resin reaches exhaustion. After the resin achieves maximum limitation, the influent water hardness increases abruptly, which necessitates resin regeneration. Figure 5.4(a and b) illustrates a softening mechanism of an SAC resin which readily accepts calcium and magnesium ions in exchange for sodium ions in their structure. Further, when the resin has exhausted its exchange capacity, a high concentration of Na ions is applied to the regenerated resin by replacing calcium and magnesium using a 10% sodium chloride solution; the resin activity regeneration proceeds following the reaction shown in Figure 5.4(c).

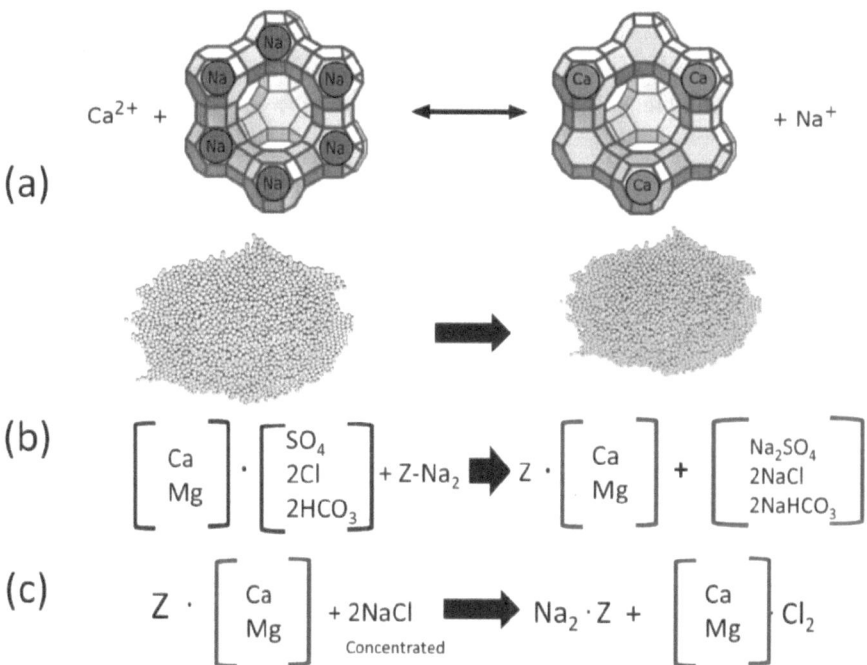

FIGURE 5.4 Mechanism of water softening using zeolitic ion-exchange clay.

Generally, a large and excessive resin regenerant with nearly three times the amount of calcium and magnesium in the resin is used to regenerate the resin's IEC. The eluted hardness in a softening system or unit is removed as waste brine by rinsing. Upon regenerating the resin, a small trace amount of calcium and magnesium will remain. In this condition, if the resin is left in the water containers for a long period, some hardness may rediffuse into the bulk of the water source. Therefore, one can expect a low concentration of hardness in the water effluent from a zeolite softener even if it has been recycled and regenerated following appropriate protocols. This is one of the limitations of zeolitic ion-exchange resins.

On the other hand, natural clay minerals are certainly the most copious inorganic nanomaterial sourced in the lithosphere. These natural clay nanomaterials possess unique properties which are very sensitive at the microscopic level due to their structural and physical arrangements. These natural clays can be easily handled and applied for specific ion-exchange applications as they offer structural alteration upon exposure to external stimuli. The change to the structural properties of clay minerals can achieved with desired environmental-scale impacts due to their swelling properties (smectites) largely found in natural clay like montmorillonite (MMT). This phenomenon in clays can be used to control the rheological and mechanical properties of sediments, soils as well as the transport of ionic solutes through them. The structures of smectites ascend from molecular-scale exchanges between a 2:1 aluminosilicate layer in MMT, the counterions that balance charges arising from isomorphic substitutions inside the layers, and associated water molecules of hydration. In natural clays, the ion-exchange occurs with variation in the types and proportions of inter-layer species causing face-to-face stacks of a 2:1 layer of MMT to swell or collapse reversibly. In spite of having such a significance, the ion-exchange mechanism is found in several environmental and geological systems that contain smectites; the presently followed models fail to describe or predict the reactions between solution composition, ion-exchange and the inter-layer swelling state that determine pore structure and macroscopic ion transportation accurately [11, 12].

There needs to be greater attention given to understanding the ion-exchange mechanism in clay nanomaterials. Therefore, this is a challenge to understand ion exchange in the confinement of inter-layers and on the surface. In this direction, there are several efforts underway following computational and experimental approaches to predict and validate how the exchange and swelling/collapse occur in complex ion-exchange systems. However, recent studies on molecular dynamic simulations have predicted free-energy relationships for a narrow range of solution conditions; nonetheless simulations are limited to a small and insignificant range of structures, sizes and timescales.

The earliest references to and applications of ion-exchange resins were made in ancient times and predominantly used in the form of sand and natural clays. However, in the early 20th century there was exploration of naturally sourced ion-exchange medias obtained from coals with a fuming sulfuric acid which resulted in the first organic ion-exchange system known as carbaneous zeolite. Further, polymer-based ion-exchange systems came into use in the early 1930s, which were synthesized via a condensation polymerization procedure to yield a phenol-formaldehyde cation-exchange resin. Further, in 1944 commercially successful styrene-based ion-exchange resins were discovered and widely used for water purification applications, which opened a new era for polymeric resins. Researchers were quick to understand the ion-exchange mechanism in polymeric systems and these systems were prepared for use both as cationic and anionic exchange resins. Further, advanced functionalized and highly stable polymeric ion-exchange resins were prepared through chloromethylation and amination that resulted in robust copolymer-based resins. This copolymer ion-exchange resin was attached to a chloromethyl functional group in a chloromethylation step which was introduced into the ethenyl-benzene nuclei. More importantly, these copolymers facilitated further functionalization opportunities for attaching various alkyl substituted aliphatic amines to polymerics back in the amination step, making them a versatile class of intermediate water treatments [13].

Interestingly, conventional ion-exchange beads (IEBs) contain 3D covalent networks to which ion-exchange groups are attached. The 3D network with optimal crosslinking preserves the structural integrity of the resin during preparation and usage. These side chains or active sites bound to a backbone provide either cationic or anionic-exchange active sites. For instance, the characteristic styrene-based polymeric resin matrix is prepared through the suspension polymerization of styrene with a variable concentration and ratios of divinylbenzene monomer. Further, as the resulting copolymeric resin beads are then swollen in organic media (organic solvents), either with a sulfonated or chloromethylated functionality inducing agent, these beads were later aminated to prepare strong cationic-exchange or anionic-exchange systems, respectively. Since then, most of the ion-exchange beads have extensively been used for the hardness removal, softening, purification and demineralization of water [14].

Recent studies conclude that excessive minerals such as calcium and magnesium, along with iron and manganese, cause water hardness in potable water sourced from ground water reservoirs. Advances in material chemistry and corrosion science have evaluated and identified hardness as a series problem and the main reason for failures in the pipelines of boilers and heat-transfer equipment. Further, these hardness-causing divalent ions can easily react with anions in soap or surfactants causing severe cleaning deficiency, resulting in added maintenance costs and the consumption of high-volume cleaning agents. Interestingly, calcium bicarbonate and magnesium bicarbonate induce temporary hardness to water upon dissolution of these bicarbonate minerals. These bicarbonate minerals with higher dissolved concentrations in ground and surface water result in Ca^{2+} and Mg^{2+} cations and carbonate (CO_3^{2-}) and bicarbonate (HCO_3^-) anions. Therefore, the presence of high concentrations of cations makes the ground and surface water hard. However, carbonate and bicarbonate minerals yield temporary hardness, unlike sulfate and chloride minerals which induce permanent hardness to water. Interestingly, temporary hardness can be reduced or completely removed by adding calcium hydroxide, also known as lime, through the process of lime softening or treatment. On the other hand, boiling hard water helps the formation of carbonate from the bicarbonate dissolved mineral which can be easily precipitated as calcium carbonate out of the water source making it softer. Permanent hardness caused by sulfate and chloride-based minerals is difficult to remove by boiling. Nonetheless, permanent-hardness-causing ions such as Ca^{2+},

Mg^{2+}, CO_3^{2-} and HCO_3^- can be removed using a polymer-based ion-exchange resin in a water softener [15].

Calcium and magnesium are the most abundant minerals in the human body and natural food supplies, respectively. Calcium and magnesium play important roles in bone structure and muscle and tissue functioning in humans. Nevertheless, high levels of calcium in drinking water causes colorectal, gastric and breast cancer. On the other hand, high concentrations of magnesium might cause symptoms of toxicity leading to muscle weakness, low blood pressure, abnormal cardiac rhythm, difficulty breathing, confusion and the deterioration of kidney function. Recent studies have included calcium and magnesium in hard water as an EP at higher concentrations. Therefore, it is important to remove or control commonly occurring bivalent cations from water. In addition to conventional heating, distillation and sand filtration, several advanced technologies like enzyme-catalyzed electrochemical processes, electrodialysis, nanofiltration, ultrasound treatment, pulsed spark discharge and ultrafiltration have been widely adopted for the removal of Ca^{2+} and Mg^{2+} hardness. However, soda softening and ion-exchange softening are the two main methods through which Ca^{2+} and Mg^{2+} hardness can be effectively removed. Among these two methods, lime-soda softening is considered a high cost operation and is widely used for large scale municipal water treatment purposes. Another major disadvantage with lime-soda softening treatment is attributed to the generation of secondary pollutants like large volumes of sludge. This necessitates a post-treatment step and the use of excessive chemicals like lime soda ash, caustic soda and large volumes of acids to adjust the pH of water sources.

Therefore, the ion-exchange process remains the best option for treating Ca^{2+} and Mg^{2+} hardness. This process can be used in a wide range of capacities from residential water softening to municipal treatment plants. However, the ion-exchange process also induces higher sodium levels in softened water which again requires a secondary step to reduce the sodium concentrations in potable water. Supplementary adsorption has been recently adopted as a promising step for this purpose using low-cost sorbets. For this purpose, a polymer-based 3D super-absorption agent has been adopted as an absorbent for capturing excessive sodium and traces of heavy metal ions from water. Hydrophilic polymers have a high affinity to cation, such as polyacrylamide polyacrylic acid; its derivatives that carry functional groups like carboxylic, hydroxyl and amide have been effectively used for metal ion removal through functional group interaction with contaminant metal (sodium) ions and other trace heavy metals.

The hypothesis and notion behind ion exchange is to swap positive and negative (contaminants) ions in water with hydronium (H^+) and hydroxide ions (OH^-), respectively. In order to achieve large scale transformation via this process, a porous permeable bed compromising of appropriately crosslinked insoluble polymers and countless ion-exchangeable sites needs to be used in an engineered module. When contaminated water is allowed to pass through the porous bed, positive metallic cations such as sodium, calcium and aluminium are exchanged with H^+, whereas anions, chloride, nitrate and sulfate will be replaced by OH^-. Following this principle, polymer-based ion-exchange resins are designed to remove nitrates, heavy metals like cobalt and nickel, and toxic metals like arsenic, copper and dissolved organic carbon (DOC). Polymer-based ion-exchange resins are considered a cost-effective solution in removing a large range of EPs both in microscale and nanoscale ion particles. The main limitation, however, is that polymer-based resin does not remove bacteria, pathogens or pyrogens. Further, once all the ion-exchange sites are exhausted the resin needs to be chemically reversed to regenerate the active ion-exchangeable sites.

Recent advancements in polymer chemistry have led to the discovery of innovative macromolecular ion-exchange resins and membranes that are widely adopted for water treatment and purification. Anion-exchange resins have been used for removing nitrates, perchlorate, phosphates, arsenic in the form of oxo-anions, nickel and vanadium. Anion exchangers have also been used for the removal of known anionic organic pollutants like natural organic matter (NOM). Even though known NOMs like fulvic acid and humic have been identified for their fouling propensity on polymeric matrices, recent advances in nanocomposite resin preparations have led to the production of anti-fouling resins. In situ disinfection of foulants may yield the production of several unknown secondary pollutants originated from the disintegration of NOMs in raw water, the decay of algal cells and algae-derived materials. This may prompt either the discarding of the resin or necessitate a secondary treatment step. One such category of secondary pollutants originates from algo-genic matter (AOM) and algal blooms in water sources. In many cases, it is recorded that AOM has been a major foulant which damages several techniques like MF, UF and ion-exchange resin during drinking water treatment [16].

Recent studies have found a suitable riposte for organic and bio-foulants in ion-exchange resins. Researchers have prepared magnetic polymer-based ion-exchange resins to eliminate AOM in a pre-treatment step which would itself greatly contribute to further processes used in hybrid systems. In this direction, AOM was completely removed in the pre-treatment step of an ion-exchange process prior to which pre-treated effluent was further fed to a UF module. In a recent approach, ion-exchange resins are used for potential foulant removal from organic and biological EP-contaminated water. As a pre-treatment step, resins are continuously submerged in contaminated water to absorb organic pollutants. However, this approach is often proved inadequate to control the algal and biofouling.

Nevertheless, several alternative approaches have been adopted to control the fouling and settlement of undesired organic matter and the growth of algal cells on filtration/resin systems. For this reason, a variety of polymers have been explored to inhibit the interaction of microorganisms and minimize the interfacial interaction of extracellular biomaterials with active material. As a result, modified polymeric resins enable the cleaning of organic and bio-foulants. For instance, poly(ethylene glycol) (PEG)-modified ion-exchange resin surfaces have shown resistance to algal and extracellular biomacromolecule contaminants. Likewise, an amphiphilic polymer-layer coating with active polar and non-polar functionalities on the surface has shown highly effective anti-fouling properties in reducing and hindering macroalga deposition on resins. For this purpose, researchers have used amphiphilic block co-polymers with PEGylated fluoroalkyl side chains. The fluoroalkyl groups on a copolymer backbone showed anti-adhesive and antifouling properties against organic and bio-foulants.

Further, several studies followed similar strategies with a series of ionic polymers and hyperbranched polymers comprising fluorinated and PEGylated groups to inhibit bio-fouling due to green algae. Among these, the use of zwitterionic polymers like chitosan have been widely used to make the resin surface super-hydrophilic, which inevitably enhances the ion-exchange property. Similarly, poly(sulfobetaine methacrylate) grafted brushes, and anionic and zwitterionic polyelectrolytes have been used following different strategies to increase ion-exchange properties and anti-fouling properties in various resins and separation media.

On the other hand, a large number of heavy metals (HMs) have been characterized for high concentrations in various natural, domestic and industrial wastewater sources. Most of the HMs even in trace amounts are toxic to aquatic life, humans and the environment at large. Also, HMs are non-biodegradable in nature under any applied external stimuli. Instead, HMs accumulate in nature and biota, and if left untreated for a long time cause various diseases and disorders. Several conventional methods like electrolytic methods, adsorption, precipitation, ion exchange and evaporation have shown varying levels of success. Among these, the ion-exchange method has shown significantly high efficiencies at low operational costs. The major advantages in using ion exchange over other methods are selectivity, recovery of metal value, low sludge production and resins that can provide high quality product water meeting all discharge norms. In this direction, synthetic ion-exchange resins are frequently preferred as an inexpensive and high performing solution for HM removal. A detailed literature survey on effective HM removal methods highlights the importance of ion-exchange resins. However, the existence of various co-pollutants and complex agents in wastewater makes it hard to remove HMs like lead, cobalt, cadmium, copper, zinc and nickel [17].

5.4 POLYMER/CARBON-BASED ADSORBENTS AND CHELATING AGENTS FOR REMOVING EPS

The ever growing risk of EPs is forcing researcher to find novel solutions to recover and eliminate a series of listed pollutants from various sources. On the other hand, fresh water resources are largely contaminated due to the discharge of various industrial wastewaters from industries like textiles, food and dairy, pulp and paper, cosmetics and pharmaceuticals. Therefore, the adequate treatment of these effluents before discharge into natural water bodies or the environment may prevent potential destruction. Based on the type of effluent and level of contamination, several treatment methods have been adopted, namely physical, chemical and biological methods [18, 19].

All physico-chemical wastewater treatment methods involving organic and inorganic derived adsorbents play a very important role in reducing the severity of the effluent. Using this route, some of the potential EPs (like Cibacron Blue 3G-A (CB) dye which originates from a synthetic class of reactive dyes widely used in the textile industry for colouring cotton yarns/fabrics and discharged in copious amounts into the environment) can be effectively treated using the adsorption technique. The CB dye has shown resistance to degradation with other conventional techniques like biological treatment; and chemical methods have been ineffective at removing it from wastewater. Also, only chemical or biological methods yield large amounts of sludge and/or other secondary pollutants like toxic matters as by-products. In such situations, a physico-chemical adsorption process can be effectively used for the complete removal of CB dye from textile wastewater. In this direction, a high surface area carbon-based adsorbent is carefully designed to selectively remove it from wastewater. Similarly, organic emerging pollutants like dyes, pigments and pharmaceutical discharges have been retained or removed from various wastewater sources using carbaneous adsorbents. This physico-chemical technique is widely applied for its simplicity, efficiency and ability to remove any class of EPs from wastewater. On the other hand, the adsorption capacity of a carbaneous adsorbent can be easily enhanced through a physical or chemical activation process. In addition to this, micro-porosity and its distribution, which contributes greatly to the high surface area in absorbents, can be easily controlled. The high surface area characteristics in activated carbon make them outstanding adsorbents for the treatment of dye/metal/API-contaminated wastewater.

High surface area activated carbon (HSAC) adsorbent produced from various plants and polymer sources have been widely used for the treatment of wastewater. However, initial processes used cost-intensive precursors to produce HSAC, which made researchers explore low cost agro-based wastes or agricultural residues for the preparation of activated carbon. Following this route, several researchers have attempted adsorption of various emerging pollutants from wastewater using various carbonaceous materials and found them to be efficient. Some of these HSAC materials discovered are originated from sawdust, rice husk, wheat shell, orange peels, pineapple stems, coconut coir dust, peanut shells, coconut husk, moringa seed, groundnut shell, residues of oil extracts, and toxic weeds. In addition to this, researchers have also used oil-palm-based HSAC for the treatment of wastewater [20, 21].

Further, various terrestrial biomasses have been explored as high-surface-area carbon sources, namely coconut shell, human hair, chicken eggshell membranes, auricularia, silk, cornstalk, lotus pollen, and so on. In general, high-surface-area carbaneous materials from biomass can be prepared through one-step or two-step processes. In one-step pyrolysis methods the biomass is directly heat treated to carbonize the biomass under a controlled atmosphere, such as 700–900 °C carbonization under nitrogen or argon, physical activation under a CO_2 atmosphere at high temperature, and a chemical activation using KOH as the activating agent. Following this method, naturally abundant weed biomass like *Parthenium hysterophorous* has been converted to a high surface area adsorbent. The resulting carbaneous powder was activated using KOH to generate highly mesoporous structures with an enhanced specific surface area and induce -OH functionalities on the surface of the carbon materials. This high HSAC (4014 m^2/g and a large porous volume of 2.0419 cm^3/g) produced from toxic weed with enriched microporosity was used for treating EPs like organic dyes and pharmaceutical EPs in a batch adsorption method. Further, a membrane was prepared using HSAC for the continuous flow removal of diverse organic EPs, such as organic dyes like methylene blue, Congo red, malachite green, eriochrome black T, and pharmaceutical active ingredients like paracetamol and ciprofloxacin from aqueous media. Thus, HSAC absorbency in membrane form is also highly effective in rejecting the EPs of cationic dyes up to 93–99.5% and anionic dyes in the range of

~50% with a high flux rate of 820–840 $Lm^{-2} h^{-1}$. Similarly, HSAC was also effective against cationic pharmaceutical ingredients with up to 90–95% rejection and an average flux of 830 $Lm^{-2} h^{-1}$. Another major advantage of using carbaneous adsorbents is recyclability. Recyclability studies of HSAC-based membranes showed over 90% retention for several EPs even after ten consecutive cycles without compromising the flux rates. This approach also affirms the possibility of converting abundant weed biomasses into a sustainable alternative for water purification in a continuous flow method. Therefore, carbon-based adsorbents are rapidly changing the treatment to a robust and scalable approach [22].

Chelation is a kind of bonding of ions and molecules to metal ions. Chelation proceeds with the formation or presence of two or more separate coordinate bonds between a polydentate ligand and a single central metal atom. These ligands are called chelating agents, chelators, chelants or sequestering agents. Chelation is a very efficient technique to treat HM contaminated waste and wastewater. Every living body has several active chelating agents in its constitution which bind to metals or heavy metals through two or more coordination bonds. Metal ions with an oxidation state of more than one (2+ or more) are mainly bound in tissues through ionic or coordination bonds. For example, multivalent metal ions readily bind to albumin, enzymes, small peptides and amino acids like methionine, selenomethionine and cysteine [23].

Recent studies show that domestic and industrial wastewater is increasingly characterized by high concentrations of HMs such as cadmium, mercury, arsenic, lead and chromium. Lead, cadmium and mercury perform no vital or specific biochemical roles in living beings; however, these HMs exercise various acute toxicities in multiple organ systems as they bind in tissues, affect endocrine function, block aquaporins, create oxidative stress and interfere with the functions of crucial cations like magnesium and zinc. Chelation is the dominant mechanism of natural detoxification of HMs through a complex formation with various chelating agents like series of biomacromolecules. Natural polymers like polysaccharides, alginate, chitosan and chlorella algal have been widely considered as potential adsorbents of HMs. Similarly, improved citrus pectin and alginate products have been applied to remove lead and mercury from various contaminated sources.

Industrial activities such as refinishing, electroplating, munitions, mining, metal working and automobile accessory manufacturing produce varying concentrations of metal-contaminated by-products that are responsible for a large amount of metal-polluted sites in soils, water and the environment at large. Industrial activities significantly pollute with HM contamination through their wastewater discharges. The most dangerous HMs in wastewater (like cadmium, lead, mercury, copper, arsenic, chromium and zinc) lead to waterborne diseases like diarrhoea or a subsequent death from various diseases. This summons urgent consideration of efficient treatment techniques. For this purpose, chelating agents like ethylenediaminetetraacetate (EDTA) and citric acid (CA) have been used for detecting and removing HM ions. These chelating agents show high sensitivity and selectivity towards several HM ions like Cu and Hg in wastewater at varying concentrations and pHs [24].

Advances are being made at preparing new and advanced classes of chelating agents in different branches of industry and agriculture. Also, chelating substances are categorized based on their complex formation capacity which plays an important role in the distribution and interaction of transition metals and HMs in the aquatic environment. Furthermore, a series of chelating agents like DTPA (diethylenetriaminepentaacetic acid), EDTA or HEDTA ((hydroxyethyl)ethylenediaminetri-acetic acid) have been successfully for HM removal in various wastewater sources.

Table 5.1 gives a list of chelating agents used for converting different HM metal ions into a chemically and biochemically active one through the complexation process. The aminocarboxylates are specially characterized by one or more tertiary or secondary amines and two or more carboxylic acid groups which readily interact with HMs. The table also provides an important list of breakdown products as a result of complexation. These are now considered as secondary pollutants which limit the use of chelating agents in water and wastewater treatment.

However, a major disadvantage of using chelating agents such as EDTA, DTPA or HEDTA is their low biodegradability and potential to produce complex secondary pollutants. On the other hand, the ecological fate of these chelating agents has raised serious concerns and received significant attention. For instance, EDTA at a higher concentration likely disturbs the natural speciation of metals and so affects metal bioavailability. A higher concentration of residual or unreacted chelating agents may lead to the remobilization of HMs in sediments and aquifers, subsequently posing a risk to groundwater and drinking water. Another main concern with agents like EDTA or phosphonates are their non-biodegradable nature and therefore their remaining in the environment for a very long time [26].

TABLE 5.1
List of Aminopolycarboxylate Chelating Agents and Their Secondary Pollutants or By-Products [25]

Abbreviations	Full Name
ADA	B-alanine diacetic acid
DTPA	Diethylenetriaminepentaacetic acid
EDDHA	Ethylenediaminedi(o-hydroxyphenylacetic) acid
EDDS	Ethylenediaminedisuccinic acid
EDTA	Ethylenediaminetetraacetic acid
HEDTA	N-(hydroxyethyl)-ethylenediaminetriacetic acid
IDS A	Iminodisuccinic acid
L-GLD A	Glutamate-N,N-diacetic acid
MGD A	Methylglycinediacetic acid
N T A	Nitrilotriacetic acid
PDT A	1,2-diaminopropanetetraacetic acid
Important breakdown products or secondary pollutants of APCs	
DTTA	Diethylenetriamine-N,N,N',N,-tetra-acetic acid
ED3A	Ethylenediaminetriacetic acid
EDD A	Ethylenediaminediacetic acid
EDM A	Ethylenediaminemonoacetic acid
IDA	Iminodiacetic acid
3KP	3-ketopiperazinediacetic acid

5.5   POLYMER COMPLEXES FOR THE REMOVAL OF EPS

Polymers are long chain organic macromolecules which offer several varying characteristics, like remarkable flexibility, high mechanical strength, chemical stability and high surface area, upon functionalization. Another major feature of polymers is that they can host a series of organic and inorganic moieties, fillers and functionalities in their bulk structure as well as on their surfaces. Therefore, polymers can be processed or tuned to develop a composite or functional material with specific targeted applications like water treatment, wastewater treatment and desalination. Polymers with polymer–polymer (blends and composites) and polymer–inorganic (nanomaterials and complexes) interactions offer several possibilities for tuning the adsorptive and chelating properties via blending, composite creation, crosslinking and surface functionalization. The main advantages with polymer complexes is that simple procedures offer highly stable and functional end products ready for use in water treatment applications immediately. Polymer-based systems are also known for their chemical stability, mechanical stability as well as their ability to withstand harsh operating conditions. Furthermore, polymer-based systems offer high adsorption capacities for a wide range of emerging pollutants and good recyclability [27, 28].

In this direction, several polymer composites have been prepared and used for the removal of emerging organic pollutants like dyes from wastewater. Recently, researchers have developed a series of copolymer systems such as styrene-*co*-acrylonitrile following a solution polymerization process. A further copolymer was electrospun to produce a stable non-woven nanofibre mat which was functionalized for carboxylic acid groups to improve organic dye pollutant adsorption. A low surface area copolymer nanofibre mat showed a high adsorption capacity of 67.11 mg/g for the basic dye, violet, in less then 30 min. Here the charged functional groups induced during the modification process imparted an ion-exchange or electrostatic interaction with dye molecules retained, making it a good adsorbent.

In another approach, a biomaterial-based, low-cost, multi-functional adsorbent, synthesized by ternary carboxylic acid and acrylamide units, was explored as a polymer adsorbent for EP removal. In a simple procedure, a parent straw was imparted with abundant carboxyl and amino groups to create an efficient complex polymer composite for dye removal. Through this process, organic EPs like methylene blue and methyl orange were removed with as high as 120 and 3053 mg/g, respectively. On the other hand, cyclodextrin-based complex systems have also gained wide attention as a new class of polymer adsorbents for dye removal from wastewater due to their remarkable physico-chemical properties and pore structures in the form of cavities. A polymer composite was prepared with crosslinked polydopamine to result in an environmentally benign complex adsorbent suitable for the removal of organic molecules. These polymer complexes have shown high adsorption capacities due to their structural characteristics and the functional groups available for interaction with organic pollutants like dyes. On the other hand, a β-cyclodextrin-induced starch complex was also adopted as a biopolymer complex for EP removal from wastewater sources. The β-cyclodextrin–starch composite with an irregular surface and COOH, NH_2 and OH functional groups played an important role in capturing organic EPs like dye. This composite polymer complex showed a higher removal efficiency compared to individual β-cyclodextrin and starch complexes.

Further to this, the creation of porosity in complex polymer structures and their effect on EP adsorption capacity was carefully studied [29].

Interestingly, porous polymeric complexes showed improved pollutant adsorption capacity as a combination of porosity and functionality helps EPs adsorption via inter-molecular and inter-particle diffusion and the electrostatic interaction of dyes. Further, draft polymers were prepared onto cellulosic cotton fibres using cyclodextrin and an amino hyperbranched polymer. This draft polymer yielded a uniquely functional adsorbent with high active sites with remarkable adsorption capacities for organic EPs like dye molecules. Similarly, agro-waste-derived cellulose was graft polymerized using 2-acrylamido-2-methylpropane sulfonic acid and acrylic acid monomers in the presence of a crosslinker to obtain highly efficient polymer composites for dye removal. Therefore, surface-functionalized waste-derived cellulose with a unique rough surface morphology showed high adsorption performance for both cationic as well as anionic emerging pollutants following Langmuir's isothermal pseudo-second-order model. Detailed experimental outcomes showed a waste-derived cellulose graft polymer efficiently removed the cationic pollutant malachite green and crystal violet as well as anionic dyes like Congo red dye from aqueous wastewater under different stimuli and process conditions. Some of these studies proved that a cellulose-based polymer complex or copolymer (branched and grafted) adsorbent was an economical alternative for the removal of EPs of both a cationic and anionic nature in industrial wastewater.

Recent studies have attempted to functionalize a natural cellulosic polymer using a hyperbranched polyethylenimine (PEI), an ionic polymer for the selective removal and separation of different organic dye molecules from water and wastewater sources. In this case, a cellulose backbone was modified via graft polymerization using polyethylenimine where a covalent linkage was established between NH_2 groups of polyethylenimine and the CHO groups on oxidized cellulose through a Schiff-base formation. Further, this copolymer complex composite was used for the very high adsorptive separation of both anionic and cationic organic pollutants like reactive red, brilliant blue and Congo red (2100 mg/g) as well as yellow dyes (1860 mg/g) in water and wastewater samples, respectively. However, these copolymer composites were ineffective against eosin and bright yellow dyes which recorded a low adsorption capacity. The main reason for the high adsorptive capacity in all copolymers (graft and branched copolymers) has been attributed to their hyperbranched structure which acts as a brush in composite polymers that improves the inter-molecular and inter-particle diffusion of EPs. However, this phenomenon is dependent on the size of the pollutant and its diffusivity. In another approach, an electrospun copolymer nanofibre adsorbent mat was prepared using diethylenetriamine and polyacrylonitrile for the adsorptive removal of EPs like dyes.

On the other hand, polymer-based adsorbents were also used for the removal of toxic HM pollutants from various sources. Conventionally, several adsorbents in various forms and formats have been successfully tested as efficient and cost-effective industrial methods for HM removal. One such class of materials are polymer adsorbents. Polymer and polymer-composite-based adsorbents are increasingly gaining industrial importance due to their chemical stability, mechanical flexibility, potential absorptivity and tunability. Interestingly, polymer adsorbents can be functionalized in various conformations like homopolymer copolymerization (graft, branched); polymer composites offer a wide variety of opportunities for water treatment application. In this direction, polystyrene sulfonate/3,4-ethylenedioxythiophene in combination with lignin, which resulted in a composite polymer complex, was used for the removal of HMs. Interestingly, a polystyrene block in a composite polymer system acted as a cation exchange moiety to adsorb HMs; on the other hand, the sulfonyl groups and the lignin blocks improved the adsorption capacity (>250 mg/g for lead) of the toxic metals through their various binding sites. In this system, a neutral pH condition best suited for water treatment application facilitated the binding of lead ions to a polymer composite. When the polymer composite was tested as having lignin as a constituent, the adsorptive capacity for lead almost doubled (450 mg/g). In another system, poly(styrene-co-maleic anhydride) was modified with the incorporation of cyclopropane and tetramethylenediamine to produce a highly active composite adsorbent for the selective removal of HMs from wastewater sources. These composite polymers were used for the selective adsorption of HM ions such as Cu, Zn and Pb in various process conditions. Polymer composites showed a high tendency for HMs in the affinity order of Cu>Zn>Pb through pseudo-second-order kinetics and the Langmuir adsorption isotherm.

Further, a novel composite copolymer was prepared via glycidyl methacrylate grafting on starch following a cyclo-addition between the NH_2 from diethylenetri-amine and the epoxy groups of glycidyl methacrylate. Interestingly, this composite copolymer showed high adsorption capacity for HMs like Cu and Pb ions which is attributed to chelation interaction between polymer composites and metal ions. On the other hand, porous polymeric systems prepared using covalently linked cyclo-dextrin and EDTA-modified chitosan showed promise of an HM-capturing capacity from wastewater sample treatment. The porous cyclodextrin/EDTA-chitosan prepared using the environmentally friendly solvent pentafluoropyridine showed efficient adsorption (>90%) for a series of HMs via host–guest inclusion and the chelating effects, which also helps in adsorbent recovery and reuse. On the other hand, conductive polymers like polypyrrole and polyaniline-based composites have been developed for metal ion removal from wastewater. Specially, a combination of imine and amine groups in polypyrrole-polyaniline, respectively, have been utilized to remove HMs from water sources via chelating through an electrostatic interaction and hydrogen bonding.

5.6   POLYMER-CARBON NANOCOMPOSITES FOR THE REMOVAL OF EPS

In the previous section, we discussed the effectiveness and efficiency of carbon-based adsorbents and polymer-based systems individually. However, recent advancements have shown the greater potential of polymer/carbon composites in treating EP-contaminated wastewater. Understanding surface functionality in both polymers and carbaneous nanomaterials led to the development of several novel composite systems. In one such attempt, a pine-sawdust-based biosorbent composite was prepared with the help of citric acid for the removal of the copper ions from waste-water. In this system, hydroxyl and carboxyl functionality from biowaste was behind the high adsorption capacity of the Cu ions. Biosorbency showed an optimal adsorption capacity at pH > 3 due to an electrostatic interaction between the positive copper ions and the negative charges (COO^-) of the composite. Furthermore, residual hydroxyl groups in the biosorbent also contributed to the adsorption of Cu ions via a hydrogen bonding interaction. More significantly, carbon is an atom with a unique electronic structure that contributes to the formation of covalent bonds with many cations, especially metals and non-metals which make carbon-based composites a potential alternative for the removal of EPs. On the other hand, polymer composites prepared by incorporating carbon nanomaterials are receiving increased attention in different applications, especially in water treatment. In this strategy, polymers provide excellent dispersion and solubility space for carbon composites to retain high surface area characteristics. These two factors impart polymer/carbon nanocomposites with a favourable characteristic for the elimination of organic and inorganic EPs from wastewater.

Carbon nanotubes (CNTs) are carbon allotropes widely recognized for their application in electronics, and energy and environmental applications. CNTs can be easily prepared at the nanoscale with tunable electronic and surface characteristics and functional properties in a simple modification process. CNTs are generally produced in a 1D cylindrical nanostructure form. CNTs possess excellent electronic and high surface area properties, making them a promising candidate for water treatment application. In spite of all these exceptional properties, CNTs suffer from poor surface functional groups and dispersibility in aqueous media. A lack of surface functionalities necessitates the formation of nanocomposites with polymers to induce exchangeable moieties to improve the adsorption properties. In this direction polymer/CNT composites have been prepared in combination with poly(N-isopropylacrylamide)-co-(acrylamide) copolymer for complex oil/water mixtures. Oil/water mixtures have been recently considered as persistent or emerging pollutants due to their potential risk to the environment. On the other hand, CNTs are known for their oleophilic properties. Therefore, a CNT-copolymer filter has been successfully used to remove oil from water. Another study used multi-walled CNTs (MWCNTs) in a polysulfone-polyether copolymer as an oil/water separating medium. In this filter medium, MWCNTs retained oil from the oil/water mixture, resulting in over 99% separation. These studies showed efficient oil/water separation strategies with a high permeate flux, rejection (>99%) and recyclability possibilities. These strategies also offer promising filter media with various types of recyclability via simple self-cleaning, washing and antifouling functions [30].

Functional polymers have been widely recognized for their metal chelating or selectivity properties. Similarly, the unique electronic structures in CNTs also potentially offer a metal binding property. Considering these factors, polymer/CNT nanocomposites have been applied to a series of HM removal. Here, the use of a hydrophilic polymer provides faster water transportation passage and surface functionality in both polymers and CNTs which retain hazardous HMs. In one such instance, CNTs were grafted with different polymers to produce effective nano-adsorbents for HM removal from contaminated water sources. A poly(hydroxybutyrate-grafted-CNT) nanocomposite prepared via a simple blend technique produced a highly efficient nano-adsorbent for the removal of HMs like Cr, Pb, As, Ni, Cr, Cu, Fe and Zn metal ions, commonly found in industrial wastewaters. Interestingly, highly dispersed CNTs in a polymer matrix provided an excellent HM capture network which led to the removal of a series of HMs. These nanocomposites retained most of the HMs in the wastewater through ion exchange and electrostatic forces.

Polymer coated CNTs were prepared in a different strategy to adsorb HMs from various sources. In this strategy, polyamidoamine-coated CNT composites were used for capturing As (432 mg/g), Co (494 mg/g) and Zn (470 mg/g) metal ions from wastewater. In another strategy, composite hydrogels were prepared using CNT nanofillers as a nanogel adsorbent. Composites of polyacrylamide/sodium alginate were incorporated with CNTs for preparing a highly flexible 3D network structure to capture HMs from wastewater. Interestingly, polyacrylamide offers a flexible, hydrophilic and porous network while sodium alginate provided mechanical strength to a nanocomposite gel comprising a polymer hydrogel. On the other hand, CNTs in hydrogel not only enhanced mechanical strength but also induced elasticity and adsorption properties. This nanocomposite upon optimal crosslinking displayed a macroporous structure which helped in enhancing hydrophilicity, product flux, low density and adsorption capacity. Further, polyaniline (PANI)-grafted MWCNT nanocomposites prepared via oxidation polymerization were applied to Cr removal. The PANI block doped with para-toluene-sulfonic acid hydrophilic groups induced enhanced permeation and HM rejection functionality. Therefore, PANI/CNTs in their characteristic morphology provided interactive groups to attract chromium ions via its amine, imine and hydroxyl moieties. Similarly, several CNT/polymer nanocomposites were prepared through various strategies and applied to the removal of organic EPs like dyes (MO, MB, CR) from various wastewater sources [31, 32].

Recent interests have explored fascinating 2D nanomaterials like graphene for various applications due to their exciting physico-chemical properties. In addition to this, graphene-based nanocomposites are also gaining a lot of interest as cost-effective adsorbents in water treatment applications. Furthermore, polymer/graphene-based adsorbents result in a porous 3D network structure providing a large surface area, macro/meso-porosity and light weight, which makes them a suitable choice for EP treatment media. Polymer/graphene composites are being prepared in various forms and formats like membranes, aerogels, hydrogels, powder beds and filter cakes and applied directly for water treatment purposes. These combinations of materials provide excellent pollutant capture efficiencies, anti-fouling properties, adsorption capacities and high permeability. These properties make polymer/carbon nanocomposite materials a promising alternative for conventional material stock.

5.7   RESIN-BASED FIXED BEDS FOR THE REMOVAL OF EPS

Industrial scale demineralization or water treatment requires an engineered module or plant set up to perform the defined contaminant removal functions. Most of the emerging pollutants in industrial wastewater, domestic wastewater or process streams are in dissolved form. It is well known that the ion-exchange resin method contributes immensely to water treatment technology for the retaining and removal of various EPs from different contaminated sources. While ion-exchange resins are generally used to treat hardness in water, in recent years they have been used for the removal of mineral and emerging pollutants. Today, ion-exchange resin-based processes are used for the complete removal of almost all ionic hazardous species. These industrial scale ion exchanges in the form of packed beds or column forms in plants, modules or units are available for both cation and anion-exchange purposes in the same column or bed, making them complete water treatment units. These engineering modules provide high quality treated water as an end product at a comparatively low cost.

However, the specific module or plant or unit design can vary, depending on the column or bed type and size and may also depend on the specific application. Nevertheless, the performance of ion-exchange bed also depends on the process condition and the composition of the wastewater. These ion-exchange units may further include components like: (a) one or more ion-exchange columns, (b) recycling and washing solution/dosing system, (c) feed storage tanks and (d) control or automatic operating systems. These systems also come in flexible configuration in order to satisfy the various end users, plant or system footprint, requirement and purity goals. Engineers, researchers and innovators design ion-exchange modules keeping these three factors in mind.

It is well known that conventional wastewater treatments are ineffective in removing merging pollutants, pharmaceutical active ingredients (APIs) (such as diclofenac, ibuprofen and their metabolites), HMs, semi-volatile organic compounds, dye molecules and nuclear waste constituents. However, these compounds have been recently characterized in high concentrations in both domestic and industrial wastewater along with less harmful co-pollutants like nitrogen and phosphorus. Among several treatment methods reported and discussed in previous sections, the membrane process, carbon adsorbents, advanced oxidation process, mercury amalgamation and electrolytic methods have been proven to have only limited success in removing these EPs. Also, some of the processes need constant washing and maintenance which generates secondary pollutants or by-products creating environmental problems in the long run. Some of the recent innovations like a solvent extraction process use extractants and an environmentally hazardous long-chain carboxylic acid as a reaction accelerator. In this situation, the ion-exchange resins discussed in the previous section have proven highly efficient; however, to make them technologically advanced and cost effective, these resins/adsorbents/ion-exchangers need to be assessed in continuous engineering modules or units or beds. For this process, several researchers have designed effective ion-exchange beds or columns using a combination of anion and cation exchange resins and co-adsorbents as a cost-effective and environmentally friendly alternative. Nevertheless, ion-exchange resin beds having a combination of functional groups like SO_3, $-NOH$, $-NH_2$, $-OH$, $-SH$ or $-NH$, exhibit excellent reactive properties with EPs like APIs, Dyes, HM and organic moieties through various interactions. In a similar approach, loosely crosslinked poly(hydroxamic acid) ion-exchange resin was prepared following suspension polymerization of acrylonitrile and divinyl benzene. These polymer-based ion-exchange materials were tested in bed filtration form for high efficiency retention of metal ions by passing contaminated feed in a continuous flow method.

It is important to understand that adsorbency plays a vital role in determining the performance of packed bed or column adsorption systems; however, the selection of a combination of ion-exchange resins may induce an optimal adsorption capacity for both anionic and cationic pollutants. In one such approach, a new class of EPs like nuclear waste constituents have been successfully separated from nuclear power plant water treatment, such as corrosive products. In a nuclear reactor, pressurized water reactors or primary coolant sections suffer technological failure due to the presence of activated corrosion products and heavy metals like Ni, Co and Ag. This necessitates the removal of these EPs to sustain plant operation, which also prevents them entering the environment upon discharge. For this purpose, industrial-scale mixed ion-exchange resin beds are used to remove these EPs from the coolant. Generally, these mixed bed systems consist of a mixture of both cation-exchange and anion-exchange resins, which simultaneously facilitates the removal of both cationic (110 mAg, 58Co and 60Co) and anionic EPs from the contaminated tanks. The main advantage of using a bed or column is that a continuous flow of feed can be controlled to maintain high performance as well as fouling. Further, mixed-bed ion-exchange beds have been widely used to produce ultra-pure water, removing all the trace chemicals required for the microelectronics and pharmaceutical industries. Also, ion-exchange mixed-bed systems have been extensively applied to the recovery of precious metals like Li, Cr, Ni and Pb from various water and wastewater sources and further as recovered metal ions used in energy conversion and storage applications.

Interestingly, several studies have also attempted to understand the parameters affecting selective separation, retention and removal efficiency of the resin for various contaminants in water and wastewater sources. These studies have considered process conditions, the nature of waste or contaminated water and the mechanism for purification in the resin beds so as to understand and predict their lifespan. More significantly, an ionic-exchange site, also referred to sometimes as cages in adsorbents, possess a 'memory effect' which determines the rate kinetics and interactive intensity for extracting target ionic pollutants with an extremely high selectivity. In this direction several reactive and transport models conclude that there are several factors based on the nature of the resin, the hydrodynamics in chemical reactions and the ionic/resin complexation mechanism. Studies also conclude that: (a) low solute concentrations in feed water and high flow rates obscure the numerical operation of such modelling approaches; (b) to produce ultra-high purity product water requires the continuous replacement of the resin before it reaches saturation; (c) there is a need for a fine simulation to understand the mechanism of resin and ionic pollutant interaction at their solid (resin bead)/liquid interfaces; and (d) the rate-limiting step in ion-exchange resin is the diffusion of ionic species between the bulk of the column and the resin beads, which determine the robustness of the ion-exchange mechanism. Unfortunately, very few studies have considered both the kinetics of the hydrodynamic as well as the multicomponent chemistry parameter to assess the specific resin systems in long term operations. Although several modelling studies considered ion-exchange modelling with specific ionic species, these models fail to outline any general observations. This is important as the further insight into the practical application of ion-exchange technologies may also shed light on evaluating EPs with a high ecological risk, as well as the quantification of EPs below detection limits, which present studies do not consider regarding an endpoint ecotoxicity assessment, nor the contribution to mixture toxicity of the saturated resin upon being discharged or discarded into the environment [33, 34].

5.8 COMPOSITE RESINS AND THEIR LIMITATION IN REMOVING EPS

Designing and developing composite ion-exchange materials is gaining increasing attention as the individual or homopolymer or pristine precursors fail to retain high performance. On the other hand, hardness is the undesirable characteristic associated with surface and groundwater generally and which is found in almost every part of the world. The dissolved multivalent ions (mainly Ca and Mg) in more than 200 ppm in water cause hardness which makes it unusable unless treated for domestic applications. A high concentration of hardness triggering ionic species can cause problems such as kidney stones, clogging water supply pipelines, scaling in water tanks and containers up to 1 mm in thickness and operational complications which drastically increase energy consumption for treatment by up to 12% and impart bad taste to drinking water.

These cumulative problems force researchers and innovators to devise industrial scale treatment techniques to remove hardness from water. Every method discussed in earlier sections have shown but limited success at removing ionic pollutants from water and wastewater including membrane separation, adsorption, ion-exchange resins, chemical methods like AOPs and thermal processes. However, residual pollutant removal for long-term sustainability still continues to instigate the search for new combination materials as cost-effective alternatives.

On the other hand, ion-exchange resin has shown high efficiency in removing ionic pollutants; however, several characteristics associated with polymeric components are causing technological challenges. Therefore, researchers have designed nanocomposite-based adsorbents or resins following cost-effective and easily implemented strategies. One strategy made use of bio-origin materials such as biopolymers and natural clays due to their abundance and easy adaptability. Natural clays are characterized by layer structures with a high surface area tunable surface charge and ion-exchange capacity. Since natural clay is originated from volcanic ashes this makes them a highly stable system for harsh chemicals and mechanical abrasions. In particular, natural clay like bentonite possesses a sandwich structure with great potential to exchange (IEC = ~98 meq/100 g) and adsorb ionic pollutants through chemical and physical interactions. Similarly, magnetic nanoparticles (MNPs) have also been explored to remove a series of EPs. Interestingly, MNPs comprise a high surface area due to their small particle size and magnetic properties which make them a better choice for easy separation post-EP adsorption.

Nanocomposites of both bentonite clay and MNPs have been prepared using several polymer matrices to enhance sorption properties. This combination of materials with a specific individual adsorption property and carrier forming abilities provide nanocomposite resins or adsorbents with an additional advantage. However, several polymer/nanomaterial combinations provide several synergic properties, making them sustainable alternatives. Still, large scale production of nanocomposite resins or adsorbents faces several compatibility issues. Among these, incompatibility induces low efficiency, aggregation formation, scale-up and operational difficulty, which limit their applications on a large scale. Another major limitation in using discrete nanoparticles or nanoclay is their potential harmful effects upon leaching or permeating into filtrate water causing a residual in the purified water.

On the other hand, high concentrations of HM ions in the aquatic ecosystem are an enormous risk for public health in industrialized countries. HMs have been characterized by their non-biodegradable toxicity even in trace amounts which cause severe physiological damage in living beings and humans upon accumulation and consumption, respectively. In particular, HMs like cadmium (Cd) cause lung and kidney damage, lead (Pb) causes physiological poisoning, chromium (Cr) causes cancer; HMs even in trace levels cause damage to the gastrointestinal track, kidney failure and liver damage. Most HMs bind to cell membranes; thiol and phosphate groups in proteins and nucleic acids cause severe neurological and hematological disfunction in humans.

Hence, these serious issues associated with HMs necessitates efficient, ion selective and inexpensive HM adsorbents. As discussed in earlier sections polymer/nanocomposite-based adsorbents and ion-exchange resins are used for HM-ion removal from water and wastewater sources. In particular, natural nanoclay, Al/Fe-oxyhydroxides, Zr-nanoparticles, MNPs and Mn-based mixed metal oxides have been used in combination with a series of polymers, such as polyvinyl alcohol, polyacrylamide and polyacrylonitrile, to prepare nanocomposite super-adsorbents. In each case, a nanomaterial or nanoclay has been tuned to selectively adsorb cationic pollutants. In addition to this, advances in nanoscience and nanotechnology procedures have facilitated tuning the unique surface and bulk properties of nanomaterials. However, to protect these properties polymers have been widely used as carriers or meshes for dispersal, which protects their active sites. In this direction, ionic polymers which complement the ion-exchange capacity of clay/nanomaterial/nanoadsorbent/feldspar (alumina-based silicate mineral) are combined in composite form. As natural clays and feldspar carry high ion-exchange capacities in particular HM exchange abilities, these pristine properties have been protected with a suitable choice of macromolecular carrier. Obviously, the chemistries of synthetic nanomaterials, feldspar minerals and clay minerals require different strategies that need to be followed to transform them into efficient nanocomposite adsorbents. Therefore, several polymers like chitosan, polysulfone and polyacrylonitrile have been used as both pristine polymers and a functional version to create effective nanocomposite treatment media. Further, polymers have also been functionalized to induce complementary ion-exchange capacities which also contribute in increasing the overall capacity of nanocomposite adsorbents or resins for the efficient removal of HMs like Cd^{2+}, Zn^{2+} and Pb^{2+} from water and wastewaters. For this, sulfuric acid assisted crosslinking for chitosan and sulfonated aniline-modified poly(vinyl alcohol) (PVA)/feldspar composite resins were applied. Through this strategy, the surface and bulk functionality of PVA increases the ion-exchange ability in PVA, chitosan and their blend systems which complement metal ion adsorption. Sulfonic acid groups on polysulfone, PVA, polyether sulfone and their nanocomposites have also induced a high cation-exchange capability, even under very acidic conditions [35–37].

Nevertheless, nanotoxicity has been considered a major future concern associated with health risks as the increased use of nanocomposite-based treatment systems are gaining commercial importance and are in use for various applications. To overcome these leaching issues, several studies have been designed using speciality polymeric materials to carry nanoparticles, clay minerals and ion-exchange moieties. Therefore, the design and application of polymer/nanocomposite resins generate a promising alternative. Recently explored nanocomposite resins used chitosan, alginate, acrylic polymers, polyacrylonitrile (PAN), agarose and a combination of these polymers in a blend and composite form.

5.9   RECYCLING AND REGENERATION OF POLYMERS AND RESINS

Ion-exchange resins and polymer materials have several advantages over conventional adsorbents and absorbents. These advantages include easy fabrication procedures to yield the desired share, size and format. Polymer-based resins also offer easy scalability to small to medium filtration plants. Another major advantage is that polymers/resins do not require a pre-treatment step or an additional secondary stage to operate. Interestingly, most of the composites such as graphene, high surface area activated carbon, CNT-induced polymers, polymeric nanocomposites, nanostructured metal oxides/hydroxides, hydrogels and natural nano-clay-type adsorbents or adsorbents can be reasonably recycled and reused.

In this direction, nanosized manganese oxides with mixed phases were produced via a facile synthetic method using green solvents at room temperature. Mn oxides prepared in a robust synthetic procedure using choline chloride and ethylene glycol as a solvent and a templating agent showed a high pollutant retention property. These mixed metal oxides have been first tested for the removal of organic EPs like dye molecules from water samples. Interestingly, metal oxide-based ion-exchange materials showed higher flux (>5000 $Lm^{-2}\,h^{-1}$) and rejection of 99.8% for cationic methylene blue dye. Furthermore, these nanomaterial-based adsorbents showed excellent recyclability of up to ten cycles without significant decrease of dye rejection performance.

Following a similar important strategy, hydrothermal carbon composite nanomaterials were prepared and used for organic pollutant removal as cost effective, easily scalable and a feasible alternative to cost-intensive conventional processes. As discussed in previous sections, several carbaneous nanocomposite adsorbents and adsorbates in individuals and composites have been prepared and tested for their efficiencies and capacities against several EPs. Further, the surface and bulk functionalization of these carbonaceous materials have been carried out using Fe, Al and Zr-based nanocomposites following a low temperature (120 °C) hydrothermal procedure. These hybrid materials were tested for their ion-exchange capacities, adsorption and filtration performances in different devices for the abatement of several emerging anionic pollutants (anionic and cationic dyes, APIs and biomacromolecules). These nanocomposite materials have shown remarkable separation potential without any leaching in order to avoid problems associated without showing any nanomaterial and agglomeration behaviour.

Further, the nanocomposite-based aerogels have been prepared by incorporating various nanoparticles, both on the surface and bulk of ultra-high porous filters. Both hydrothermally prepared carbaneous microcleaner and aerogel based macroporous filters exhibited a maximum adsorption capacity for a series of emerging pollutants with high efficiency from contaminated water sources. The extended lab-scale studies have also showed scalable potentials of these nanocomposites even at higher concentrations found in industrial effluents. Most importantly, Fe-oxide nanoparticle-induced hydrothermal carbon and aerogels have produced high quality water at varying applied conditions. These magnetically active nanocomposites can be easily removed, regenerated and recycled for further use, making them overall low-cost alternatives.

In a similar approach, biopolymer-based aerogels were designed to separate an oil/water mixture and a thermodynamically stable oil/water emulsion. These oil/water mixtures and emulsions were collected from real industrial discharges and oil spills. Superhydrophilic adsorbent systems prepared using biopolymers have shown significantly high oil separation efficiency from the contaminated mixture. Most importantly, biopolymer-based polymer systems showed long term stability and reusable capabilities. Detailed recycling and regeneration studies have shown promising aerogel recyclability after recovering dye molecules, HMs and oil from various feed mixtures. Recycled polymer systems showed easy surface recovery by retaining a high flux and rejection for various pollutants.

Overall, recyclability studies on nanocomposites, polymer composites and ion-exchange resins have shown promising results. Recently reported studies also unravelled a unique sustainable methodology to obtain recyclable carbon materials, metal-oxide/hydroxide resin composites, polymer-based nanocomposites and ion-exchange systems. Interestingly, in most of the cases, surface and bulk functionality resulted in inducing important ion-exchange, regenerative and performance retaining properties [38–40].

5.10   LIMITATIONS OF INORGANIC/POLYMER RESINS AND POLYMER COMPOSITES IN REMOVING EPS

Historically popular natural zeolitic and synthetic aluminosilicate-based softeners have been proven to be efficient media to treat residual hardness in domestic water sources and industrial effluent. Age old technologies have shown several advantages in their lifespan. Nevertheless, recent studies have revisited the use of sodium zeolite softeners for the removal of trace emerging pollutants like organics and heavy metal ions from contaminated sources. Sodium zeolitic softeners have been effective in controlling scaling due to the excessive presence of conventional pollutants like multivalent ions below 2 ppm in most domestic water supplies. These ion-exchange systems offer simple and reliable operating and maintenance possibilities. Ion-exchange capacity can be easily regenerated in a simple recyclability process using relatively low-cost regeneration agents like salts. Regeneration salt is not only inexpensive but also environmentally friendly in nature. A regeneration process does not yield waste sludge. These inorganic ion-exchange systems can be adopted in various plant footprints to suit the end application from simple household to industrial demineralization.

Although zeolitic and aluminosilicate-based softeners efficiently decrease the metal ion concentrations in water, secondary parameters characteristic of polluted water like total dissolved solids, alkalinity and increasing nanomaterial amounts remain unaltered. Another major disadvantage is the replaced ionic species remain in the bulk of treated water. Ion-exchange resins are also ineffective against influent turbidity. Resins experience severe surface fouling if used in the long run for turbid water (>1.0 JTU) which may influence scale formation following which surface fouling damages the activity of the resins. Unlike in polymeric resins, inorganic ion-exchange systems are vulnerable to heavy metal fouling. Higher concentrations of HMs like Fe and Al create irreversible scaling on the resin surface. At present there are no suitable cleaning methods or washing solutions recorded for HM scaling in resins. In addition to this, the presence of an excessive combination of Fe and Mn in water demands periodical and rigorous cleaning to regenerate the functionality of ion-exchange resins. Inorganic ion-exchange systems also need to be carefully operated to avoid aluminium coagulation which is generally used ahead of zeolite softeners to avoid the infiltration of Al contamination to a plant by controlling the pH and other operational parameters. On the other hand, the use of strong oxidizing agents to create chemical intermediates and to facilitate the easy separation of contaminates may damage or degrade the resin. Therefore, chlorine, a commonly found pollutant in domestic or municipal water, should be removed in a pre-treatment step. The pre-treatment step generally comprises the use of either high surface activated carbon or sodium sulphite treatment.

On the other hand, a demineralization process is adopted as a supplementary or complementary technique to remove TDS, traces of HMs and biominerals. Resin-based softners have a proven limited success in retaining a large number of dissolved solids such as anionic minerals (Cl^-, SO_4^{2-}, NO_3^-) and cationic pollutants like silica, sodium, alkalinity, Fe, Al and Mn.

Generally, the presence of strong acid cation resin in the H-form converts and dissolves the various salts in the softner tank into their corresponding acids, and the strong base anion resin in the OH-form removes these acids. However, several intermediate by-products and traces of contaminants percolate in to the treated water. These intermediates have also been considered as EPs. A closer understanding of the nature of EPs, their origin and possible transformative routes show a worrying trend concerning the quality of municipal water. At present, complete demineralization is performed following a cost-effective and conventional distillation technique. However, future approaches demand a prior or post-softner step for the complete removal of EPs from municipal water. Nevertheless, demineralizers conventionally used can produce high-purity drinking water from ground and contaminated water sources. Demineralizers can also produce a high quality comparable to distilled water at a fraction of the cost. However, the demineralization process requires pre-filtered water as an influent. This is due to the demineralizer having various resins beds and packs susceptible to fouling and which can interact with strong ionic species such as Fe and Cl in the influent and undergo degradation. In many cases, anion-exchange resins are highly susceptible to organic and biofouling as the surfaces are generally characterized by high organic and biological contents like humic acid and bacteria, respectively. On the other hand, modern pollutants like nanomaterial discharges percolated in various water sources pose a serious threat to the stability of resins. Modern emerging pollutants like colloidal silica and other nanomaterial adsorbents used excessively remain unchanged and pose a serious threat to treated water. Conventionally, colloidal silica is removed from the scalant layer or from the containers with the help of hot alkaline boiling water which converts it to simple silicates.

On the other hand, both inorganic and organic resins suffer from severe surface fouling and from various undissolved or suspended minerals. This organic and biofouling can be managed to some extent with suitable preventive techniques in ion-exchange resins, but most of the inorganic fouling is irreversible. The resin generally susceptible to fouling and degradation from washing solutions or the regenerative process or major inorganic fouling originates in ion-exchange resins from the high presence of iron and manganese. Interestingly, most of the recent studies have classified these metal ions as EPs. Medium to high concentrations of Fe-metal ionic contamination may originate from ferrous or ferric inorganic salt or from the sequestered organic ion-exchange complex itself. Ferrous ions from resin exchanges into water, which may be the main source of the high concentration (ferric-iron is insoluble). Therefore, ferric iron forms a uniform coating layer on a cation resin which further prevents or drastically reduces the ion-exchange ability. In such a case, an acid or a strong reducing agent is used to react with the Fe to remove it from the surface of the resin. However, this process leads to the washing of cationic resin but which fouls the anionic resin as the Fe-metal ion, upon washing, moves from the cationic surface to the anionic resin surface. On the other hand, manganese originates from the high concentration in the treatment tank from influent sources like surface and well water which follows the Fe-metal ion which causes resin fouling. Similarly, aluminium in the form of aluminium hydroxide precipitates out from alum or sodium aluminate treatment further in high concentration deposits or creates a layer on the zeolitic resin. Calcium sulphate has one of the lowest solubilities in water, which can easily precipitate on to a strong acid cation resin. Here, calcium forms a thick coating layer upon precipitation on the resin bed drastically reducing the ion-exchange property of the system. Barium sulphate is even less soluble in water than calcium sulphate which causes severe damage to the resin surface. Surface water contaminated with oil spills, upon passing through the resin bed, causes a surface layer deposition on ion-exchange resins. Simple surfactant washing can removal oil depositions, however it may create a source of secondary pollutants and by-products. These secondary pollutants may foul resin. Similarly, ground water and surface water are characterized by a high concentration of microbial or pathogenic contaminations. Both inorganic and organic resins exposed to these biological contaminants may face serious surface interactive depositions. The increased deposition of a microbial layer on the resin bed and individual resin beads may cause severe clogging, resulting in an excessive pressure drop across the fouled resin.

On the other hand, organic fouling in both polymeric and nanocomposite resin systems is a major cause of concern which severely damages overall performance.

Organic and biofouling are the most common and technologically challenging intrusions in resin which is gaining a lot of attention from researchers, technologists and innovators. Even though ground water and municipal water sources are generally characterized with traces or low levels of organic matter like humic acid, fulvic acid, tannic acid and other industrial organics, surface water generally sourced for drinking water production or household applications are characterized by medium to high concentrations (several tens to hundreds ppm) of organic contaminants which are now designated as EPs. These aromatic substances can also easily interact with heavy metals in water sources like Fe, Al and Mn to result in complex insoluble and risky secondary pollutants.

Also, organic strong base sites or moieties in polymeric and nanocomposite resin are more susceptible to interaction with various EPs, even in trace amounts. This potential risk may cause blockage in resin beds in the long run which reduces the salt splitting capacity which is essential for ion exchange to take place. When contaminated water is fed through the resin bed (HM or API or organic substance) in a continuous flow, the pollutant may turn into a foulant and remain on the resin surface via various interactions. In particular, inorganic pollutants on strong base sites severely restrict their activity, causing a drastic decrease in ion-exchange capacity. Washing or cleaning with harsh chemicals leads to polymer degradation or erosion. These phenomena in anionic resin can be easily figured out as the resin changes colour due to surface interaction during regeneration in which the resin turns from a tea-coloured surface to dark brown.

However, researchers have invested a lot of time in proposing prevention measures for various types of fouling and surface deposition on inorganic, polymeric and nanocomposite resins. One of the simple strategies followed for the pre-treatment of contaminated water is at the source itself. Conventionally, (a) organic or biologically contaminated water is perchlorinated at the source and clarified with the help of a suitable reagent. (b) Pre-treatment has been another strategy followed with the help of ultra-high active surface area carbon filters which can easily retain organics in traces, meta-ions and biological contaminants to some extent. This approach can significantly reduce the potential risk to the resin bed or bead upon which it is allowed to pass through. (c) The use of macroporous and loosely packed resin beds prevents potential clogging during emerging pollutant treatment or retention. (d) The use of macroporous and weak base resin prior to strong base resin reduces the risk of counter-pollutant or secondary pollutant deposition due to the presence of organics on the resin surface. Similarly, inorganic pollutants can be prevented from precipitating on the resin bed by the use of weak base or macroporous resin. (e) Developing speciality organic moieties or nanomaterial or a polymeric backbone induces anti-fouling, chemical resistant and mechanical stability to polymeric and nanocomposite resins. For this purpose, acrylic-based polymers, which are well known for their hydrophilicity and use of anti-bacterial nanoparticles, have been proposed. These strategies make resins less susceptible to organic and bio-fouling.

Although ion-exchange resins or beads and/or polymer-based nanocomposites offer an effective alternative to all conventional water treatment techniques for EPs in various contaminated sources, it should be noted that they also suffer from a number of drawbacks. These functional polymers and ion-exchange resins inherently carry several disadvantages to them which are induced by the functionalization step's vital physical properties like a high surface area and active sites which are sacrificed to impart mechanical strength to the system via chemical crosslinking. The use of organic and relatively hazardous solvents in manufacturing processes make them cost intensive and environmentally non-friendly.

Nonetheless, the shelf life of an inorganic, organic or nanocomposite ion-exchange resin or system can be greatly enhanced following a simple but important strategy like constant inspection and cleaning. Conventionally, manufacturers recommend a set of cleaning and maintenance procedures based on the standard users' protocols. However, a rise in the number of EPs on the list of potentially hazardous compounds or substances demands regular inspection and cleaning of the ion-exchange system. This simple strategy would certainly help to preserve the life of the resin bead or system at large. In this direction, several recent studies recommend the following cleaning procedures: (a) The use of warm (48 °C) brine and caustic soda; these mild oxidants or solubilizing mediators are known to improve the cleaning efficiencies in anionic resins. (b) Metal-ion scaling can be washed with the use of hydrochloric acid which can easily remove and prevent EPs like Fe, Al and Mn on resin surfaces. (c) Regular circulation of a sodium hypochlorite (0.25–0.5%) solution can prevent organic fouling; however, a high concentration of sodium hypochlorite may degrade polymer resins to a lower degree of crosslinking. (f) Simultaneous testing and analysis also improve the life span of ion-exchange resin systems. The health of a water treatment system is highly desired. Therefore, to track the condition of the ion-exchange resin and to devise the best cleaning solution and timing, the ion-exchange resins and the resin system, and the unit or plant should be subjected to periodical sampling, analysis and evaluation for their chemical changes, physical robustness, fouling conditions and levels, as well as subject to examinations of their regenerative properties.

REFERENCES

1. C. Calmon, Recent developments in water treatment by ion exchange, *Reactive Polymers, Ion Exchangers, Sorbents*, 1986, 4(2), 131–146.
2. C. J. Johnson and P. C. Singer, Impact of a magnetic ion exchange resin on ozone demand and bromate formation during drinking water treatment, *Water Research*, 2004, 38, 3738–3750.
3. T. H. Boyer and P. C. Singer, Bench-scale testing of a magnetic ion exchange resin for removal of disinfection by-product precursors, *Water Research*, 2005, 39, 1265–1276.
4. L. V. Morozova, A. E. Lapshin, and I. A. Drozdova, Preparation and investigation of porous aluminosilicate ceramic materials, *Glass Physics and Chemistry*, 2008, 34(4), 443–448.
5. S. Babel and T. A. Kurniawan, Low-cost adsorbents for heavy metals uptake from contaminated water: a review, *Journal of Hazardous Materials*, 2003, 97(1–3), 219–243.
6. Mahmoud Fathy, Th. Abdel Moghny, Ahmed E. Awadallah, and Abdel-Hameed A-A. El-Bellihi, Study the adsorption of sulfates by high cross-linked polystyrene divinylbenzene anion-exchange resin, *Applied Water Science*, 2017, 7, 309–313.
7. Sanjeev Kumar and Sapna Jain, History, introduction, and kinetics of ion exchange materials, *Journal of Chemistry*, 2013, 2013, 957647.

8. Greg T. Hermanson, *Bioconjugate Techniques*, ISBN 978-0-12-382239-0, 2013, Amsterdam, Netherlands, Elsevier Inc.

9. M. Naushad, Inorganic and composite ion exchange materials and their applications, *Ion Exchange Letters*, 2009, 2, 1–14. View at: Google Scholar

10. K. G. Varshney and A. M. Khan, Amorphous inorganic ion exchangers, in *Inorganic Ion Exchangers in Chemical Analysis*, Edited by M. Qureshi and K. G. Varshney (eds.), 1991 (pp. 177–270), Boca Raton, FL, CRC Press.

11. M. Shanika Fernandoa, A. K. D. V. K. Wimalasiria, S.P. Ratnayakeb, J. M. A. R. B. Jayasinghe, Gareth R. William, D. P. Dissanayake, K. M. Nalin de Silva, and Rohini M. de Silva, Improved nanocomposite of montmorillonite and hydroxyapatite for defluoridation of water, *RSC Advances*, 2019, 9, 35588–35598.

12. Julie Salvé, Brian Grégoire, Leslie Imbert, Fabien Hubert, Nathalie Karpel, Vel Leitner, and Maud Leloup, Design of hybrid Chitosan-Montmorillonite materials for water treatment: study of the performance and stability, *Chemical Engineering Journal Advances*, 2021, 6, 100087.

13. B. A. Adams and E. L. Holmes, Adsorptive properties of synthetic resins, *Journal of Chemical Society*, 1935, 54, 1–6.

14. F. Helfferich, *Ion Exchange*, 1962, New York, NY, McGraw Hill.

15. M. H. Entezari and M. Tahmasbi, Water softening by combination of ultrasound and ion exchange, *Ultrasonics Sonochemistry*, 2009, 16(3), 356–360.

16. Brian Bolto, David Dixon, Rob Eldridge, Simon King, and Kathryn Linge, Removal of natural organic matter by ion exchange, *Water Research*, 2002, 36(20), 5057–5065.

17. G. Al-Enezi, M.F. Hamoda, and N. Fawzi, Ion exchange extraction of heavy metals from wastewater sludges, *Journal of Environmental Science and Health, Part A*, 2004, 39(2), 455–464.

18. I. A. Abbas, A. M. Al-Amer, T. Laoui, et al., Heavy metal removal from aqueous solution by advanced carbon nanotubes: critical review of adsorption applications, *Separation and Purification Technology*, 2016, 157, 141–161.

19. T. A. Kurniawan, G. Y. S. Chan, W.-H. Lo, and S. Babel, Physico-chemical treatment techniques for wastewater laden with heavy metals, *Chemical Engineering Journal*, 2006, 118(1–2), 83–98.

20. V.T. Sharma, S.V. Kamath, D. Mondal, and S. K. Nataraj, Fe–Al based nanocomposite reinforced hydrothermal carbon: efficient and robust absorbent for anionic dyes, *Chemosphere*, 2020, 259, 127421.

21. H.M. Manohara, K. Aruchamy, S. Chakraborty, N. Radha, M.R. Nidhi, D. Ghosh, and S. K. Nataraj, Sustainable water purification using an engineered solvothermal carbon based membrane derived from a eutectic system, *ACS Sustainable Chemistry & Engineering*, 2019, 7(11), 10143–10153.

22. K. Aruchamy, D. Kalpana, D. Mondal, and S.K. Nataraj, Creating ultrahigh surface area functional carbon from biomass for high performance supercapacitor and facile removal of emerging pollutants, *Chemical Engineering Journal*, 2022, 427, 131477.

23. L. Sartore, M. Barbaglio, L. Borgese, and E. Bontempi, Polymer-grafted QCM chemical sensor and application to heavy metal ions real time detection, *Sensors and Actuators B: Chemical*, 2011, 155(2), 538–544.

24. Z. Ezzeddine, I. Batonneau-Gener, Y. Pouilloux, H. Hamad, Z. Saad, and V. Kazpard, Divalent heavy metals adsorption onto different types of EDTA-modified mesoporous materials: effectiveness and complexation rate, *Microporous and Mesoporous Materials*, 2015, 212, 125–136.

25. Bernd Nowack, Environmental chemistry of aminopolycarboxylate chelating agents, *Environmental Science & Technology*, 2002, 36(19), 4009–4016.

26. Simon Gluhar, Anela Kaurin, and Domen Lestan, Soil washing with biodegradable chelating agents and EDTA: technological feasibility, remediation efficiency and environmental sustainability, *Chemosphere*, 2020, 257, 127226.

27. Muzammil Anjum, R. Miandad, Muhammad Waqas, F. Gehany, and M. A. Barakat, Remediation of wastewater using various nano-materials, *Arabian Journal of Chemistry*, 2019, 12, 4897–4919.

28. Al Arsh Basheer, New generation nano-adsorbents for the removal of emerging contaminants in water, *Journal of Molecular Liquids*, 2018, 261, 583–593.

29. Bingren Tian, Shiyao Hua, Yu Tian, and Jiayue Liu, Cyclodextrin-based adsorbents for the removal of pollutants from wastewater: a review, *Environmental Science and Pollution Research*, 2021, 28, 1317–1340.

30. Chaudhery Mustansar Hussain and Ajay Kumar Mishra, *New Polymer Nanocomposites for Environmental Remediation*, ISBN 978-0-12-811033-1, 2018, Amsterdam, Netherlands, Elsevier Inc.

31. L. Hu, S. Gao, X. Ding, et al., Photothermal-responsive single-walled carbon nanotube-based ultrathin membranes for on/off switchable separation of oil-in-water nanoemulsions, *ACS Nano*, 2015, 9(5), 4835–4842.

32. R. Kumar, M. O. Ansari, A. Alshahrie, et al., Adsorption modeling and mechanistic insight of hazardous chromium on para toluene sulfonic acid immobilized-polyaniline@ CNTs nanocomposites, *Journal of Saudi Chemical Society*, 2019, 23(2), 188–197.

33. Kelly A. Landry and Treavor H. Boyer, Fixed bed modeling of nonsteroidal anti-inflammatory drug removal by ion-exchange in synthetic urine: mass removal or toxicity reduction?, *Environmental Science & Technology*, 2017, 51(17), 10072–10080.

34. Mateus Gustavo Sausen, Fabiano Bisinella Scheufele, Helton José Alves, Melissa Gurgel Adeodato Vieira, Meuris Gurgel Carlosda Silva, Fernando Henrique Borba, and Carlos Eduardo Borba, Efficiency, maximization of fixed-bed adsorption by applying hybrid statistical-phenomenological modelling, *Separation and Purification Technology*, 2018, 207, 477–488.

35. Ira Yudovin-Farber, Nurit Beyth, Abraham Nyska, Ervin I. Weiss, Jacob Golenser, and Abraham J. Domb, Surface characterization and biocompatibility of restorative resin containing nanoparticles, *Biomacromolecules*, 2008, 9, 3044–3050.

36. Majed M. Alghamdia, Adel A. El-Zahhara, Abubakr M. Idrisa, Tarek O. Saida, Taher Sahlabjia, and Ahmed El Nemr, Synthesis, characterization, and application of a novel polymeric-bentonitemagnetite composite resin for water softening, *Separation and Purification Technology*, 2019, 224, 356–365.

37. Emmanuel I. Unuabonah, Bamidele I. Olu-Owolabi, Andreas Taubert, Elizabeth B. Omolehin, and Kayode O. Adebowale, SAPK: a novel composite resin for water treatment with very high Zn^{2+}, Cd^{2+}, and Pb^{2+} adsorption capacity, *Industrial & Engineering Chemistry Research*, 2013, 52, 578–585.

38. M.Y. Haddad and H.F. Alharbi, Enhancement of heavy metal ion adsorption using electrospun polyacrylonitrile nanofibers loaded with ZnO nanoparticles, *Journal of Applied Polymer Science*, 2019, 136, 47209.

39. Q. Liu, L. B. Zhong, Q. B. Zhao, C. Frear, and Y. M. Zheng, Synthesis of Fe3O4/polyacrylonitrile composite electrospun nanofiber mat for effective adsorption of tetracycline, *ACS Applied Materials & Interfaces*, 2015, 7, 14573–14583.

40. N.D. Gultekin, N. Ucar, and Tanger, The effect of sepiolite clay on the properties of polyacrylonitrile composite nanofibers, In: *Nanocon 2014, 6th International Conference*, 5–7 November 2014, Brno, Czech Republic, 2015, pp. 363–367.

6 Functional Materials-Based Microcleaners and Adsorbents

6.1 INTRODUCTION

One of the major challenges the world is facing today is the provision of an adequate supply of sustainable and safe drinking water. In the recent past, several attempts have been made to adopt greener and more sustainable water treatment technologies. Water plays an important role in creating a healthy society and nation and which directly impacts the quality of life and economic growth of a country. On the other hand, two-thirds of the global population face medium to severe water scarcity and many are suffering due to inadequate access to clean potable water. Moreover, increased industrialization, urbanization, expansion and mining processes have worsened the quality of fresh water resources through disproportionate contamination by various conventional and emerging pollutants (EPs).

Presently, it is estimated that over 600 million people lack access to safe and clean drinking water globally. Now, the presence of a new class of pollutants is posing a serious threat to the environment and human beings. In addition to this, EPs of different physical, chemical and biological natures are causing and propagating serious waterborne diseases. On the other hand, EPs leach into fresh water sources and have an origin in many industrial effluents such as those from distilleries, the pulp and paper industry, the dyeing and textile industry, leather processing units and automobile discharges. It is estimated that the paper industry alone releases >50,000 tons of unutilized or excessive dyes into the aquatic ecosystem annually and causes considerable fresh water pollution which results in high COD and BOD contamination. In addition to this, discharges of huge amounts of pharmaceutical and personal care products, also classified as persistent pollutants or EPs such as APIs, surfactants, organics, dye molecules, nanomaterials and synthetic hormones, are making conventional treatment techniques ineffective. On the other hand, novel toxic organic pollutants like plasticizers, pesticides, phenols, polynuclear aromatic hydrocarbons (PAHs), polybrominated diphenyl ethers (PBDEs) and polychlorinated biphenyls (PCBs) have also been categorized as EPs under new norms for which effective treatment needs to be devised. Even though earlier known inorganic heavy metal pollutants, including lead, arsenic, chromium, cadmium, mercury and many other (toxic) metals, have been considered primary pollutants, their emerging ecological toxicity studies have reorganized them into EP categories. These EPs are not only causing a shortage of fresh water sources but continuously adulterate the water bodies and thus threaten life in the ecological environment gradually.

Nevertheless, several environmental monitoring and regulatory authorities are keenly observing the developments in the discharge of polluted waters from industrial sources, the influence of novel nanomaterial-based products entering consumer markets and subsequent discharges into the environment. According to the WHO and European regulatory authorities, the aquatic environment has already been contaminated with more than 700 EPs. Furthermore, the absence of suitable detection and analytical techniques in place makes it increasingly difficult to track these EPs, their metabolites and the transformative products in the ecosystem. These scenarios compel researchers to adopt quick detection as well as removal or treatment solutions to address EPs in the environment on an urgent basis. More worryingly, domestic or municipal wastewater discharges are increasingly characterized by their EPs being above permissible limits, even in treated product water samples. Therefore, the removal of such EPs by a conventional method is proving ineffective and again this is posing newer challenges. On the other hand, several countries including India, America, Middle Eastern nations and African countries have raised serious concerns about the deteriorating ground water quality. These regions have seen high concentrations of EPs like arsenic, fluoride, heavy metals and micropollutants. According to WHO guidelines, these EPs are known to cause deadly diseases when consumed above their permissible limit (arsenic 50 µg L^{-1} and fluoride above 1.5 mg L^{-1}) [1–5].

Nevertheless, recent times have seen considerable effort being made to treat these EPs found in several fresh water sources in order to minimize the environmental impact. The severity of the problem also forces both individual countries and world regulatory authorities to frame revised guidelines to effectively address these impending EP-related issues. In this direction, researchers and innovators have so far devised and tested both conventional and advanced water treatment techniques only with limited success. Also, recent efforts in addressing these global water problems of obtaining high-quality freshwater from polluted water have shown promising results. In this direction, many improved conventional methods have been proposed for the removal or retention of EPs like HMs, APIs, dyes, toxic ionic and non-ionic pollutants like arsenic, fluoride and surfactants. For this purpose, advanced processes like ion-exchange systems have been used in both resin form and membranes for electrodialysis, membrane-based separation processes, photothermal-assisted degradation technologies, adsorption, precipitation coagulation, advanced oxidation processes and improved biological treatment techniques.

Among these, membrane-based separation processes and advanced oxidation technologies dominate the EP-contaminated water treatment technologies. However, conventional membranes prepared using polymeric precursors suffer from a number of issues, like fouling and productivity decline in the long run, which makes systems economically unrealistic. In contrast, adsorption techniques have emerged as the most promising technology due to the use of low-cost precursors and simple preparation methods, high pollutant removal efficiency and operational simplicity. Further, advances in nanotechnology procedures have improved the adsorbents' properties in retaining EPs of a different nature from water sources. In particular, carbon nanomaterials have shown promising results in retaining several pollutants with high adsorption capacities, as discussed in the previous chapter. On the other hand, ion-exchange resin and polymers have also shown higher water disinfection efficiencies due to their unique properties, such as high ion-exchange capacities, and improved and tunable functionalities.

Nevertheless, nanotechnology-assisted products promise to deliver smarter and effective technological solutions at both conventional lab-scale and device-level operations through immobilized active nano-sites. Lower dimensions of nanomaterials also provide a large surface area and greater contact points with the pollutants, which makes them robust and sustainable alternatives to real water treatment applications. In this category, inorganic nanomaterials like metal oxides and hydroxides, engineered biomaterials in gel and aerogel form, and a combination of organic/inorganic systems have also been scrutinized for their pollutant abatement abilities. On the other hand, biopolymers can be extracted from abundant biomass resources. In particular, seaweed biomass is the abundant source for several biopolymers that are used in numerous applications including water treatment. Biopolymers possess exceptional hydrophilicity, anti-bacterial properties, a pore formatting capacity and easy processability that makes them an exceptional choice of material for devising advanced water purification systems. These modern times also require such a robust material stock that can be adapted easily to prepare technologies. Today, biopolymers are extensively used in the form of adsorbents, ion-exchange systems, membranes and hydrogels for conventional as well as emerging pollutants. Biopolymers are also extensively used for oil/water separation owing to their exceptional super-hydrophilic properties. Among them, chitosan, agarose, gelatin and cellulose-derivatives, in individual capacities as well in composite form, show exceptional pollutant adsorption behaviours [6, 7].

Ever since their discovery, petroleum based synthetic polymers have been widely used in various industrially important applications. On the other hand, nanomaterials such as carbaneous end products, biopolymer-based nanogels, aerogels and metal-oxide/hydroxide/mixed oxide materials have also been considered as smart and high surface area materials with tunable physico-chemical properties. However, the inert nature of most synthetic polymers and nanomaterials limits their development for specific applications in various industries. Thus, surface modification of such diverse material stock must be carried out to improve their physico-chemical interaction, adhesion and surface wettability by attaching more polar functional groups and easy to access surface functional groups on both polymers and nanomaterials. In the past, several conventional surface functionalization methods and techniques were developed following (i) attaching one or more primary reactive functional groups both to the bulk as well onto the surface of polymer chains, (ii) tuning the reactivity of reactive groups, (iii) inducing extreme hydrophobic and hydrophilic monomers and (iv) enhancing specific surface characteristics to suit the applications by attaching oligomers or polymers in the form of grafting or blending. Similar strategies have also been followed for other host materials to make them target specific material stocks [8, 9].

In another strategy, the immobilization of surface active or bioactive functional groups on a polymeric surface is commonly accomplished through electrostatic interactions, covalent bonding and a ligand–receptor pairing mechanism. On the other hand, soft polymers like biopolymers have been functionalized through non-covalent physical adsorption to induce an affinity towards specific EPs like APIs and surfactants in water and wastewater samples. This strategy also allows them to retain regenerative properties as well as antimicrobial characteristics. Nevertheless, the covalent immobilizations of specific functional groups on both polymers and nanomaterials offer added advantages such as mechanical and chemical stability and longer shelf life. Among other strategies, polymer grafting is used to achieve polyfunctional moiety immobilization on a polymer backbone. However, selection of the functional group, process and functionalization protocols always depends on the nature of the polymer and nanomaterial for targeted application.

Interestingly, both polymers and nanomaterials have been widely used in membrane technology. There are several advantages to polymers and nanomaterials using membranes in water purification. Specific pore size membranes can be prepared using a suitable polymer dope solution and the desired selectivity can be attached to a membrane either by polymer functionalization or composite formation using designed nanomaterials. Nevertheless, fundamental membrane characteristics like flux and selectivity in a particular process greatly depend on the nature of the polymer, the surface characteristics of the membrane, wettability, pore size and their distribution. Similarly, different water treatment materials require specific surface and bulk properties to perform pollutant rejection and retention abilities. These properties are generally tuned into various stock materials following different strategies which determine the overall performance of the water treatment media and those characteristics are generally determined by considering selectivity and the chemical and mechanical stability of water treatment media in an accelerated chemical, mechanical or thermal environment. Nonetheless, the emerging pollutant removal behaviour of a functional material also greatly depends on operating conditions, fouling propensity, engineering modules and the contaminant nature and functional properties of filters or adsorbents in recycling.

6.2 POLYMER-BASED FUNCTIONAL MATERIALS FOR REMOVAL OF EPS

As discussed, in the previous section, the overall performance of the functionalized polymer or nanomaterial depends on two important factors that essentially need to be taken into consideration before designing a final application, namely (i) mechanical robustness: the physical state of the material in the water and (ii) chemical stability: the chemical nature of the functional moieties attached to the polymer and nanomaterial. In the past few decades, advances in polymer science and chemistry have led to the discovery of versatility in the polymerization methods following which a wide range of polymeric materials have been synthesized and explored for various applications. Polymers have been broadly classified based on their solubility: as either soluble (broadly as hydrophilic) or insoluble in water (broadly as hydrophobic). The significant characteristic of a hydrophilic polymer is that it induces several water lovable properties which are essential for sustainable applications. This helps in the removal of pollutant hydrophilic interaction of water-soluble polymers which are widely used for liquid-phase mixture separations. These approaches have been popularly used in membrane separation techniques where a polymer-based separation medium determines the retention of specific pollutants in processes such as ultrafiltration, nanofiltration and reverse osmosis [10, 11].

On the other hand, processes like solid-phase extraction consisting of fixed-bed or ion-exchange resin made of a functional polymeric system in a column have been popularly used for the removal of both conventional and EPs. This process generally uses a water insoluble polymer with functional active sites both in the bulk and on surfaces. Interestingly, the hydrophilic nature of polymer materials facilitates the organic moiety to have a direct interaction with a polymer backbone and to attach functional groups which will be used for the capturing of pollutants from a contaminated mixture. This process also induces overall enhancement in the hydrophilicity and interaction ability in the polymer, enhancing the efficiency of the separation media. However, the conventional functionalization process is time consuming and cost intensive as the process occurs in multiple steps and uses expensive precursors. This also makes overall polymer–pollutant complex formation during wastewater treatment an expensive process. Moreover, conventional functional polymer systems are difficult to regenerate or difficult to recycle for further use which limits their implementation at an industrial scale. On the contrary, the pollutant removal in solid-phase extraction like an ion-exchange process is easy in which the slow removal or capture or ion-exchange kinetics and low removal rate in polymeric media is due to poor solubility or low interaction of the polymer matrix which makes ion-exchange reversal difficult during washing. Furthermore, the flexibility of hydrophilic polymers enables their application in elution processes like membrane forms, fixed beds or resin columns with high productivity and retention rates. Thus, conventionally the selection of water-soluble polymers has been the choice for the preparation of functionalized materials, which facilitates pollutant retention and regeneration that can be achieved simultaneously. Nevertheless, faster separation adsorption kinetics and high flux rates can still be achieved using water-soluble polymers with suitable regeneration and recycle procedures in place [12].

Recent scares in the form of EPs, like HMs, APIs, organic-micropollutants, pesticides, biologicals, microplastics and nanomaterial residues, have found their way into the food chain and drinking water. Therefore, to weaken their presence in natural sources, various techniques have been developed using precursors of both natural and synthetic origin. These precursors have been transformed into efficient water treatment media through a simple extraction, conversion and functionalization process. Recent studies have also reported the utilization of discarded natural waste into an efficient adsorbent with a high number of active functional groups via renewable engineering processes. These natural wasted resources have shown remarkable potential in retaining a series of EPs from waste sources. Simultaneously, this has also led to the adoption of nanotechnology in producing novel nanomaterials for many industrial and eco-friendly water treatment applications. Therefore, the paradigm has shifted toward the application of sustainable nanomaterials utilizing the bulk and surface properties of polymer systems. Now, attention has been diverted toward utilizing bulk properties to activate every possible potential site with the help of nanotechnology procedures. Over the period, researchers have been dedicated to decoding the advantages of functionalized polymers compared to bare polymer systems by following the physico-chemical properties and nature of a polymer. This is being carried out using nanomaterials as active sites in the bulk of the polymer matrix and attaching pollutant-specific active moieties to the backbone of a polymer. However, prior knowledge of the polymer properties which are to be functionalized helps in selecting appropriate pathways linking the functional groups, polymer backbones and pollutants. Besides the choice of functional groups and the polymer's chemical nature, process parameters also play an important role in for example (i) the physical state, (ii) physical features such as shape, size and dimensions and (iii) morphologies toward the selectivity of polymer-functionalized materials [13].

Nevertheless, functionalization of an organic compound or substance is not new in the literature. However, emerging technological challenges in the field of EPs requires special attention regarding attaching target specific functional groups to a polymer backbone, namely N-donors, like amides, amines and O-donors, from sources such as ethers and alcohol which both have shown greater affinity toward EP attraction. Figure 6.1 shows a list of commonly used functional groups on a polymeric backbone to capture or retain EPs from water and wastewater samples. The design, fabrication and optimization of polymer-functionalized materials is also carried out by attaching common functional moieties such as sulfonyl groups, carboxylic groups and amino groups. However, the removal and capturing of a reaction pathway largely depends on the combination of polymers and functional groups carried by the parent materials in addition to the effluent's acidity. For instance, the deletion and removal of metal cations require that the anionic functional moieties via ion exchange in which metal ion-functional group complexation interactions are likely to occur is as a result of uncharged functional moieties. Nevertheless,

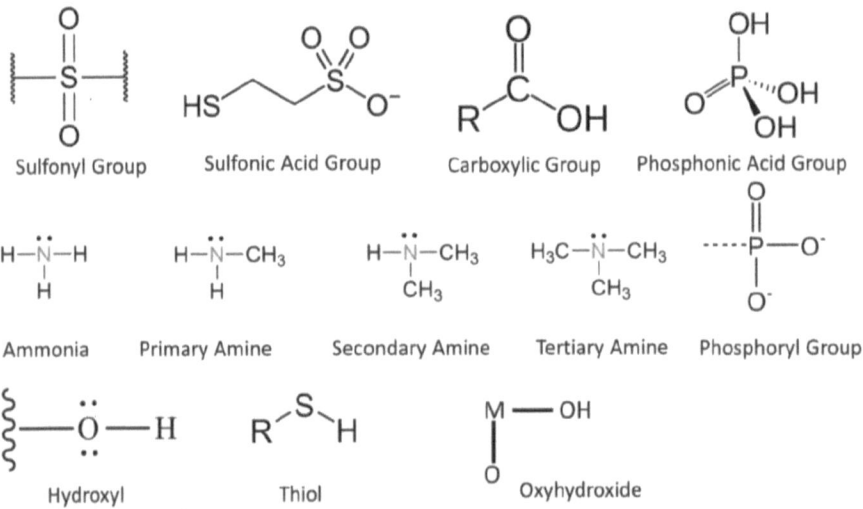

FIGURE 6.1 Commonly used functional groups on polymeric materials useful for EP capture in water and wastewater samples.

functional polymeric composites offer a wide range of property control when specifically applied to, for example, EP capture or removal from wastewater. For instance, the pollutant selectivity and removal mechanism of a functional polymeric material can easily be controlled by adjusting the pH of an effluent solution or wastewater.

On the other hand, the influence of the physical nature or state of polymeric composites and the properties of functional moieties attached on a backbone during application to a water or wastewater treatment process is crucial. Since ground, surface or wastewater samples are characterized by a mixture of pollutants, this necessitates designing a combination of two or more functional groups, such as hydroxyl, phosphonic, sulfonic, carboxylic and amines to enhance performance towards the specific selectivity of functionalized composite materials bearing various targeted pollutants. Further, nanomaterials bearing fascinating interaction properties via various mechanisms have been adopted to induce special EPs retention properties in polymeric composites.

In this direction, researchers have used superhydrophilic polymeric backbones to prepare functionalized bio-nanomaterial (FBN)-based universal filters for the effective removal of fluoride, arsenic and other EPs. As FBN was developed in combination with varying process parameters and material compositions, devices were developed to work at ultra-low applied pressures to filter EP contaminated water through. Further, researchers have developed easy to use and affordable FBNs in tea-bag like pouches that can be supplied in single or multi-packet models to remote inhabitants or for individual or community usage, as shown in Figure 6.2. Here in this procedure, the functionalizing of FBN proceeded with AlOOH and FeOOH activation on to composite biopolymer blends. Further, FBN filter material, packed in a

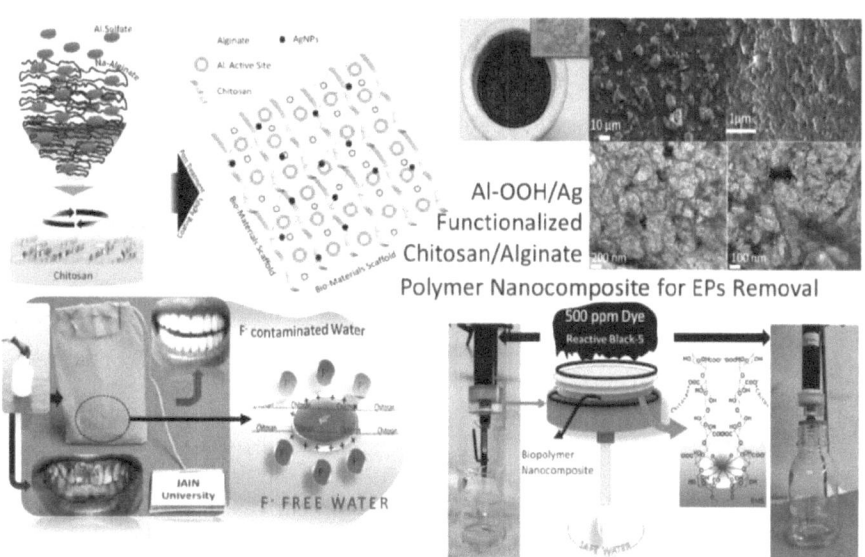

FIGURE 6.2 Preparation, characterization and application of biopolymer-based functional nanocomposite materials used for emerging pollutant removal in different application formats.

layer-by-layer porous bed in membrane form, is optimized for the flux and rejection. Using this user-friendly device, simultaneous removal of fluoride, arsenic (III/V) ions, heavy metals and other EPs from aqueous media can be achieved. These simple approaches enable preparing large scale easy-to-use community kits that can be used for a longer period as the uptake capacity is much higher and performance can be reactivated easily by washing in a simple alcoholic solution for a few minutes [14].

Here, in situ functionalized alumina hydroxide resulted in the FBN's scaffold-like structure and showed high levels of fluoride (F^-), chromium (Cr(VI)) and dye from contaminated water samples. Further, the FBN was surface modified with silver nanoparticles (Ag NPs) to induce anti-bacterial properties which enhances the shelf life of the nanocomposite. The Ag NP-coated on the FBN exhibited a very high EP retention capacity for fluoride at ~168 and ~60 mg g^{-1} at pH 4 and 7, respectively, whereas for Cr(VI) the FBN showed an ~8.5 mg g^{-1} uptake capacity at room temperature. Further, nanomaterial packed in the filter kit form showed >99% organic EPs like Reactive Black 5 (RB-5) dye with outstanding surface regeneration properties. Similarly, several other combinations of biopolymers and synthetic polymers have been adopted to functionalize the surface with appropriate active sites. In each case, the end-user affordability, in different formats (tea-bag, powdered adsorbent form, fixed bed filter and filter cake form), has been tested for EP removal successfully. These strategies have demonstrated a viable and sustainable alternative to the removal of HMs, toxic pollutants like arsenic, fluoride and chromium, and organic contaminants from water sources. Moreover, polymeric backbones provided highly stable and water tolerant network structures which helped in maintaining the mechanical robustness of the adsorbent or filter media. Further, to explore the multipollutant removal approaches, chitosan-based iron, aluminium and a mixture of two metal composites were explored as efficient adsorbents for the removal of arsenic, APIs, organic molecules like dyes, fluoride and biological contaminants from an aqueous medium, though there are limitations to discrete water systems.

Likewise, several superhydrophilic biopolymers have been converted into functional water filters with the help of simple preparation routes. However, currently used adsorption-based techniques or methods have shown limited success in powder or filter form. Therefore, hybrid methods and other methods are being adopted in the form of filter membranes which are emerging as a remedy to EP-contaminated water-related problems and are showing efficient removal of organic EPs and HM contaminated water with high a retention capacity and product flux. However, the high-cost material precursor, limited flexibility for diversified or multiple pollutants, and low water permeability restrict their large-scale utility. Thus, the development of a cost-effective protocol for rapid and versatile water purification is of the utmost importance. Alternatively, highly porous polymeric systems like sponges, possessing enhanced affinity toward multiple pollutants, have been shown to be new-generation adsorbents for the purification of water. Nevertheless, this functional polymer nanocomposite approach demonstrates the super-adsorbent capacities for several EPs from water and wastewater samples. Extensive characterization-supported results reveal that: (i) a functional polymer nanocomposite shows a remarkable As, F^-, Cr(VI), API, biological contaminant and organic EP uptake efficiency; (ii) in many cases functional polymer nanocomposite material can be easily recycled and reused with high efficiency of regeneration; and (iii) a super-hydrophilic polymeric adsorbent also offers after-use value addition. For instance, a fluoride-absorbed functional polymer nanocomposite forms a ralstonite-like permanent strong bonded structure because of the strong binding of F^- with a functional polymer nanocomposite. In such a situation, a super-hydrophilic biopolymer-based functional polymer nanocomposite can be used for the separation of oil/water emulsions. On the other hand, a stable powder or granular end product can be easily packed to create a filtration device. Some of the approaches have shown the use of a functional polymer nanocomposite in user-friendly modules with different configurations, namely a column filter, a cake filter, a powder adsorbent and tea-bag-like pouches. Therefore, a functional polymer nanocomposite offers an appropriate, sustainable and environmentally benign alternative solution to producing reusable and in some cases safe drinking water from contaminated underground, surface water and industrial wastewater streams. Hence, functional polymer nanocomposites immensely contribute to solving the endemic problems of EP-contaminated water, this approach in particular proposes an alternative solution to industrial water cycling and producing safe drinking water for much needed rural and semi-urban inhabitants.

6.3 POLYELECTROLYTE COMPLEXES IN THE REMOVAL OF EPS

In addition to functional nanocomposites, polymers have been considered in various forms and compositions using state-of-the-art techniques for wastewater treatment applications. The advanced methods and techniques are capable of controlling surface characteristics and functionality to sustain pollutant retention and polymeric active sites during and after capturing EPs from wastewater. Recent synthetic tools and techniques have led to the development of several water treatment systems, transforming the use of polymers and methods to convert them into active treatment media with several advantages. Recent studies have shown that the polymers with a specific composition can respond in some desired direction and way to changes in external stimuli like pH, temperature, electrical or magnetic field, mechanical responses and physico-chemical variations. However, the driving force behind these transformations and changes are external stimuli like pH shifts which alters the surface charge from anionic or cationic to neutral. These changes may also happen due either to a pH shift or the addition/altering of an oppositely charged polymer. This influences changes in possible interactions and initiates possible new interactions. Generally, in such situations the efficiency of hydrogen bonding alters, with an increase in ionic strength or temperature which may lead to the collapse of hydrogels and the interpenetration of the polymer network [15, 16].

Therefore, charged polymers can not only transform active substances into highly interactive ones, but also can be charge distributions that can be controlled to yield biocompatible and environmentally friendly complexes. On the other hand, ionic polymers play a vital role in various applications that are either organic or inorganic in nature in which both covalent and ionic bonds make up their molecular structure. These are the fundamental characteristics that distinguish the ionic polymers from other conventional polymeric systems. Therefore, in such systems polymers are held together by primarily covalent bonds and then van der Waals and hydrogen bonds.

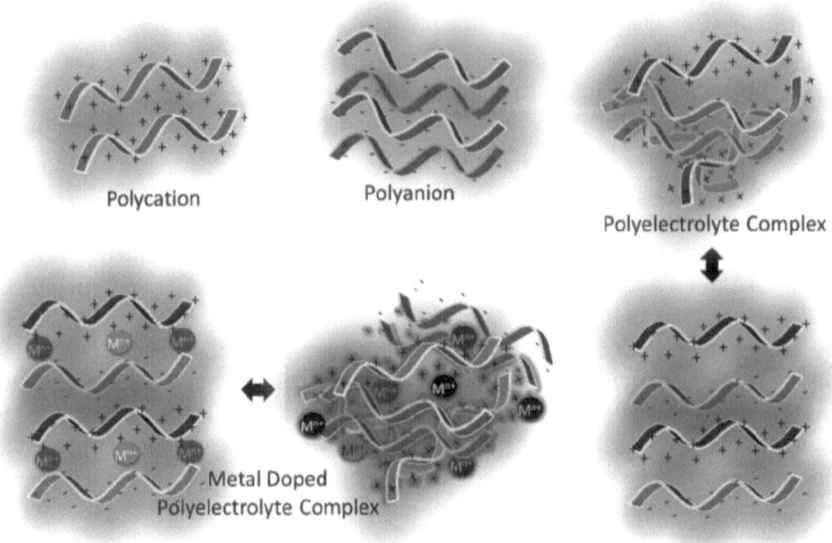

FIGURE 6.3   Schematic illustration of polyelectrolyte complex (PEC) formation.

Therefore, when blended or mixed in composition, interaction between two oppositely charged polymers results in the formation of a complex, termed a polyelectrolyte complex (PEC) as shown in Figure 6.3. Also, PECs are defined as the associated complexes formed between oppositely charged anionic and cationic polymers. These PECs have been designed and optimized to actively participate in pollutant removal and retention, and can be adapted to meet various requirements like mechanical and chemical robustness and high EP uptake efficiency.

PECs are formed due to electrostatic interaction between oppositely charged polyion, polyanionic and polycationic systems. Unlike in crosslinked or interpenetrating networks, PECs do not require any use of chemical crosslinking agents, which make them non-toxic and environmentally friendly. The PECs once formed attain highly stable complexes even under varied pH conditions. For instance, a biomacromolecule system like DNA is combined with polycation like chitosan to result in PECs generally used for biomedical applications. Generally, the polymers that contain a net negative (polyanion) or positive (polycation) charge at a near-neutral pH are also called polyelectrolytes. PECs are generally soluble in water and their solubility is influenced by the electrostatic interactions between water molecules and the charged monomeric unit. Examples of such polymers include DNA, protein, certain derivatives of cellulose polymers, chitosan, acrylic polymers and carrageenan [17].

The polyelectrolytic systems are mainly classified into two types: natural and synthetic polyelectrolytes. However, PECs are further divided into several categories based on their origin and composition as natural, synthetic and chemically modified biopolymers as shown in Table 6.1. Further, based on the molecular architecture, PECs are also sub-divided into linear, branched, cross-linked and based on the

TABLE 6.1

Natural, Synthetic and Modified Polyelectrolytic Systems

Natural Polyelectrolytes

	Nucleic acids
	Poly (L-lysine)
	Poly (L-glutamic acid)
Polyanion	Carrageenan
	Alginates
	Hyaluronic acid
Modified Biopolymers	
Polyanion	Pectin
Polycation	Chitosan (deacetylation of chitin)
	Cellulose-based
Polyanion/Polycation	Starch-based
	Dextran-based
Synthetic polyelectrolytes	
	Poly (vinylbenzyl trialkyl ammonium)
	Poly (4-vinyl-N-alkyl-pyridimiun)
Polycation	Poly (acryloyl-oxyalkyl-trialkyl ammonium)
	Poly (acryamidoalkyl-trialkyl ammonium)
	Poly (diallydimethyl-ammonium)
	Poly (styrenesulfonic acid)
	Poly (vinylsulfonic acid)
Polyanion	Poly (acrylic or methacrylic acid)
	Poly (itaconic acid)
Polyampholytes	Maleic acid/diallyamine copolymer

electrochemistry of the constituent polyacids or polyanions, polybases or polycations, and polyampholytes.

On the other hand, PECs of both natural and synthetic origin have been used as coagulation and flocculation agents in a variety of wastewater treatments. PECs are known for their ability to interact with pollutants in wastewater via electrostatic interactions and the hydrophilic properties make them easy to interact with water soluble EPs. In water treatment applications, the presence of charge–charge interactions in a PEC ionic and pollutant plays an important role in removal efficiency. In these systems retention of a charged pollutant will be strongly affected by the existence of charge–charge interactions in the PEC. Therefore, recently, these interactions have been widely explored regarding the removal of EPs following PEC/EP interactions. These interactions have earlier been utilized in pervaporation and fuel cell applications. However, lately, PECs have been adopted for nanofiltration and ion-exchange purposes in battery applications. Further, a dip-coating method was followed to modify sodium carboxymethyl cellulose (Na-CMC) to prepare NF membranes using a water-soluble quaternary ammonium cellulose ether-based PEC. Thus, hydrophilicity and NA-CMC/PEC-NF membranes with surface characteristics can be easily altered using types of polyelectrolytes and their mixing ratios.

Similarly, a layer-by-layer (LBL) method was followed to prepare poly(diallyldimethylammonium) (PDDA) and sulfonated polyether ether ketone (SPEEK) multi-layered PEC membranes on a hydrolyzed polyacrylonitrile support.

Further, this multi-layer PEC membrane was used for the filtration of organic solvents like DMF, THF and isopropanol from various mixtures. In another simple facial PEC preparation method, a nanoporous PEC was prepared using poly[1-cyanomethyl-3-vinylimidazolium bis(trifluoromethanesulfonyl)-imide] (PCMVImTf2N) and poly-(acrylic acid) (PAA) as a polycation and polyanion, respectively. Similar to the LBL approach, a nanoporous PEC was produced via the alternate dipping of oppositely charged solutions, consecutively. In this procedure, an LBL-coated multi-layer film was immersed in an aqueous NH_3 solution; the PEC turned into a porous film via electrostatic complexation. This simple scalable process resulted in a highly stable and porous NF membrane which was later used for several contaminant removals from wastewater sources. Further, the mechanically stable polyacrylonitrile porous support was surface modified to result in a negatively charged PEC membrane. This novel membrane was achieved using a blended mixture of negatively charged copolymers from acrylic acid and acrylonitrile. In this PEC system, the acrylic acid part acted on active and functional sites, whereas the acrylonitrile provided a film-forming property to the complex. Further, similar PEC systems were prepared and applied to various wastewater treatments.

In addition to the list provided in Table 6.1, several synthetic polyelectrolyte systems have been in use for water treatment application with varying levels of competency. The most commonly used synthetic systems in water treatment include poly(allylamine hydrochloride) (PAH), poly(ammonium 13 acrylate) (PAC), poly-ethylenimine (PEI), polyacrylonitrile (PAN), poly (vinyl alcohol) (PVA), poly(styrenesulfonate) (PSS) and polyacrylic acid (PAA), which have been studied for heavy metal removal. Interestingly, these polymers have been used in both film and electrospun fibre mat form. Even though polyelectrolytes are generally water soluble, to electrospin the polymeric system into a fibre mat requires non-aqueous media. Nevertheless, advantages in electrospinning technology have enabled the fabrication of water soluble polyelectrolytes or PECs to be used in water treatment applications. In general, synthetic polymers possess hydrophobicity which make them susceptible to surface fouling when used in membrane filtration applications. Therefore, increased efforts have been made to prepare, modify and utilize hydrophilic polyelectrolyte or PECs for water filtration application which inherently prevents the fouling phenomenon. Nevertheless, a combination of synthetic polymers such as chitosan (extracted from chitin), PAH and PAA can improve the hydrophilicity of the membranes. In one such effort, a layer-by-layer deposited PAA/PAH polyelectrolyte complex has been converted into fibre membranes for EP removal. Here, a PAA/PAH PEC solution was electrospun into fibre mats and used for the removal of heavy metals from water sources. Further, this approach was extended to modify the surface of an ultra-porous UF support with PAA/PAH nanofibre mats and used for HM removal from landfill leachate, highly heavy metal contaminated surface water and natural organic matter (NOM) removal [18].

Similarly, charge functionality in polyelectrolytes has been widely utilized to attract heavy metals to bind to and be removed from the contaminated sources. Here, functional groups such as carboxylates on a PEC matrix establishes strong interactions with metal ions which facilitates the retention of such EPs from the wastewater sources. In addition to this, the PECs containing ionizable repeating units such as amine, carboxylate and sulfonate play a crucial role in solubility and pollutant retention properties. However, several detailed studies reveal the mechanical robustness of electrospun PECs as having polyelectrolytes such as PAA/PAH or PAA and chitosan (CS) in aqueous solutions. Interestingly, electrospinning techniques can yield ultra-thin nanofibres with diameters ranging from microns to sub-microns. These nanofibre characteristics induce a high surface area and activity to PEC mats. Another technological advantage with nanofibre PEC mats is the scalability which makes them a cost-effective solution to EP treatment in medium to large scale operations. Also, the high porosity and aspect ratio, surface charges and physical nature helps in immobilizing active nanoparticles and functional groups on to the electrospun PEC fibres in both in situ as well as post-modification procedures. Further, prepared PEC systems have been used for the removal of organic EPs like dyes, oil/water separation, heavy metal removal and the treatment of oily wastewater. Nevertheless, electrospun PECs and membranes have been highly efficient in retaining heavy metal contaminants from wastewater sources. Specifically, electrospun PEC fibre mats produced using PAA and PAH have shown excellent heavy metal uptake properties with a highly stable mechanical integrity in aqueous solutions. Further studies on PAA/PAH have shown high EP uptake efficiencies, such as cadmium (Cd), lead (Pb) and copper (Cu), in different pH conditions. Also, the effect of a co-pollutant such as NOM on PAA/PAH PEC HM removal efficiencies have been evaluated. Interestingly, this approach has shown the mechanical robustness of PECs in the presence of NOM and HM-interacted NOM.

Nevertheless, the determining effect of a water matrix and the level of contamination on EP removal using such PEC fibers is essential when considering practical water treatment applications. There needs to be further studies to address the above adsorption/removal mechanism of PECs for different EPs. This critical knowledge gap will also depend on water quality indexes like concentration, severity, pH, temperature and origin of the wastewater.

6.4   CARBON-BASED FUNCTIONAL MATERIALS FOR REMOVAL OF EPS

Carbaneous materials in different forms such as graphite, graphene, fullerenes, carbon nanotubes and graphene oxides are generally prepared from fossil-fuel-based precursors. Recently, advancement in nanomaterials and physico-chemical methods like chemical vapour deposition have also been explored to prepare carbaneous materials. Even though these methods are tedious and energy-consuming synthetic processes, advanced applications in the field of energy conversion, storage, sensing and electronic applications require a huge supply of these materials. On the other hand, in a commercially successful form, activated carbon (AC) is one of the most frequently used adsorbents in the removal process of industrially important EPs such as heavy metals, organic compounds, herbicides, dyes, APIs and many other toxic compounds. To produce a high surface area AC, there are two types of activation procedures, namely physical or thermal, and chemical. In a physical activation process, the ligno-cellulosic precursors, such as biomass or partially carbonized materials, can undergo gasification with water vapour and/or carbon dioxide. A chemical activation process proceeds with the impregnation of chemicals such as KOH, $ZnCl_2$ or H_3PO_4 on lignocellulosic or carbonaceous raw materials, which later undergo carbonization or pyrolysis. This process is subsequently followed by the washing of chemical precursors to remove the residual activating agent [19, 20].

Generally, carbaneous materials are prepared based on the end-user specifications. The controlled pyrolysis is generally carried out under air or nitrogen to generate a highly porous carbon nanomaterial. Commercially, most of the ACs are prepared following oxidative pyrolysis using biomasses of soft and hardwoods, mineral carbon, lignite, peat, bones and coconut shells. Initially, AC was considered to be an expensive material due to the processing conditions involved. Further, the chemical and physical treatment methods add to the final product cost. However, these processes yield low productivity at the expense of high energy consumption. Nevertheless, recent studies and technologies have proven cost-effective for the production of functional activated carbon. Also, functionalized carbon involving nanomaterials are being prepared using various strategies. Several strategies have been adopted to drastically reduce the cost of functional-activated carbon production in which wasted resources from industry and agricultural residual wastes are used as precursors.

At present, there are various types of small, medium and large scale sophisticated activated carbon filters that are available on the market for industrial wastewater treatment. Porosity, the nature of precursors and the desired high surface area play important roles for targeting specific EP removal in contaminated sources. On the other hand, EPs are largely characterized by their unique surface functionality which requires a designed functional moiety in AC to match up the uptake efficiency in both powder or bed. However, most of the activated carbon systems limit their uptake efficiency for conventional pollutants. Therefore, it is very important to design EP-specific surface functionality in functional carbon, which is also the type-specific porosity and level of uptake capacity to match the high EP concentration of contaminants. With this strategy it is expected that the carbon-based filters will last for several years. However, EPs still require a specific functionality on an AC surface to effectively remove contaminates from wastewater for which high temperature (800–1000 °C) activation using either steam or chemical activation with the help of activating agents is followed as discussed earlier.

Nevertheless, conventional activated carbon filters have also shown limited but very important removal efficiencies for EPs like disinfection by-products, micro-organic contaminants like pesticides, APIs, dyes and heavy metals. A high surface area AC is conventionally used for the removal of chlorine and disinfection by-products. However, as dichlorination occurs quickly, the carbon experiences surface erosion. This necessitates the replacement of the carbon filter to maintain the desired dichlorination efficiencies in the long run. Among many disadvantages, AC provides a suitable environment for bacterial accumulation, growth and proliferation. On the other hand, the removal of micro-organic pollutants from wastewater is a highly desired. AC has the capacity to retain micro-organic pollutants like dyes, APIs, pesticide and herbicides which are now classified as EPs from surface as well as wastewater samples. AC adsorption of micro-organic pollutants depends on several factors: (i) physical properties like surface area and pore size and their distribution; (ii) the chemical functionality on the surface, like active hydrogen and oxygen moieties; (iii) the chemical nature of the pollutant in the wastewater; (iv) the pH and temperature of the wastewater; (v) EP and AC exposure or interaction timing; and (vi) the chemical and mechanical stability of the ACs. The major drawback of conventional AC is the absence of target specific functionality, cost intensive AC conversion methods and the process control in tuning the surface morphologies in carbaneous end products.

Among many precursors, agro-bio-waste or agriculture residues have been considered sustainable, green and cost-effective carbon sources. In this direction, toxic weed like *Parthenium hysterophorous* (PH), which is considered an agricultural hazard, has been effectively converted into an ultra-high surface area, highly porous and functional carbon for the removal of dyes and heavy metal ions. The PH also offers a simple conversion into a highly functional, biomimetic helical carbon with varying degrees of oxygenated functionalities on the surface. These end products have been used for attracting bio-catalysts and further used for the removal of EPs from water sources. Further, to produce sustainable and greener end products, researchers are investing in finding natural-origin material stocks which can be converted through an environmentally friendly synthetic route. This automatically considers the drastic reduction of production costs, and renewable precursors and processes leave behind a low carbon footprint. Consequently, carbaneous materials derived from biomass have attracted remarkable interest due to their easy availability, abundance, tunable properties and inherent chemical functionalities. In this direction, different biomasses have been utilized as carbon sources and successfully explored for diverse applications, namely coconut shells, human hair, chicken eggshell membranes, lotus pollen, auricularia, silk, cornstalk, toxic weeds and seaweed. Overall, high surface area carbaneous materials can be produced from biomass through a one-step or two-step process. In the one-step approach, (i) the high temperature (700–900 °C) carbonization or pyrolysis is carried out under an argon or nitrogen atmosphere; (ii) the high porosity in pyrolyzed carbon is induced via physical activation under a CO_2 atmosphere at high temperature; and (iii) chemical activation proceeds with the use of activating agents such as KOH. In the two-step method, both pyrolysis and physical activation/chemical activation are followed simultaneously. The chemical activation is carried out using activating agents like NaOH, KOH, $ZnCl_2$, NH_3, H_3PO_4, $CaCl_2$/urea and $K_4 [Fe(CN)_6]$, which induce high porosity in the resulting carbon materials. Among these, KOH and NaOH are extensively used due to their low cost and facilitating highly microporous structures with an enhanced specific surface area. More significantly, these activating agents induce high –OH functionalities on the surface of the carbon materials.

On the other hand, hydrothermal (HTC), solvothermal (STC) and iono-thermal-based (ITC) functional carbon materials possess rich surface functional groups. These functional groups have been widely explored in various applications such as water purification, catalysis and for energy conversion and energy storage. Conventionally, HTC, STC and ITC-based carbon materials are produced via low temperature in the range of 150–250 °C. Here, the carbonization occurs at a low vapour pressure with the precursors of both synthetic and natural origin. HC has been widely used in the past as one of the cost-effective methods and is widely used for the conversion of biomass which can be converted into hydrothermal carbon through a simple protocol. As a result hydrothermal carbon is frequently used for energy and environment related applications. Nevertheless, HTC, STC and ITC processes induce a relatively low surface area and limited functionality. However, the surface area and functionality can be further enhanced by the incorporation of (i) templating agents in the form of soft/hard templates during the carbonization process and (ii) the use of pore formation agents, namely hyper-saline eutectic salt mixtures such as $ZnCl_2$ and LiCl [21, 22].

However, these low temperature carbonization techniques have several disadvantages, such as the use of hazardous chemical reagents to remove the hard template and the requirement of an additional post-activation process to obtain a high surface area. Also, the preparation of carbon using these techniques does not provide tunable properties in the precursor carbon and final product properties, such as shape, functionality and surface acidity, which cannot be controlled. Nevertheless, these methods offer simple, sustainable and cost-effective process routes to produce highly functional carbon which can be easily scaled up via eco-friendly approaches. These methods also offer varying morphologies in the resulting carbaneous product, with a highly diverse heterogeneous functional carbon. Hence, several ionic liquids (ILs) and green solvents like deep eutectic solvents (DESs) have been used as a template and structure directing agents for carbon products. These template-assisted STC and ITC carbonization processes can be used to produce controlled morphologies with several optimization routes. These functional carbons have been widely used for the removal of a series of EPs like heavy metals, APIs, artificial hormones, biological contaminants and organic pollutants such as pesticides and herbicides from contaminated wastewater sources [23].

Further, the functional group attached to carbon in different templating agents and experimental environments yields outstanding properties which interestingly induce significantly improved carbonization with attractive surface properties. However, issues related to low carbon yield, non-porous structures, high cost and the toxicity of IL-based STCs limit their use on the industrial scale. On the other hand, due to their attractive attributes, such as ease of preparation, cheapness, scalability and low toxicity, DESs have been employed as a potential substitute for ILs for the solvothermal conversion of biomass to carbon helices, as the reaction medium for material synthesis, and as the precursors for carbon materials prepared via high temperature carbonization (500–900 °C). DESs, also for some time referred to as green solvents, are known for their natural origin and are gaining increased interest due to several advantages they possess, such as ease of preparation, high thermal stability, low vapour pressure, high conductivity, moderate viscosity, cost-effectiveness and scalability. Interestingly, recent advances in green material development have adopted a series of DESs in multi-functional product synthesis where they play important roles. Remarkably, DESs are used simultaneously as a structure directing template, catalyst, precursor and reaction medium/solvent. In one such attempt, a eutectic solvent system was designed using ethylene glycol (EG), glucose (Glu) and choline chloride (ChoCl) to both act as solvent and templating agent. These eutectic systems are also known for their non-hazardous and non-toxic components in the synthesis of STCs. Also, metal-based ternary DESs also have been widely used for inducing functionality to hydro/solve/iono-thermal carbon. Nevertheless, activated carbon in granular form has been widely used for drinking water treatment and as a hybrid step in wastewater treatment. Activated carbon is generally produced in bulk from naturally abundant organic materials in the absence of oxygen, which activates the surface, creating a highly porous functional carbon system.

Further, metal assisted ternary templates have been used to functionalize carbon surfaces. Novel approaches have used a series of metal salts like $FeSO_4$, $FeCl_3$, $ZnCl_2$ and $KMnO_4$ as template agents to prepare highly oxygenated, metal-doped, task-specific and multi-functional carbon adsorbents. These strategies well serve the supply of on-demand applications. On the other hand, innovations in nanotechnology procedures have also led to the understanding of the recyclability and regeneration of carbon-based adsorbents. In this direction, metal-doped, surface functionalized, hydro/solve/iono-thermal carbons have been further used as new surface regeneration catalysts during and after the uptake of EPs. Among these, metal-doped carbon such as Fe-STC has emerged as Fenton catalysts (Fe^{2+} + H_2O_2) to simultaneously uptake and degrade micro-organic EPs from wastewater. Even though some of the precursors and procedures are currently considered expensive, efforts are being made to drastically reduce the production cost of functional carbon by following new strategies. One of these efforts is an iono-thermal procedure which has produced highly surface functionalized carbon that is an effective EP removal medium at low-cost and scalable potential. These active microcleaners have been extensively tested for their micro-organic pollutant removal from water sources. In all cases, carbon attains highly oxygenated polar and acidic surface functionalities which play a key role in adsorbing cationic pollutants; the presence of metal traces supports the degradation of adsorbed organic pollutants through a Fenton reaction. Similarly, hydroxyl and carbonyl-functionalized carbon has shown high carbo-catalytic-behaviour for the removal of nitrobenzene.

The industrial importance of carbocatalysts was realized via the selective reduction of nitroarene which is a vital process for producing respective aromatic amine. These approaches play a vital role from both an industrial and economical point of view. Therefore, green-solvent-assisted functional carbon has proven to be a highly effective adsorbent for EP-contaminated waters. On the other hand, a major feature of functionalized carbon filters or adsorbents is their recyclability and surface regeneration. Functional carbon materials have been successfully regenerated at the surface to retain >90% activity in a repeated cycle without compromising the permeate flux and high adsorption capacities. These features make hydro/solvo/iono-thermal carbon a sustainable alternative for EP-contaminated water purification in both batch and continuous flow methods.

Further, carbon precursors have been surface modified with nanomaterials which induce a desired functionality to the filter media. In this approach, carbon precursors were mixed either in situ or via post-treatment routes with a series of metal precursors like Mn, Zr, Al, Fe and a combination of multi-metals. Upon hydro/solve/iono-thermal treatment, highly surface functionalized active carbon was produced. In one such study, a CC-EG-based deep eutectic solvent system was used to synthesize active Mn_3O_4 nanomaterial. Here, DES acted as both a solvent and reactant in the metal oxide conversion from a $KMnO_4$ precursor at different temperature ranges; however, optimal 140 °C for 2 h showed high performing active materials as end products [24].

Thus, functional carbon materials prepared with the help of green solvents like water, deep eutectic solvents or ionic liquids yield highly stable, robust and easily recyclable adsorbents. These functional carbons can be easily applied to EP-contaminated water treatments such as (i) an adsorbent or filter for the selective retention and removal of ionic pollutants like APIs, HMs and micro-organic pollutants; (ii) a Photo-Fenton catalyst for the degradation of organic dyes and adsorption of secondary degraded pollutants; and (iii) a heterogeneous carbocatalyst for the reduction of hazardous organics like nitrobenzene. Furthermore, activated carbon and functional carbon adsorbents have been also extended to the adsorption of synthetic dyes and pharmaceutical drugs, synthetic hormones, biological contaminants and secondary pollutants. Among many functional wastewater treatment approaches, the carbon-based adsorption method has been found to be an efficient and economic process for EP extraction from contaminated sources.

6.5   2D MATERIAL-BASED FUNCTIONAL MATERIALS FOR REMOVAL OF EPS

Recent times have seen the use of nanostructured materials in various applications including water treatment and environmental applications. Generally, nanostructure materials are classified into the several following general dimension rules, namely 0D nanoparticles or nanodots; 1D nanowires, nano-strands or nanotubes; and 2D nanosheets or nanoplates. This categorization mainly considers at least one of the three dimensions of nanomaterials smaller than 100 nm in order to be considered a nanomaterial structure. Advancements in nanotechnology are opening up new avenues in utilizing next-generation water purification and treatment systems. Nanomaterials, in particular the 2D nanomaterials shown in Figure 6.4, are also considered as nano-engineered materials which have been widely considered for a series of conventional as well as EP treatments in different water samples. Ever since their discovery and successful understanding, single-layer graphenes, 2D nanomaterials, layered nanostructures such as boron nitride (BN), graphite-carbon nitride (g-C_3N_4) nanosheets and various transition metal dichalcogenides (TMDs) have received increasingly significant attention due to their unique physical, chemical and physicochemical properties over recent decades. These 2D nanomaterials typically have a large surface area, electrical conductivity, chemical stability, easy functionalization and mechanical strength. Besides these, other 2D nanomaterials composed of two or more elements possess more novel properties [25, 26].

Engineered 2D nanomaterials have been particularly effective in adsorbing or degrading or retaining various EPs like micro-organics, such as APIs, pesticides, herbicides, metal ions, dyes and other organic pollutants. Interestingly, 2D nanomaterials have been successfully characterized for their antimicrobial, anti-viral and anti-bacterial properties against waterborne microbes. 2D nanomaterials offer extraordinary features, supporting their potential as efficient remedial solutions with unique material features like a high surface area, tunable porosity, excellent mechanical properties, ease of fabrication, lower cost and energy requirements, greater chemical reactivity, ease of functionalization and recyclability. This section focuses on

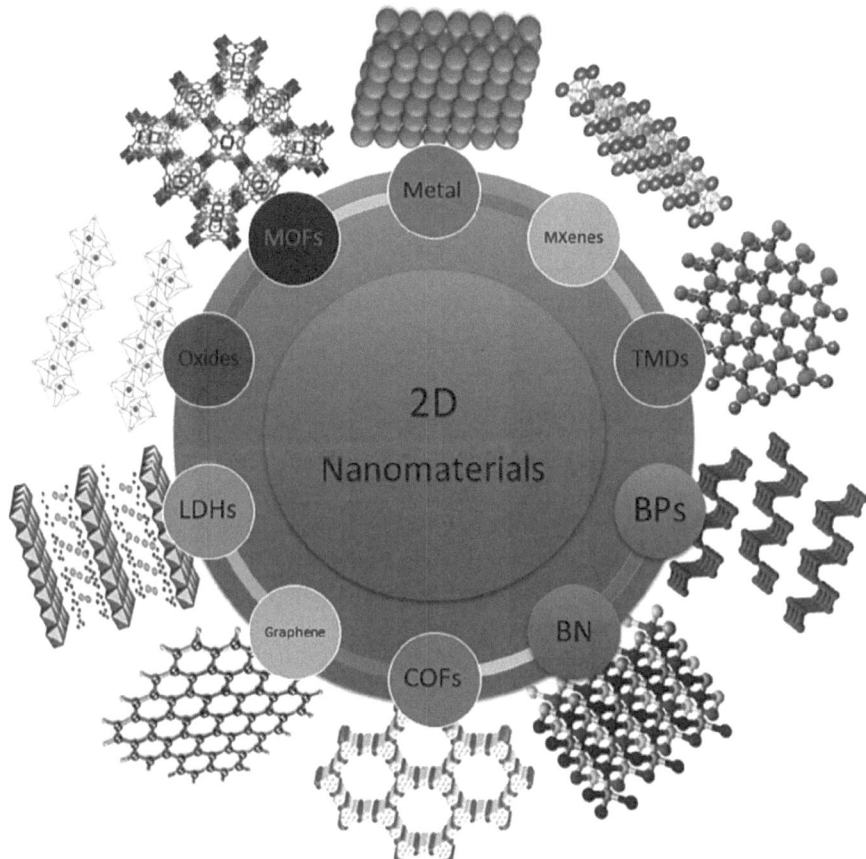

FIGURE 6.4 Ultra-thin 2D nanomaterials available for various applications, now typically considered as potential material stock for water treatment applications.

providing an informative description of the features of 2D nanomaterials used for water treatment applications. This section also briefly gives the advantages and limitations of 2D nanomaterials in water treatment applications, and the definitive risks involved. However, today, the nanomaterials area is considered a new class of materials having several advantages over conventional technologies and being generally regarded as more beneficial than harmful. Many different 2D nanomaterials have found environmental applications, such as nanoscale natural clays, zeolites, metal oxides, hydroxides, mixed oxides and carbaneous materials, carbon nanotubes, graphene-based materials, nanofibres, enzymes and various metal-based systems. All the nanomaterials found in 2D architecture and morphologies are generally prepared and utilized in the form of graphene-based composites, carbon nanotubes, metal organic frameworks, graphitic carbon nitride composites, metal oxides and composites, and commercially available nanomaterials. Nonetheless, these 2D nanomaterials have also been used in composites with polymers as membrane materials.

Membrane technology has been a highly effective means to purification, separation and treatment of various types of conventional and emerging pollutants from water sources. Among several advantages, membrane technology offers a clean and direct separation of contaminants from water sources. However, it also suffers from some drawbacks such as membrane fouling, permeability and selectivity trade-off, long-term membrane stability, fouled membrane waste and concentrated solid waste, which have limited the development and implementation of membranes for EP treatment. Membrane fouling refers to an accumulation or deposition of insoluble pollutants or solutes on membrane surfaces that are susceptible to adsorption, cake formation and pore blocking, which further leads to the concentration polarization of contaminants. These unwanted interactions between membranes and pollutants affect the overall performance of the membranes and the unit in operation. In particular, membrane pore blocking results in trans-membrane pressure increase which leads to a decline in pollutant rejection, flux and membrane lifetime. Fouling in membranes mainly occurs due to the relatively high hydrophobic nature of the membrane material which attracts constituents that are categorized as organic, inorganic or biological in nature.

Therefore, in order to improve anti-fouling properties in membranes, several 2D nanomaterials have been used as composites in membrane preparation. As discussed in the membrane section, several strategies have been adopted to increase membrane life such as chemical, physical, biological and hybrid physico-chemical routes. Among these, physical cleaning methods have been considered advantages for operational membranes. The use of nanomaterials has increased many times as a composite to improve membrane performance. Although these methods lessen fouling, these cleaning techniques involve harsh chemicals and large pre-treatment approaches which result in a drastic increase in operational cost and the generation of secondary pollutants. Therefore, improving membrane properties can be achieved by developing composite structures involving intrinsic anti-fouling characteristics. However, the use of nanoparticles such as TiO_2, zirconia, alumina and clay nanomaterials in membranes has shown some drawbacks, including pore-blocking, agglomeration, degradation of the membrane when used as a photoactive nanomaterial creating defects, and the leaching of the nanoparticles. On the other hand, 2D nanomaterials have shown remarkable potential as a membrane constituent for building blocks or filler components which also offer flexible membrane structure, mechanical strength, chemical inertness and separation performance. Among several nanomaterials, 2D architectures have received significant importance for their superior anti-fouling properties, photo-degradation and pollutant retention properties. Membranes prepared using graphene-based 2D nanomaterials have shown greater stability, permeability, rejection and anti-fouling [27].

On the other hand, 2D nanomaterials have a unique nanosheet architecture, high surface areas and an atomic thickness that offers a separation barrier for all of the ionic pollutants exposed to the surface via electronic coupling. In addition to this, the separation behaviour of 2D nanomaterials differs from that of bulk or conventional nanomaterials: due to their large surface area, nanosheet architecture improves both the surface coarseness and hydrophilicity in addressing inactive organic and biological pollutant accumulation. Moreover, the presence of a 2D nanomaterial sheet structure in thin-film composite membranes has shown great improvements in chlorine resistance, inducing higher mechanical robustness and chemical resistance. Nevertheless, in all membrane applications, 2D nanomaterials such as graphene-based systems have been mainly used to induce anti-fouling properties which limits the scope of 2D nanomaterials in membranes.

On the other hand, the combination of inorganic 2D nanomaterials with polymers allows hybrid adsorbents or membranes to combine the cost-effectiveness of polymers with the superior perm-selectivity and outstanding mechanical, thermal and chemical stability of inorganic 2D nanofillers. However, in the case of 2D nanomaterials, high concentrations on low specific surfaces tend to form aggregates which undermine the advantageous effects of 2D materials. Therefore, novel composite preparation routes have been adopted to maximize the utilization of 2D nanomaterials modified by filtration media. With these advantages, 2D nanostructured materials stand out for their water filtration applications.

On the other hand, ever since their discovery in the mid-2000s, graphene-based nanomaterials in multi-layer and single layer forms have been considered promising molecular separation media in water treatment systems. Based on their nanoporous, promising molecular sieving properties, graphene oxide (GO) nanostructures in nanosheet forms are continuously being discovered for water purification applications. The structure of GO resembles a honeycomb lattice with functional groups on the boundaries or around holes. Oxidized sp^3 and pristine graphitic sp^2 regions are arbitrarily distributed in the basal plane of the nanosheet structure. GO nanosheets also carry various defects that originate from their preparation or during their modification procedures. Nevertheless, there are significant surface functionalities like hydroxyl and epoxy groups, while carbonyls are located at the edges and in holes to help in retaining various EPs from water samples. However, a major disadvantage with GO comes from its weak Van der Waals attractive forces, which makes it unstable in water. Nevertheless, GO as free-standing membranes have been widely tested for their potential in colloidal, dye, micro-organic and pharmaceutical contaminant separation in various simulated as well as real wastewaters. Several researchers have proved the potential of GO nanosheets converted into moderately large-scale sheet membranes following various physical and physico-chemical routes like chemical vapour deposition techniques. However, in all these cases, efforts have fallen short as the mechanical stability of free-standing GO-based membranes have shown limited scope for large scale applications. However, GO in combination with a polymer matrix is still considered as an effective and cost-effective material stock for EP treatment.

In addition, given the large surface area of 2D nanomaterials, it is important to induce more active sites and greater affinity for targeting EPs, which make them highly significant for improving photo-catalytic efficiency. The main features that contribute water treatment properties to 2D nanomaterials are: (i) 2D nanomaterials have a higher surface-to-volume ratio in combination with a high number of active sites that exist at their surface; (ii) conductive 2D nanomaterials provide excellent electron mobility which can induce an ultra-fast intra-particle transfer of electrons from the bulk to the surface/interface which effectively separates charge carriers in photo-degradation; (iii) the dispersibility of photo-catalysts can be improved by 2D nanomaterials due to their unique dimension features such as their small size and nanosheet structure which is helpful for light-harvesting and photo-catalysis; (iv) 2D structures offer compatible, easier and convenient features to form hybrids, blends or composites with other functional species, in particular 2D standalone and 2D/nano-composites and 2D/polymer composites which exhibit a large contact area and a more efficient interfacial charge transfer. Therefore, these extraordinary features bestow 2D nanomaterials with promising applications in water treatment applications [28–30].

Recently, a new class of 2D nanomaterials, including molybdenum disulfide (MoS_2), graphitic carbon nitride (g-C_3N_4), bismuth oxide (Bi_2O_3), tungsten oxide (WO_3) and bismuth oxyhalide (BiOx), has been widely explored for the photo-catalytic decontamination of EPs in water and wastewater samples. Among these, MoS_2 nanosheets owning a band gap at ≈1.8 eV are frequently used as photo-catalysts or co-photo-catalysts for removing organic EPs. Similarly, g-C_3N_4 with a band gap of ≈2.6 eV, WO_3 with a band gap of ≈2.7 eV, Bi_2O_3 with a band gap of ≈2.5 eV and BiOx with a band gap of ≈3.0 eV have widely been used as visible light-activated photo-catalysts for the decontamination of various EPs.

6.6   MOFS, COFS AND ZIFS IN REMOVAL OF EPS

Inorganic and metal-organic nanomaterials have also received significant attention due to their simple preparation techniques and scalability. One such class of 2D nanomaterials used for water purification applications are oxides, layered double hydroxide (LDH), covalent organic frameworks (COF), metal organic frameworks (MOF), zeolitic imidazole-frameworks (ZIFs), transition metal dichalcogenides (TMD) and graphitic carbon nitride. Among other water remediation techniques, the photocatalytic degradation of pollutants using metal-oxides (MOs) and composites of graphene/MOs, GO/MOs, LDHs, MOFs and BNs have been frequently considered as one of the most cost-effective and environmentally friendly strategies. These 2D nanomaterials utilize light energy to generate free radical groups that subsequently oxidize organic EPs into less-harmful substances. Based on the photo-catalytic process, most of these 2D nanomaterials are effective in breaking up various EPs like dyes, pharmaceuticals, pesticides, micro-organic compounds like endocrine disrupting compounds, and semi-volatile organic compounds from water and wastewater sources. Interestingly, these 2D nanomaterials offer unique catalytic degradation features for treating EPs via simple and cost-effective means. Some 2D nanomaterials are highly efficient in utilizing solar energy to initiate photo-catalytic degradation and the complete elimination of EPs to make them environmentally friendly methods. Nevertheless, research efforts have been devoted to the developments of visible-light-driven photocatalysts possessing a band gap in 2D nanomaterials to reduce it to the range of 1.65–3.10 eV.

MOFs are typical porous and crystalline nanomaterials that are built using metal ions/clusters and organic ligands via coordination bonds. MOFs feature well-defined and highly tunable porous nanostructures, as they carry secondary building units in the form of organic moiety which can be easily altered for their ligand topology, connectivity and chemical functionality. Therefore, MOFs are promising structural nanomaterials for molecular sieving membranes which can be potentially used for the separation of liquid mixtures based on molecular size and shape of specific components. In recent years, several researchers have dedicated their time to exploring MOF-based polycrystalline nanomaterials, or composite membranes and other hybrid materials, for gas as well as liquid mixture separations. Initial studies were dedicated to understanding the stability of MOFs in water; based on initial findings, it was observed that MOFs possess poor stability in water which restricted their liquid separation applications. Initially explored polycrystalline MOF-5-based membranes were stable in organic solvents, however they failed to sustain structural integrity in moist conditions. Lately, advancements in MOF preparation techniques have induced structural and chemical stability to nanomaterials. On the other hand, polycrystalline ZIF-8 in which zinc ions are coordinated by four imidazolate rings have been explored to retain ionic species and pollutants from various water samples. However, ZIF-8 components were unstable and susceptible to dynamic degradation in aqueous media through continuous Zn^{2+} leaching which restricted the wider use of MOFs and ZIF-8 separation and purification applications involving water. This promoted the early exposure of MOFs and ZIF-8 to gas separation applications in combination with highly stable polymers like polyimides.

Recently, it was noticed that the water stability of MOFs is particularly determined by the strength of metal-ligand bonds. Earlier, researchers struggled to achieve the chemical and structural stability of MOFs in aqueous media. Water treatment applications require highly water-stable MOF systems. Therefore, finding new material combinations involving metal-ligand systems to prepare water-stable MOFs was the most desired aspect for the treatment of aqueous pollutants.

Some of the very important materials in this category like HKUST-1 and MOF-5 have moisture sensitivity. However, later studies noticed that the stability of MOFs in water was particularly dominated by the strength of metal-ligand bonds. Thus, researchers designed novel approaches for the preparation of water stable, robust coordination bonds such as Hf-O (802 kJ mol^{-1}) or Zr-O (776 kJ mol^{-1}) to construct MOF-based membranes. In this direction hydrothermally grown polycrystalline UiO-66(Zr)-membranes have been explored for their application to the desalination of seawater via pervaporation dehydration. Further, innovations in the material preparation technique have led to the preparation of a series of water-stable MOFs, like Cr and Fe-based MIL-100, Cr and Al-based MIL 53, Cr, Fe and Al-based MOF-74, Cr, Fe and Al-based MIL-101, and ZIF-8 and Zr-based MOFs. The metal and organic linkers are susceptible to breaking in MOFs when they come in contact with water due to weak coordination. However, MOFs like Cr-MIL-101 with a Cr-O bond, MOF74 with an Mg-O bond, Al-MIL-53 with an Al-O bond, ZIF-8 with a Zn-N bond and MIL-100(Fe) with a ZIF-8 bond were observed to be more stable in aquatic environments. Another study also revealed that they were water stable during direct exposure to water. Further, recent studies have shown the extensive use of UiO-66($-NH_2$), HKUST-1, DMOF-1-NH_2, Mg MOF-74, DMOF-1 and UMCM-1 for water treatment applications. Among these, zirconium-based MOFs such as UiO-66 and UiO-66($-NH_2$) have been proven to possess highly stable characteristics in water and have been widely considered for water purification applications. On the other hand, aluminum hydroxide isophthalate-based MOF CAU-10-H has been tested for its stability in various accelerated hydrothermal conditions.

Nevertheless, a significantly improved MOF structure has been in use in recent times with varying morphologies, sizes and chemical compositions. To achieve this, both metal and organic ligand chemistry has been explored following an ingenious design and synthetic route to tune MOF properties. These design strategies have helped researchers to develop products with accurate control over the pore architecture. Using these advantages, initially MOF nanofillers were used as structurally flexible materials in membrane construction for water purification applications. These crystalline MOFs with zeolitic porous structures provide better compatibility and affinity for polymer matrices, resulting in flexible and controlled pore size membranes. Subsequently, polymer/MOF NF membranes have been extensively used for a series of pollutant removal applications. However, the hydrophobic nature of MOFs still restricts their use for the large scale and wider use in water purification systems.

These developments and limitations with MOFs have motivated researchers to disperse these nanofillers in an organic phase having monomer diamine followed by polymerization to yield in a randomly distributed nanocomposite polyamide layer. However, MOFs have failed to appear uniformly in the polyamide layer due to the hydrophobic interaction in the aqueous phase, making the overall polyamide layer an underperforming barrier. However, recent studies reported good dispersion of UiO-66(Zr)-NH_2 MOFs in a polymer matrix which resulted in a high performance membrane being used for seawater desalination [31–34].

In spite of these several advantageous characteristics, when MOFs were added to a polymer matrix there resulted mixed matrix membranes (MMMs) or a thin-film nanocomposite (TFNC), and most of the time a synergic effect was missing. In many cases, the performance of MMMs was restricted, due to the formation of MOF aggregation which may also block useful porosity and restrict their distribution. These drawbacks have led to the adoption of alternative membrane preparation strategies, placing MOFs in the top active layer which helps in rejecting high concentration EPs and retaining traces of EPs. This strategy also helped in post-surface modification of polymers as well as MOFs in the active layer to induce a useful surface functionality such as hydrophilicity. Therefore, MOFs induced surface functionalization which helped in tuning permeation and rejection properties in membranes. However, there are three main factors that govern the functionalization of membranes with MOFs, namely (i) nanofiller loading and physical parameters like type, size, surface charge and functional groups on MOFs; (ii) the continuous polymer matrix is used for membrane preparation which induces surface charge and roughness, and offers functional group modification; and (iii) the functionalization process, both on MOFs and the polymer matrix. For instance, dopamine has the capability to self-polymerize in an alkaline solution at pH ~8.5 in the presence of oxygen, resulting in an adhesive polydopamine (pDA) end product. In addition, pDA also carries catechol moieties that are available for secondary reaction. These special features of MOFs and pDA helps in the functionalization of copper-MOF (Cu-MOF) induced in both the bulk and surface of NF membranes. Here, dopamine was used as the functionalizing agent which also facilitated a binding and stabilization effect between the Cu-MOFs and the polyamide layer of TFC membranes; this induced enhanced permeation and rejection. The filtration tests on Cu-MOF/pDA showed high permeation with high rejections for organic EPs like methylene blue, a positively charged and methyl orange, and a negatively charged pollutant through a charge interaction. Further, several MOF/polymer-based nanocomposite membranes have been successfully used for dye, textile, pharmaceutical and automobile industry wastewaters.

Recently, MOFs have also gained considerable interest as efficient adsorbents for various pollutants due to their high porosity, and chemical and structural tunability. Hydrophilic MOFs have displayed strong adsorption affinities following the fast adsorption kinetics of organic EPs like dyes, pesticides, herbicides, pharmaceuticals, perfluoro substances like perfluorooctane sulfonate (PFOS), and perfluorooctanoic acid (PFOA). Interestingly, MIL101(Cr) with a high adsorption capacity has been explored for organic pollutants as well as for PFOA. A high specific surface area and mesoporous structure in MIL101(Cr) have enabled higher adsorption efficiencies for EPs. Even though MIL101(Cr) has been used in a restricted form for drinking water applications owing to their potential toxic Cr leaching vulnerabilities, they have been widely considered for wastewater treatment applications. On the other hand, Zr-based MOFs containing carboxylate-based organic ligands have been an interesting choice for practical use in large scale adsorption applications due to their exceptional stability which arises from the strong Zr–O bond. Therefore, specifically, UiO-66 and Zr-MOF have proven to be promising alternatives for water treatment applications. There have been reports showing both Zr-MOFs and UiO-66 to be effective in removing PFAS and PFOS from contaminated water sources.

Interestingly, adsorption capacities in MOFs can be significantly improved by introducing defects in the structures. With this hypothesis a series of MOFs was prepared for the adsorption of organic and perfluoro compounds. In particular, UiO-66 has been highly successful in adsorbing PFOS and PFBS from contaminated water sources. Further, this approach was followed to generate different MOFs with higher porosity and a larger number of coordinatively unsaturated sites such as (CUS) Zr^{4+}; they have been utilized for adsorption applications.

Further, industrial effluents consisting of various organic and inorganic substances like phthalates exhibit endocrine disrupting properties. In one such attempt, MIL-53(Al) has been used to successfully remove organic EPs like dimethyl phthalate (DMP) from industrial wastewater. In addition, various forms of MIL-53(Al) were prepared using different Al sources like aluminium hydroxide, boehmite, alumina and aluminium nitrate, and showed a high adsorption capacity and affinity for various organic EPs. In this series, the adsorption capacities of UiO-66(Zr), UiO-66(Zr), ZIF-8, NH^{2-} for EPs like diethyl phthalate (DEP) and phthalic acid (PA) were successfully carried out. Interestingly, among these, ZIF-8 showed the maximum adsorption capacity of >650 mg g^{-1} for PA at different pH values. This enhanced the adsorption capacity of ZIF-8 and was attributed to its electrostatic and acid-base interactions. Another main feature of MOF or ZIF or COFs is that, upon adsorption, active materials can be easily regenerated and recycled with a simple methanol wash.

In another study, phenolic compounds such as bisphenol-A (BPA), a well-known endocrine disruptor, have been successfully removed from contaminated wastewater using MIL-53(Cr). In this MOF, the adsorption of BPA was proceeded with *p–p* interactions and hydrogen bonding between the pollutant BPA and hydroxyl groups of MIL-53(Cr). Similarly, BPA was successfully removed from contaminated sources using MIL-101(Cr) and MIL-100(Fe). Among these, MIL-101(Cr) has proven to be the best MOF in terms of high adsorption capacity and kinetics which is attributed to its higher surface area and pore size characteristics. Similarly, COFs with their unique structural features like permanent porosity, pore size and distribution, are highly flexible as well as possessing a controllable molecular design, and various accessible building blocks make them a potential choice as a precise molecular sepa-rating media. COF-based water treatment systems in both forms as adsorbent and composite membranes with a polymer matrix have been shown to be highly efficient EP-separating media. Even though COFs are relatively new compared to MOFs and ZIFs for water treatment applications, COF-based separation systems have shown greater potential for water and wastewater treatment applications [35–37].

6.7   GELS AND AEROGELS IN REMOVAL OF EPS

Advancements in polymer science and technology have led to the invention of new forms of polymeric end products that are being widely used in various applications. Hydrogels are one such class of highly crosslinked three-dimensional polymer net-work structures emerging as an effective adsorbent for water treatment applications. A hydrogel is a moderate to highly crosslinked 3D network of hydrophilic poly-mers that can swell in water and retain a large volume of water while sustaining the structure. The hydrogel can be stabilized against external stimuli and high solvent conditions through chemical or physical crosslinking of specific polymer chains. Hydrogels are generally prepared using hydrophilic polymers of both synthetic and biopolymer origin that are made water insoluble via crosslinking. However, polymer swelling capacity can be easily controlled with a degree of crosslinking in different working and application conditions. This swelling property and affinity for different crosslinking agents and solvents has been utilized. This property makes hydrogels highly swellable and can absorb a large volume of solvent inside its 3D reticulated network structure. On the other hand, crosslinkable sites on the polymer back can retain EPs like heavy metals, organic pollutants, pharmaceutically active species and synthetic dyes from wastewater [38–40].

Utilizing these advantages, research emphasis has been shifted toward the design and development of new crosslinked polymeric adsorbents for different pollutants as an alternative economical and efficient treatment method. Further, hydrogels have been prepared in functional as well as composite forms to enhance the pollutant retention property. In this direction hydrogels are being engineered by incorporating various nanomaterials such as magnetic nanoparticles, 2D/3D nanoparticles and photo-active nanomaterials which can be utilized efficiently for the removal of EPs. With these advancements, hydrogels have been widely explored for highly hazardous municipal, ground as well as industrial wastewater treatment applications.

A simple route of hydrogel preparation was gradually transformed to utilize poly-mer complex systems to be an efficient charge carrying gel useful for water treat-ment. In this direction, polyelectrolyte complex systems have been adopted as superior pollutant adsorbents. Depending on the nature of the polymer constituents in polyelectrolyte complexes, the overall active charge on the product gel can be tuned and then further utilized for EP adsorption purposes. Nevertheless, as discussed in the previous section, a major drawback of polyelectrolyte complexes comes from their limited adsorption capacities; they may also stabilize pollutants in water instead of completely settling due to their excessive charge. Nevertheless, biologically active polydopamine (PD)-PAH polyelectrolyte complexes with the help of agarose were initially converted to hydrogel beads and further used in water treatment applications. Prepared PEC-loaded hydrogel beads were exposed to ionic pollutants in water to study their adsorption and transportation phenomenon. Hydrogel exposed to ionic pollutants behaved like a conventional flocculant in removing pollutants from water. Further, the adsorption capability of the PD-PAH PECs loaded in the agarose hydrogel beads were extensively studied for chemical stability and regeneration capacities. However, PEC-based hydrogel systems showed limited potential in EP removal from contaminated water samples.

However, with the proper selection of polymer and crosslinking agent combinations, hydrogels can be tuned to obtain a large scale of flexibility, fast swelling kinetics, and adjustable surface characteristics such as functionality, charge, a large area, a fast diffusion process, fast kinetics, a controllable pore structure, permeability catalysis, thermo-stability, interesting acid/base properties and hydrophilicity. Following these parameters, recent studies have applied advanced hydrogel systems for the removal of organic and inorganic pollutants including metal ions, harmful dyes and persistent pharmaceutical ingredients from various water and wastewater samples.

In hydrogels, an EP removal mechanism depends on several factors such as the nature of the polymer and functional groups. In addition to these, feed parameters like pollutant chemistry and experimental conditions like concentration, pH, temperature, ionic strength and ligands also influence EP removal efficiency. Further, hydrogel and pollutant interfaces occur via hydrogen bonding, electrostatic interaction, acid–base interactions, ion exchange, complexation, hydrophobic interaction and coordination/chelation. However, in most of the hydrogels, EP adsorption mechanisms are described by a combination of two or more interactions that occur concurrently to varying degrees. Nevertheless, most hydrogels remove EPs from contaminated water sources via physisorption and chemisorption which is stimulated to take place simultaneously via H-bonding, acid–base interactions, chelation and/or ion exchange. For instance, in chitin or chitosan-based hydrogels, heavy metal pollutants are adsorbed via one or more interactions, depending on the chemical composition and process operating conditions. Similarly, in hydrogels with amino and vicinal hydroxyl groups, pollutant adsorption happens by a coordination reaction in which protonated amino groups facilitate an ion-exchange mechanism by replacing the counter-ions with the metal anions. Further, electrostatic interactions stabilize the adsorption processes under acidic conditions, making the removal process comprehensive. In another example, cyclodextrin (CD)-based hydrogel adsorb heavy metal ions through a complex formation in which CD molecules and heavy metal contaminants are involved in host–guest interactions. Here, the hydrophobic interactions and chemical crosslinking and network structure controls and dominates the heavy metal adsorption.

On the other hand, an aerogel is an ultra-high porous, light weight, synthetic foam produced from various precursors. An aerogel is usually produced as a result of freeze-drying the hydrogel in which all the solvent component (>90%) is subjected to sublimation (removal of ice) processes. This freeze-drying process converts a highly swollen hydrogel into the highly porous macrostructure of a 3D hybrid network as shown in Figure 6.5. Recent advancements and understanding of hydrogels, which were increasingly tested for their potential application to water filtration, has led to innovations in aerogel developments. Recent years have seen extensive research efforts in design and developing various precursors-based aerogels for water treatment applications. Initially, biopolymer-based aerogels were developed for industrially important oil/water separation, dye removal from industrial wastewater, and multiple EP removal from ground and surface waters [41, 42].

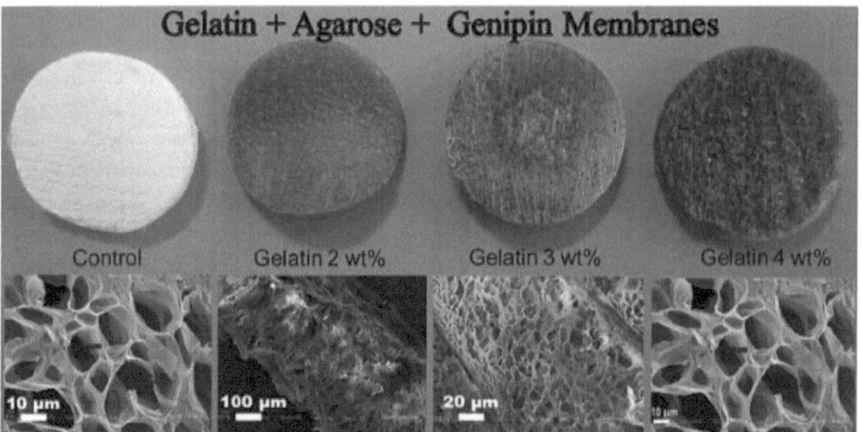

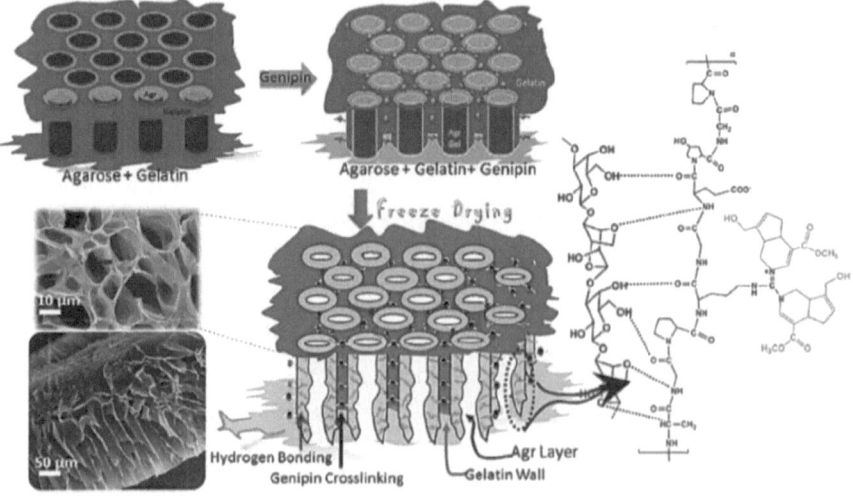

FIGURE 6.5   Actual images, SEM micrographs and schematic illustration of the pore formation mechanism in biopolymer-based aerogels.

In addition to this, recently innovations in affordable cryogenic technologies have led to the exploration of a series of synthetic polymer-based precursors such as polyvinylidene difluoride, synthetic fibres, polyvinyl alcohol and graphene-based hybrid materials for aerogel preparation. Even though initial studies focused on utilizing biopolymer precursors aimed at preparing superhydrophilic ultra-porous aerogel for oil–water separation in which a high permeate flux of water was achieved in a continuous flow method. These efforts were also extended to treat highly complex oil spills, automobile industrial oil/water wastes and organic/water mixture separations. For this, biopolymer-based systems like agarose, chitosan, gelatine, aerogels and blends and composites of chitosan/agarose, agarose/gelatin, sodium alginate/cellulose nanofibril (CNF), alginate/graphene oxide and agarose-/bovine serum albumin have been extensively used in foam, membrane and ultra-porous filter forms for various organic/water, oil/water and industrial wastewater treatments. In most of the cases, biopolymer-based hybrid aerogel membranes or foams or filters have proven to be robust separation media for water from oil/water mixtures, thermodynamically stable emulsions and ship-breaking yard spill over waters with moderate to high permeate flux rates. These oil contaminated water or wastewaters have been lately categorized as EPs and potential risks for aquatic life and the environment.

Although aerogels possess exceptional light weight, low density and ultra-high porosity, which helps in separating oil/water, emulsion and oil contaminated wastewater efficiently, they fail to retain EPs like APIs, organics and HMs. Therefore, it is more interesting to improve their EP retention properties by incorporating functional groups or active nanomaterials. In this direction, several efforts have been made to improve their EP separation. These efforts have identified and utilized remarkable properties of biopolymers such as natural abundance, biodegradability, negligible toxicity and stability under different testing conditions to create an ideal parent or base material for functionalization or nanomaterials. This approach yields highly surface active, functionally rich, active sites on a host biopolymer 3D network structure for the eco-friendly separation of various EPs. Several studies have utilized biopolymer functionality and nanomaterial active sites as EP-capturing media, while porous channels have provided easy pathways for purified water to pass through. In one such attempt, a task-specific nanomaterial was in situ decorated on a biopolymer-based aerogel membrane for the removal of multiple contaminants including both conventional and emerging pollutants from water. In this case, an agarose/chitosan biopolymer-based aerogel was prepared with in situ doping of α-FeOOH and γ-AlOOH active nanomaterials to achieve sustainable and robust water purification media. Here, a superhydrophilic biopolymer network provided large water permeation channels, whereas active α-FeOOH and γ-AlOOH nanomaterial sites captured multiple EPs like F, As, Pb, Cr, pesticides and pharmaceuticals such as paracetamol, ciprofloxacin, organic dye molecules and synthetic hormones [43].

These aerogel-based water filter media are increasingly gaining importance due to their target specific design to achieve ultrafast water permeability. These aerogel foams, filters and membranes have been increasingly proven to be advantageous over conventional polymer film-based membranes due to their high flux rates, target specific design options, easy recyclability and biodegradability after the active materials run out of their activities. In addition to this, a large biopolymer stock provides ample opportunity to tune a specific porosity and surface functionality to target specific EP treatments. Among the several biopolymers, a chitosan-based blend and nanocomposite aerogels have been extensively employed due to their sustainability and antimicrobial nature for real industrial contaminated water treatment. Similarly, a series of polysaccharide-based macromolecules and agarose have been explored for EP treatment which undergoes easy gelation through the formation of extensive intermolecular H-bonding, resulting in double-helical structures that aggregate into thick bundles. Also, the stable hybrid 3D structure of a chitosan/agarose aerogel was prepared with the help of ethylenediaminetetraacetate dianhydride (EDTAD) crosslinking as well as a pore forming agent. This recipe resulted in a highly water stable superhydrophilic aerogel with the ability to withstand harsh conditions like pH variations and applied temperature ranges. In this configuration, chitosan was design as a major component to induce an antibacterial property into the resulting aerogel. On the other hand, AlOOH and α-FeOOH nanocomposites induced excellent uptake capacity for fluoride, organic contaminants, arsenic and APIs. Further, these designed aerogel filters have shown high performance toward the successful removal of multiple pollutants like surfactants, dyes, hormones and pharmaceuticals, with high fluxes and substantial rejections of 99% for most EPs. Even though aerogel technology has still to attain large scale production, nanotechnology-assisted nanocomposite aerogels have shown greater potential for the large-scale applicability and excellent recyclability to real industrial contaminated water purification. Nevertheless, aerogel technology has proven itself to be a sustainable alternative to conventional filtration methods and technologies currently in use to mitigate, in part, global water scarcity.

6.8   LIQUID MEMBRANES IN REMOVAL OF EPS

Ever since their discovery by Li in 1968, liquid membranes (LMs) have been considered as promoting water treatment techniques. An LM is defined as a separation membrane media made of liquid. An LM system comprises a liquid phase like a thin oil film, present either in a supported or unsupported arrangement, and which serves as a membrane barrier between two phases of aqueous solutions or gas mixtures. The design of LMs mainly involves water-immiscible emulsions as suitable sources comprising an oil phase of surfactants and additives in a hydrocarbon solvent which captures the microscopic droplets of an aqueous solution of suitable reagents for removing or stripping and/or trapping contaminants from wastewater sources. For instance, the removal of ammonia from wastewater samples can be achieved using liquid membranes, in which ammonia is present in mobile equilibrium with the ammonium ion.

The large volume of different types of wastewater are increasingly posing a threat to the environment due to rapid industrialization. This has led to the development of various water treatment technique adsorptions, sand filtrations, polymeric adsorbents and chemical processes like flocculation using aluminium salts, iron salts and biological processes. However, many of these methods face issues due to the high level of toxic organic compounds which hinder the heavy metal retention capacity in various adsorption materials which leads to the collapse of the treatment technique. In such situations, liquid membrane processes offer superior pollutant retention properties in special cases through a solvent extraction route at relatively low treatment costs.

Interestingly, emulsion liquid membranes (ELMs) have been recognized as a potential alternative for the precipitation and removal of EPs like heavy metals from wastewaters. ELMs have also been effectively used for the recovery of precious platinum group metals from precious metal refinery and mining wastewaters. Similarly, the removal of Pb, Cu, Ni, Zn and Cd has been achieved from their waste mixtures using di-2-ethylhexyl phosphoric acid as the EP carrier and extractant over a wide range of pH and ionic strength values. The nanoparticle pollutant of $PbCrO_4$ has been successfully removed from various industrial wastewaters using a kerosene, sorbitan monooleate liquid membrane. Similarly, cobalt extraction in the form of $CoCl_4$ from various wastewater sources was achieved using a trialkyl amine hydrochloride LM. Similarly, trialkyl amine hydrochlorides ELM was used as an extractant for the selective removal of cobalt in excess of 99% from a hydro-metallurgical wastewater effluent. The tannery industry has used Cr(VI) extensively for leather processing purposes and which leaches out into the wastewater system and may cause detrimental impacts on human and environmental health, and can be easily pre-concentrated using an ELM process. Further, Cr(VI) can be reduced to Cr(III) with the help of $FeSO_4$ catalysis at pH 2, making it environmentally benign. Similarly, the removal and/or recovery of Zn metal ions can be removed from acidic industrial wastewater using ELMs up to 99.3% efficiency after optimization. However, several ELMs also suffer from a number of drawbacks such as stability and concentration alteration which limits their large-scale use in EP treatment. These limitations may also originate from the collective swelling of the ELM with elevated surfactant concentration as a result of their affinity to water.

Pharmaceutically active ingredients like antibiotics in wastewater such as tetracyclin and chloramphenicol pose serious risks to the environment upon their long-term accumulation. Separation of these antibiotics from production broth or reaction mixtures has been popularly carried out using UF, NF or diafiltration processes. However, the anti-microbial nature of these pollutants may pose difficulties for the materials involved in traditional wastewater treatment systems. This necessitates the use of LMs instead of cost-intensive membrane separation. For instance, an ELM-based process has been successfully achieved for the extraction of antibiotics along with alcohols and carboxylic acids from wastewater. Similarly, both ELMs and supported liquid membranes (SLMs) were utilized for the extraction of antibiotics like erythromycin.

On the other hand, protein-rich contaminants have been increasingly characterized in food industry and dairy industry wastewaters. The amino-acid-rich dairy industry wastewaters pose a serious COD and BOD induction potential upon release into aquatic systems. These proteins and counter-pollutants induce physical, chemical and biological severity to food and dairy wastewaters. Upon accumulation, pollutants like lactose, fats and proteins convert wastewater into organically rich EPs. Conventionally, protein and fat are generally precipitated out to reduce pollution loading to some extent. However, the precipitation removal method has shown a low removal efficiency and requires expensive reagents. Among several other proteins, casein accounts for 80% of the total proteins in dairy industry wastewaters, which is generally hydrolyzed to achieve biodegradation. However, the isolation and recovery of amino acids from food and dairy wastewaters may be recovered as a value-added product of high commercial importance using an LM process. On the other hand, purified amino acids can be extracted and utilized as nutritional supplements in the diet of livestock and human nutrition. Therefore, an LM process can be utilized to recovery amino acids from dairy and food industry proteinaceous wastewaters which can add high industrial value in the long run. Reverse micelles have been used to retain amino acids and proteins at optimal pH conditions from wastewater sources via balanced electrostatic forces and hydrophobic interactions.

Further, PTFE-supported LMs like polypropylene glycol and polybutylene have been used to remove organic EPs like p-nitrophenol from industrial wastewater. Similarly, a PVDF microporous-supported thin non-porous polydimethylsiloxane (PDMS)-based SLM was used for the removal of phenol from wastewater. In addition to this, a highly efficient poly(dimethyl)siloxane and poly(methylvinyl)siloxane-supported LM like linear monoalkyl cyclohexane, dibenzo-18-crown-6, N,N-di(1-methyl heptyl) acetamide, trioctylamine, dodecane and N-octanoylpyrrolidine have been used for the removal organic EPs like phenol and cresols from various wastewaters. Also, a flat-sheet polymer-supported LM has been successfully utilized for the removal of erythromycin from pharmaceutical wastewaters.

Thus, liquid membranes offer an economically alternative and easy to operate EP treatment option where conventional methods fall short. In particular, LM techniques are useful in removing heavy metals and ammonia efficiently and economically from wastewater. Therefore, considerable effort is being made to develop economic scale-up designs of liquid membrane processes to achieve sustainable commercialization in the EP-contaminated water treatment sector [44].

6.9   ION-EXCHANGE MEMBRANE METHOD FOR REMOVAL OF EPS

The previous chapter presented a detailed discussion on the use of ion-exchange-based resins and polymers for the effective removal and retention of EPs from various wastewater sources. Inorganic nanomaterials and natural clays with highly nanoporous structures have been widely explored for the purpose. However, there are limited reports on the effort to use ion-exchange membrane systems for the purpose of EP treatment from water and wastewater sources. Also, conventional wastewater treatment processes suffer several drawbacks during EPs treatment due to their complexities and high-cost operation. In particular, conventional wastewater processes suffer setbacks in treatment that are characterized by high concentrations of EPs like phenols, APIs and other PPCPs. Natural water streams have also been increasingly characterized regarding therapeutic constituents or PCPs like anti-epileptics or carbamazepine, tranquillizers or diazepam, analgesics-ibuprofen, naproxen, diclofenac, antibiotics such as roxithromycin, erythromycin, sulfamethoxazole, trimethoprim and organic micropollutants like polycyclic musk fragrances, namely galaxolide, tonalide and celestolide [45].

Ion-exchange membranes have been tested for their potential to remove these EPs from various water and wastewater sources. However, except for desalination and ionic species removal, ion-exchange membranes have shown limited success. However, a hybrid technology involving an ion-exchange membrane process, which is one of the important components, like electrodialysis, has shown greater potential in concentrating ionic pollutants beyond desalination. However, newer approaches and task-specific engineering designs may open new avenues for ion-exchange membrane-based technologies in the near future. At present, innovations in ion-exchange membrane processes are gaining pace and adopting innovative designs which particularly depend on the specific type of EP to be removed from wastewater sources. This may further lead to the faster adoption of the technology for real field applications in the near future.

REFERENCES

1. *Water Quality and Health Strategy 2013–2020*, WHO, 1 January 2013, https://www.who.int/water_sanitation_health/publications/2013/water_quality_strategy.pdf
2. *Guidelines for Drinking-Water Quality*, 4th ed., Geneva, Switzerland: World Health Organization Press, ISBN: 9789241548151, 2011.
3. M. A. Gonzalez, R. Trocoli, I. Pavlovic, C. Barriga, and F. L. Mantia, Capturing Cd(II) and Pb(II) from contaminated water sources by electro-deposition on hydrotalcite-like compounds, *Physical Chemistry Chemical Physics*, 2016, 18, 1838–1845.
4. S. Annamalai, M. Santhanam, S. Sudanthiramoorthy, K. Pandian, and M. Pazos, Greener technology for organic reactive dye degradation in textile dye-contaminated field soil and in situ formation of "electroactive species" at the anode by electrokinetics. *RSC Advances*, 2016, 6, 3552–3560.
5. M. Amini, K. Mueller, K.C. Abbaspour, T. Rosenberg, M. Afyuni, K.N. Møller, M. Sarr, and C.A. Johnson, Statistical modeling of global geogenic fluoride contamination in ground waters. *Environmental Science & Technology* 2008, 42, 3662–3668.
6. Neeta Pandey, S.K. Shukla, and N.B. Singh, Water purification by polymer nanocomposites: an overview, *Nanocomposites*, 2017, 3(2), 47.
7. M. Elkady, M. El-Aassar, and H. Hassan, Adsorption profile of basic dye onto novel fabricated carboxylated functionalized Co-polymer nanofibers, *Polymers*, 2016, 8(5), 177.
8. Kevin J. De France, Todd Hoare, and Emily D. Cranston, Review of hydrogels and aerogels containing nanocellulose, *Chemistry of Materials*, 2017, 29(11), 4609–4631.
9. Grégorio Crini, Nadia Morin-Crini, Nicolas Fatin-Rouge, Sébastien Déon, and Patrick Fievet, Metal removal from aqueous media by polymer-assisted ultrafiltration with chitosan, *Arabian Journal of Chemistry*, 2017, 10(2), S3826–S3839.
10. Bhuvaneshwari Balasubramaniam, Prateek, Sudhir Ranjan, Mohit Saraf, Prasenjit Kar, Surya Pratap Singh, Vijay Kumar Thakur, Anand Singh, and Raju Kumar Gupta, Antibacterial and antiviral functional materials: chemistry and biological activity toward tackling COVID-19-like pandemics, *ACS Pharmacology & Translational Science*, 2020, 4(1), 8–54.
11. Katarzyna Lewandowska, New functional materials based on natural polymers obtained from renewable resources, *Molecules*, (ISSN 1420-3049), EISSN 1420-3049, Published by MDPI.
12. Saurabh Joshi, Himanshu Kathuria, Sandeep Verma, and Suresh Valiyaveettil, Functional catechol–metal polymers via interfacial polymerization for applications in water purification, *ACS Applied Materials & Interfaces*, 2020, 12(16), 19044–19053.
13. Tawfik A. Saleh, Prakash Parthasarathy, Muhammad Irfan, Advanced functional polymer nanocomposites and their use in water ultra-purification, *Trends in Environmental Analytical Chemistry*, 2019, 24, e00067.

14. Anshu Kumar, Parimal Paul, and Sanna Kotrappanavar Nataraj, Bionanomaterial scaffolds for effective removal of fluoride, chromium, and dye, *ACS Sustainable Chemistry & Engineering*, 2017, 5, 895–903.

15. S. Lankalapalli and V. R. M. Kolapalli, Polyelectrolyte complexes: a review of their applicability in drug delivery technology, *Indian Journal of Pharmaceutical Sciences*, 2009, 71(5), 481–487.

16. L. Anders, W. Charlotte, W. Staffan, Flocculation of cationic polymers and nanosized particles. *Colloids and Surfaces A: Physicochemical and Engineering Aspects*, 1999, 159, 65–76.

17. M. Sai Bhargava Reddy, Deepalekshmi Ponnamma, Rajan Choudhary, and Kishor Kumar Sadasivuni, A comparative review of natural and synthetic biopolymer composite scaffolds, *Polymers*, 2021, 13, 1105.

18. Gudrun Petzold and Simona Schwarz, Polyelectrolyte complexes in flocculation applications, *Advances in Polymer Science,* Springer-Verlag, Berlin Heidelberg, 2013, DOI:10.1007/12_2012_205.

19. A. Mahto, A. Kumar, J. P. Chaudhary, M. Bhatt, A. K. Sharma, P. Paul, and S. K. Nataraj, Solvent-free production of nano-FeS anchored graphene from Ulva fasciata: A scalable synthesis of super-adsorbent for lead, chromium and dyes, *Journal of Hazardous Materials*, 2018, 353, 190–203.

20. A. Mahto, R. Gupta, K. K. Ghara, D. N. Srivastava, P. Maiti, D. Kalpana, and S. K. Nataraj, Development of high-performance supercapacitor electrode derived from sugar industry spent wash waste, *Journal of Hazardous Materials*, 2017, 340, 189–201.

21. S. M. Alatalo, E. Mäkilä, E. Repo, M. Heinonen, J. Salonen, E. Kukk, M. Sillanpää, and M. M. Titirici, *Green Chemistry*, 2016, 18, 1137–1146.

22. S.-H. Yu, X. Cui, L. Li, K. Li, B. Yu, M. Antonietti, and H. C¨olfen, *Advanced Materials*, 2004, 16, 1636–1640.

23. Manohara Halanur, Supratim Chakraborty, Kanakaraj Aruchamy, Debasis Ghosh, Nripat Singh, Kamalesh Prasad, D. Kalpana, S. K. Nataraj, and Dibyendu Mondal, Engineering Fe-doped highly oxygenated solvothermal carbon from glucose-based eutectic system as active microcleaner and efficient carbocatalyst. *Journal of Materials Chemistry A*, 2019, 7, 4988–4997.

24. K. Aruchamy, A. Mahto, R. Nagaraj, D. Kalpana, D. Ghosh, D. Mondal, and S. K. Nataraj, Ultrafast synthesis of exfoliated manganese oxides in deep eutectic solvents for water purification and energy storage, *Chemical Engineering Journal*, 2020, 379, 122327.

25. J. Safaei, P. Xiong, and G. Wang, Progress and prospects of two-dimensional materials for membrane-based water desalination, *Materials Today Advances*, 2020, 8, 100108.

26. Lara Loske, Keizo Nakagawa, Tomohisa Yoshioka, and Hideto Matsuyama, 2D nanocomposite membranes: water purification and fouling mitigation, *Membranes (Basel)*, 2020, 10(10), 295.

27. Saoirse Dervin, Dionysios D. Dionysiou, and Suresh C. Pillai, 2D nanostructures for water purification: graphene and beyond, *Nanoscale*, 2016, 8, 15115–15131.

28. Seungju Kim, Huanting Wang, and Young Moo Lee, 2D nanosheets and their composite membranes for water, gas, and ion separation, *Angewandte Chemie*, 2019, 58(49), 17512–17527.

29. Ting Liu, Xiaoyan Liu, Nigel Graham, Wenzheng Yub, and Kening Sun, Two-dimensional MXene incorporated graphene oxide composite membrane with enhanced water purification performance, *Journal of Membrane Science*, 2020, 593(1), 117431.

30. Minxiang Zeng, Mingfeng Chen, Dali Huang, Shijun Lei, Xuan Zhang, Ling Wang, and Zhengdong Cheng, Engineered two-dimensional nanomaterials: an emerging paradigm for water purification and monitoring, *Materials Horizons*, 2021, 8, 758–802.

31. Worood A. El-Mehalmey, Youssef Safwat, Mohamed Bassyouni, and Mohamed H. Alkordi, Strong interplay between polymer surface charge and MOF cage chemistry in mixed-matrix membrane for water treatment applications, *ACS Applied Materials & Interfaces,* 2020, 12(24), 27625–27631.

32. Sarita Dhaka, Rahul Kumar, Akash Deep, Mayur B. Kurade, Sang-Woo Ji, Byong-Hun Jeon, Metal–organic frameworks (MOFs) for the removal of emerging contaminants from aquatic environments, *Coordination Chemistry Reviews,* 2019, 380, 330–352.

33. Chelsea A. Clark, Kimberly N. Heck, Camilah D. Powell, and Michael S. Wong, Highly defective UiO-66 materials for the adsorptive removal of perfluorooctanesulfonate, *ACS Sustainable Chemistry & Engineering,* 2019, 7(7), 6619–6628.

34. Jian Li, Xiang Zhou, Jing Wang, and Xiufen Li, Two-dimensional covalent organic frameworks (COFs) for membrane separation: a mini review, *Industrial & Engineering Chemistry Research,* 2019, 58, 15394–15406.

35. Hongwei Fan, Manhua Peng, Ina Strauss, Alexander Mundstock, Hong Meng, and Jürgen Caro, MOF-in-COF molecular sieving membrane for selective hydrogen separation, *Nature Communications,* 2021, 12(38), 1–10.

36. Shushan Yuan, Xin Li, Junyong Zhu, Gang Zhang, Peter Van Puyvelde, and Bart Van der Bruggen, Covalent organic frameworks for membrane separation, *Chemical Society Reviews,* 2019, 48, 2665–2681.

37. Bahram Hosseini Monjezi, Ksenia Kutonova, Manuel Tsotsalas, Sebastian Henke, Alexander Knebel, Current trends in metal–organic and covalent organic framework membrane materials, *Angewandte Chemie,* 2021, 60(28), 15153.

38. Vibha Sinha, Sumedha Chakma, Advances in the preparation of hydrogel for wastewater treatment: a concise review, *Journal of Environmental Chemical Engineering,* 2019, 7, 103295

39. Li Yu, Xiaokong Liu, Weichang Yuan, Lauren Joan Brown, and Dayang Wang, Confined flocculation of ionic pollutants by poly(l-dopa)-based polyelectrolyte complexes in hydrogel beads for three-dimensional, quantitative, efficient water decontamination, *Langmuir,* 2015, 31(23), 6351–6366.

40. Yi Meng, Tanglong Liu, Shanshan Yu, Yi Cheng, Jie Lu, Xianzheng Yuan, and Haisong Wang, Biomimic-inspired and recyclable nanogel for contamination removal from water and the application in treating bleaching effluents, *Industrial & Engineering Chemistry Research,* 2020, 59, 8622–8631.

41. Jai Prakash Chaudhary, Sanna Kotrappanavar Nataraj, Azaz Gogdaa, and Ramavatar Meena, Bio-based superhydrophilic foam membranes for sustainable oil–water separation, *Green Chemistry,* 2014, 16, 4552–4558.

42. Jai Prakash Chaudhary, Nilesh Vadodariya, Sanna Kotrappanavar Nataraj, and Ramavatar Meena, Chitosan-based aerogel membrane for robust oil-in-water emulsion separation, *ACS Applied Materials & Interfaces,* 2015, 7, 24957–24962.

43. Manohara Halanur Mruthunjayappa, Vibha T. Sharma, Kalpana Dharmalingam, Nataraj Sanna Kotrappanavar, and Dibyendu Mondal, Engineering a biopolymer-based ultrafast permeable aerogel membrane decorated with task-specific Fe–Al nanocomposites for robust water purification, *ACS Applied Bio Materials,* 2020, 3(8), 5233–5243.

44. Roman Tandlich, Chapter-8: Application of Liquid Membranes in Wastewater Treatment, in *Liquid Membranes: Principles and Applications in Chemical Separations and Wastewater Treatment,* Roman Tandlich (ed.), 2010, Amsterdam, Netherlands, Elsevier B.V.

45. Getachew Dagnew Gebreeyessus, Status of hybrid membrane–ion-exchange systems for desalination: a comprehensive review, *Applied Water Science,* 2019, 9(5), 1–14.

7 Nanomaterial-Based Water Filters for Emerging Pollutants

7.1 INTRODUCTION

Water is vital for life, and providing clean and safe drinking water at an affordable cost would drastically improve the lives of many people in underdeveloped countries. The world is witnessing an exponential growth of population, urbanization and industrialization, which are the essentials to the expansion and development of the global economy. Regrettably, today, nearly 80% of the world's wastewater, produced by industrial and other human activities, is dumped unprocessed and untreated back into water bodies without proper assessment of its environmental risks, thus leading to deteriorating water quality. Wastewaters presently produced are characterized by high concentrations of EPs like toxic organic and inorganic compounds, which may undergo bio-magnification through the food chain, thereby threatening the environment and aquatic life on earth. It is also evident that they greatly affect the health of living beings by triggering various health issues such as developmental retardation, immunological disorders, kidney damage and cancer. On the other hand, less than 1% of the world's freshwater that is accessible has been recovered or recycled from wastewater. If immediate measures are not taken to reclaim water and reuse it environmental degradation will become worse by 2050, hence the global demand for freshwater is predicted to increase by 30% from what it is now. This growing demand for drinking quality water can be satisfied only by water and wastewater resources that are managed sustainably and efficiently. Thus, the management of water and the treatment of wastewater resources is a precondition [1].

To address the issue of water scarcity, water reclamation and pollution mitigation and recover a high value-added resource, several traditional materials, methods and treatment technologies are being employed. These methods and technologies include highly successful membrane-based filtration, commercially popular adsorption, conventional coagulation, precipitation, ion exchange, advanced oxidation processes, floatation, biological treatment and many others which fit the purpose of environmental remediation. Among these, membrane technology, in particular, has been shown to be a promising and widespread solution option in comparison to other methods, as membrane systems have proven to be very effective at lowering costs and achieving quick contamination separation in a variety of applications. Some of the major advantages associated with commercially successful membrane technology are: (a) it operates simply without heating or added chemicals, making it less energy intensive; (b) the separation of the contaminant from water is purely physical in nature in which both retentate and permeate fractions can be used for further application, directly; (c) it does not require post-physical or chemical treatment which makes it widely suitable for water treatment application [2].

However, the main drawbacks associated with membrane technology concerns the membrane itself which frequently undergoes scaling and fouling problems. Membrane fouling promotes the accumulation of foulants and the degradation of selective layers. This phenomenon leads to a drastically reduced lifespan for the membrane, thus requiring regular preservation and washing, which in turn increase the operational cost. Therefore, to be an effective separation medium, an ideal membrane should retain its high permeability and high contaminant rejection rate all the time. In this direction, there have been several attempts made to induce anti-fouling and anti-scaling properties to polymeric membranes towards reclaiming high quality water and separating both conventional and emerging pollutants from wastewater sources effectively. Further, the categories of anti-fouling and high flux membranes have also been developed to address the emerging issues. However, the organic nature of polymer membranes still poses vulnerabilities to micro-organic pollutants and series of EPs. Even though membrane-based separation processes have been effective in treating most of the conventional contaminants like TDS, BOD, COD and salinity, these conventional membranes pose problems like their being ineffective against EPs such as APIs, fluoride, heavy metals, pesticides and dyes. Also, it is difficulty to tune their porosity and distribution to suit EP rejection, following the conventional phase inversion technique. This widely used preparation method is still lacking in providing higher permeability and selectivity, and is susceptible to pore blocking, resulting in poor flux permeability and poor mechanical robustness and activity [3].

Another major drawback of membrane technology comes from the recovery rate. Membrane processes offer a limited water recyclability option which is currently restricted to less than 50% for the best performing membrane. In such a situation, the management of feed and its disposal is a major environmental concern. Furthermore, the presence of emerging micro-contaminants such as microplastics, nanomaterials, colloidal pollutants, semi-volatile organic compounds and endocrine disrupting chemicals (EDCs) and API ions in contaminated water sources have made practically all conventional water and wastewater treatment systems ineffective at meeting treatment and regulatory requirements.

Nevertheless, advances in functional material chemistry, nanoscience and nanotechnology-based innovations, and conventional technologies have marginally improved their performance against EPs. In particular, membrane purification technology has adopted notable nanomaterials and enabled technical advancements and functionalization processes to improve the overall performance of a unit. On the other hand, nanomaterial-based technologies are gaining focus and importance in addressing environmental-related issues. Initially, nanomaterials were used as a replacement for conventional adoption technologies. Conventional adsorption technologies such as activated carbon, ion exchange resins and polymeric systems have been widely used for retaining commonly occurring contaminants in ground and surface water. However, the invention of a new diverse class of nanomaterials and the availability of wide varieties of precursors have opened new arenas for exploring nanomaterials in wider applications and extending sustainable water treatment. Nanomaterials have redefined the adsorption process in which the principle of adsorption is gaining importance globally due to their great pollutant retention efficiency at a low cost in simple and inexpensive operations. Nanomaterials offer a large surface area to volume ratio which makes them highly active and with a surface functionality that can be utilized to enhance removal efficiency, induce antifouling and antibacterial characteristics, alter the catalytic activity, and control wettability in nano-adsorbents through various interactions. These diverse features make nanomaterial-based adsorbents highly effective in removing EPs from wastewater, thereby providing an economically feasible and effective alternative.

In this direction, various nanomaterials with a large surface area and a large number of surface functional groups have been explored in recent years for water treatment applications. However, some of these nanomaterials when added or used in water or wastewater tanks often pose regeneration or recycling issues. In some cases, nanomaterial leaching or leaking from the loosely packed testing kit or system, due to their nano sizes, pose serious threats to humans or the environment. These may lead to a generation of secondary pollutants, which has raised fears among environmental and occupational health specialists that nanomaterial-based filtrations susceptible to leaching are more hazardous than previously encountered chemicals, necessitating specialized risk and exposure evaluations to determine the level of risk to human health and safety. To these limitations associated with the nanomaterial-based water treatment system, contemporary research focuses mainly on assembling nanomaterials into macrostructure packaging, through composites and the trapping of nano-active materials in 3D sponges, hydrogels, aerogels, foams, beads, membranes and so on [4, 5].

Nevertheless, the demand for novel filtration materials that passage water molecules selectively with high contamination retention separation efficiencies has led to innovative exploring and designing of advanced nanomaterial-based filters. Currently, methods have been adopted to convert their unique functionalities into a packed nanomaterial filter to provide ultrafast permeance and enhanced rejection as a result of structural alterations and physicochemical adjustments. Evidently, several researchers and technologists have adopted skilful module designs to achieve a 'nanotechnology' solution following tunable porosity by nanomaterials packing to manipulate the flux and rejection trade-offs. Until now, many hybrid nanocomposites have been prepared in order to achieve the goal. First among these were the functional polymers and carbaneous nanomaterials. Functional polymers include both biopolymers and synthetic original macromolecules which carry a tunable functional group on their surface and backbone. Amongst these are carbon in the form of hydrothermal carbon (HC), activated carbon (AC), graphene (GO) and its derivative oxides, and carbon nanotubes (CNTs). In all cases, active nanomaterials have been either converted to filters or membranes or bead forms for the production of drinking water from contaminated sources.

7.2 POLYMER-BASED NANOCOMPOSITES FOR REMOVAL OF EPS

Despite facing several drawbacks, high-rejection membrane-based processes empower us to solve industrially important applications. Polymer-based membranes still govern and dominate the water filtration membrane industry due to their efficiency regarding the removal of contaminants and which has greatly added to their economic value and capability of restructuring new technologies for water purification. As discussed previously, to enhance the efficiency and specificity of conventional membranes, advanced membranes with a combination of unique functionalities have been discovered in the form of nanomaterials. On the other hand, membrane technology with ultra-fast permeance and enhanced rejection has been widely improved with the help of engineering innovation. With these outlooks over the years, polymer-based packings or beds or film have rapidly transformed from an academic pursuit to a commercial reality [6].

Polymers have traditionally been successful in solving water-related problems and as an efficient way to remove contaminants from wastewater with enhanced adsorption capacity and improved flux in the form of hybrid membranes or polymer ion exchange adsorption beds, fixed packs and modules and have evolved as an all-in-one solution to address EPs in water sources. However, the use of petroleum-based synthetic polymers has been characterized by their low sustainability on the basis of the hazards involved in preparation and disposal once they have served their lifespan. Also, synthetic polymer preparation and conversion, to be an effective filter medium, is becoming increasingly cost-intensive, lacking flexibility in addressing a wide variety of contaminants and with limited water permeability, which makes it a costly business proportion. However, the ever-increasing water pollution due to the unregulated release of wastewater has been the starting point for altering the use of polymer sources. This has forced researchers to consider bio-based polymers and their composites which are capable of substituting synthetic sources for water treatment applications. Thus, the design and fabrication of cost-effective polymer-nanocomposite water filters and devices using bio-based sources like seaweeds and biomass extracts are of great interest. Mainly, biopolymer-based systems offer oxidation resistance, easy processability, a high modulus, easy availability, cost-effectiveness, a renewable nature and recycling prospects. Also, bio-gen polymers are considered renewable, non-hazardous, decomposable and compatible with the environment and aquatic ecosystem. On the technology front, high functional characteristics offer both task specific and universal filtration prospects for filters.

7.2.1 CELLULOSE-BASED NANOCOMPOSITES

Cellulose is a natural, long-chain polysaccharide polymer consisting of a linear chain of several hundred to millions of β linked D-glucose units. Cellulose is a significant structural component of the primary cell wall of green plants and vegetation at large, and is also abundantly found in many forms of algae, oomycetes and seaweeds. Nevertheless, cellulose is a complex carbohydrate macromolecule that occurs naturally in various ecological environments that are highly abundant and widely distributed. Cellulose shows great affinity for water due to the presence of abundant hydroxyl groups, and these functionalities make natural cellulose a versatile medium to be converted into beads, films and resins. Conventionally, these cellulose systems have been utilized for the removal of EPs like heavy metals and hazardous organic pollutants. However, a major shortcoming of bio-gen cellulose-based adsorbents or filters is their relatively low adsorption capabilities. The surface and bulk functionality of the cellulosic membranes restricts their use in treatment to only ionic EPs like dyes and heavy metals [7–9].

Thus, composites of cellulose have been explored that have the potential to replace pristine polymers as high-performance adsorbents in wastewater treatment. In this direction, cellulose/graphene oxide nanocomposite solutions have been vacuum-filtration filtered to create reusable functionalized cellulose composite membranes (CGCMs) which were then used as adsorbent filters to remove organic contaminants from simulated wastewaters. It was evident from the series of experiments that the CGCM filters can effectively retain various EPs from wastewater. The CGCM filters showed separating adsorbents with as high as 86.4 mg/g maximum adsorption capacity for organic EPs like Rhodamine B. This demonstrates the practicality and high potential of composite cellulose filters in removing EPs, which might be significant for their use in wastewater treatment. In another study, Fe-modified cellulose (MCFBs) adsorbent membranes were fabricated following simple and environmentally friendly preparation routes. These MCFB adsorbents were further devised in the form of a filter bed for the removal of dyes and heavy metals. The MCFB beads showed an adsorption of 1186.8 and 151.8 mg/g for methylene blue and Rhodamine B, respectively, which is due to their characteristic porous structure and carboxyl functionality on a polymer surface. This functionality and the porous nature of the composite also helps in inducing chemical and thermal stability.

Thus, a cellulose-based nanocomposite provides a successful alternative to separate EPs from wastewater and also offers regeneration and recyclability of adsorbents to make them sustainable. Also, the 3D nanostructure and magnetic properties of MCFBs have been greatly characterized by their mechanical robustness which helps in the recovery and recycling of adsorbent beads. In another study, a mechanically robust 3D network structure of cellulose/cyclodextrin (CNF-CD) blend structures with higher porosity and a large surface area have been effectively used as adsorbents for organic pollutants such as phenol. In this study, functionalized cellulose nanofibril aerogels have shown a high adsorption of 148 μmol/g for phenol. Furthermore, the CNF-CD aerogels have been tested for their outstanding reusability. These characteristics make CNF-CD aerogel an ecologically friendly and promising adsorbent for the elimination of highly hazardous organic EPs like phenol pollutants, which are abundantly found in several industrial wastewaters.

On the other hand, nanocellulose, graphene oxide and their composite systems have been mixtures been designed to remove heavy metal contamination and organic pollutants from wastewater. This approach has been followed to induce unique adsorption abilities, mechanical properties, a coordination site to help attract metal ions, and a surface charge. Such a strategy was adopted to fabricate bio-hybrid adsorbents following the self-assembly of graphene oxide sheets (GOs) and 2,2,6,6-tetramethylpiperidine-1-oxylradical mediated oxidation nanofibers (TOCNFs) which were vacuum filtered to achieve nanocomposite filters. This approach has yielded a unique hydrolytic stabile, an easily recoverable composite biopolymer filter as shown in Figure 7.1, as prepared biohybrid filters have been successfully used for the

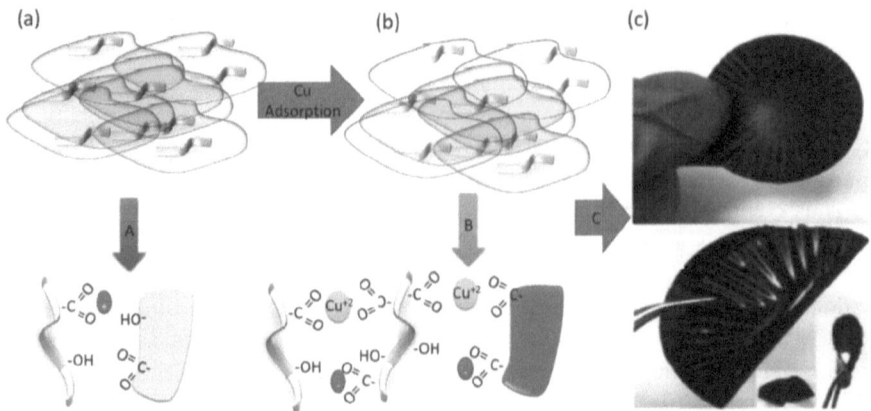

FIGURE 7.1   (a) Schematic of self-assembled bio-hybrid structure of TOCNF + GO, (b) the TOCNF+GO biohybrid following Cu(II) adsorption is represented in changed colour via functional groups having hydrogen bond interactions and electrostatic interactions during TOCNF and GO hybrid formation, (c) the freestanding biohybrid membrane used for EP removal.

selective removal of Cu(II) from water and wastewater systems. In this retention mechanism, metal ions were adsorbed via ionic crosslinking between the negatively charged biohybrid and the adsorbed contaminant Cu(II). These findings validate the potential of TOCNF/GO biohybrids to generate a new class of EP removal filters with unique features such as high flexibility, hydrolytic stability, performance and mechanical toughness [10].

Therefore, cellulose-based bio-adsorbents have proven to be relatively high-capacity adsorbent in retaining EPs from various water sources. Interestingly, biopolymers and their nanocomposites with the help of active inorganic/organic moieties have been proven to possess outstanding multifunctional properties. These cellulose composite EP treatment systems have also shown greater potential to be employed as smart and engineering materials. In this system, the incorporation of task-specific moieties has induced improved ductility and Young's modulus during organic EP removal and regeneration. In this approach, easily recyclable $NiFe_2O_4$/MWCNT engineering biomaterial has been used to functionalize cellulose yielding composites of magnetic m-$NiFe_2O_4$/MWCNTs@cellulose as an effective bio-adsorbent. These magnetic bio-nanocomposites further used abatement of Congo red (CR) from water and wastewater sources. The magnetic adsorbents exhibited a robust adsorption separation capacity of 95.70 mg g^{-1} for CR and easy recyclability.

7.2.2   CHITOSAN-BASED NANOCOMPOSITES

Unlike cellulose, chitosan (CS) originates from Chitin, a long-chain polymer of N-acetylglucosamine which is again an amide derivative of glucose. This polysaccharide is abundantly found in the cell walls of fungi, the radulae of molluscs, cephalopod beaks and the scales of fish and the skin of lissamphibians. Chitosan is also the second most abundant natural polysaccharide extracted from chitin as a deacetylated macromolecule. Chitosan has unique and useful ionic characteristics that include biodegradability, biocompatibility, non-toxicity and active surface functionality that attract EPs like heavy metals. Chitosan (CS) and chitosan-based nanocomposite adsorbents have shown incredible affinity for a series of EPs like heavy metal ions, organics and APIs due to their chelating ability owing to the $-NH_2$, $-OH$ chitosan functionalities. In this study, a significant amount of metal ions was successfully retained on chemically modified-CS bio-nanocomposite. In this approach, metal oxyhydroxide functionalized chitosan composite materials as an all-in-one drinking water purifier series of EPs like fluoride, arsenic, dyes, general TDS and biological contaminants. Similarly, chitosan/alginate/Al have been prepared in both granular and fine particle form to be used as a bio-nanocomposite to treat various EPs like F, Cr, As and dyes from contaminated water in both filter beds and easy to use tea-bag pouches as shown in Figure 7.2. These bio-nanocomposites have been tested at the pilot scale successfully as one of the cost-effective water filters of different contaminated sources [11].

Similarly, chitosan-based bio-nanocomposites have been prepared to produce affordable arsenic-free drinking water from contaminated sources. In this study, biopolymeric caged metastable 2-Line ferrihydrite (CM2LF) was prepared at room temperature with simple and reagent chemical free routes which make it an environmentally friendly adsorbent. Further, CM2LF bio-nanocomposite adsorbents were applied to field studies to produce high performance adsorbents with an exceptional adsorption ability for both As(III) and As(V). Therefore, chitosan-based nanocomposite bio-adsorbents have been successfully prepared and tested for EP removal.

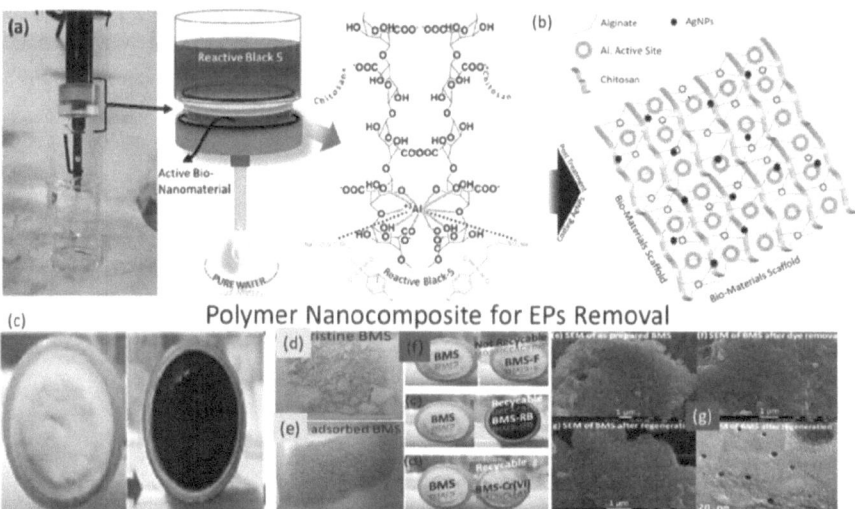

FIGURE 7.2   (a) Real time dye removal filtration set up in a continuous flow configuration, (b) schematic of the formation of an aluminium oxyhydroxide-chitosan/alginate granular bio-nanocomposite, (c–f) photographs before filtration of nanocomposite filter for EPs like F, As, dyes and Cr from an aqueous medium, (g) photograph and SEM and TEM micrographs of prepared Ag surface-modified bio-nanocomposites after a leaching study.

These CM2LF nanocomposite bio-adsorbents are eco-friendly, low-cost, water purification systems which can be successfully scaled up for EP treatment of ground,
surface as well as industrial wastewater.

Further, chitosan has been widely used for the preparation of stable aerogels for
oil/water separation and metal ion adsorbents. Interestingly, compared to other nanocomposite adsorbents, aerogel, a light-weight filter medium, has shown certain
unique features, such as large micro- and macroporous and adjustable pore volumes,
and high specific surface areas which are vital for adsorbing various EPs from water
and wastewater sources. In one such attempt, a unique biopolymer-based aerogel
membrane was functionalized with Fe-Al nanocomposites for vigorous water

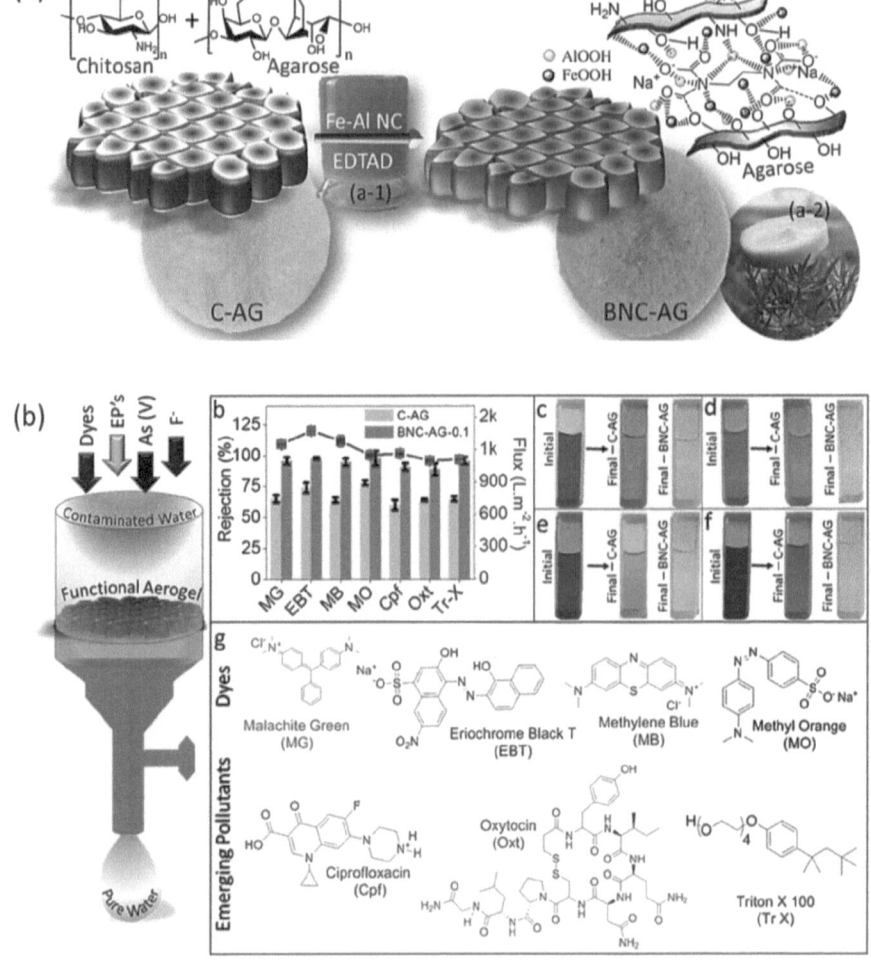

FIGURE 7.3 Schematic representation of chitosan derived (a) from bio-nanocomposite
aerogel-based membranes for (b) the separation of a series of fluoride, As, organic-like dyes
and pharmaceutical EPs, and surfactants from contaminated sources.

purification with the goal of establishing a cost-effective technique for quick and flexible water purification. The functionalized aerogel membranes had the potential to remove >95% of dyes, >89% of EPs, >93% for F^- and >99% rejection for As(V) was obtained. Overall, a novel material with outstanding adsorption capacity for a wide range of pollutants was proven to have potential for removing arsenic and fluoride with admirable reusability for point-of-use applications as shown in Figure 7.3.

Thus, a low-cost and environmentally friendly membrane with tremendous promise at dealing with real-world wastewater is being developed as a possible solution for universal water purification for both ground and wastewater sources. Similarly, a 3D ultra-high porous aerogel was prepared using polydopamine-functionalized attapulgite/chitosan and further used for the adsorptive removal of uranium(VI) from wastewater. Results showed that most of the adsorption took place within 40 min (328k) at pH 5.0 with a maximum adsorption capacity of 175.1 mg/g. Further, recyclability experiments have also shown that the removal effectiveness was retained at 78% at the end of several cycles [12].

7.3 PROTEIN-BASED FUNCTIONAL NANOMATERIALS

Natural fibres are a rich source of proteineous macromolecules originated from animal, vegetable or mineral sources and which are convertible into non-woven filter mats such as fibre cloths, fibre paper, self-assembled beds, spinning into yarns, and woven cloths. A natural fibre may be further extracted to create a defined size, shape and morphology. Although natural fibres come with a lot of impurities, methods have been developed to extract high surface-active functionality and mechanical robustness. Natural-origin fibres such as cellulosic fibres (like grain, straw, cotton and wood) can easily be adopted for environmental applications. Apart from economic considerations, the natural fibres offer the usefulness of scaling up for commercial purposes based on their technologically important properties such as abundance, abrasion resistance, elasticity, length, strength, absorbency and various surface properties.

On the other hand, the advancements in nanoscale materials with outstanding porosity and surface activity have aided the search for new generation filters, such as natural fibers, in particular protein-based fibre for water treatment. In one such attempt, protein-based fibres were used the first time for the preparation of novel filters consisting of a crosslinked biomacromolecular structure with a highly stable self-assembled structure. In this study, a tightly packed filter was prepared using silk nanofibrils for the removal of EPs like dyes with ultra-high permeation compared to conventional polymeric membranes. Further, nanomaterial-induced 2D fibre materials were designed as effective nano-adsorbents and tested for the removal of dyes, fluoride, arsenic and biological contaminants from various water sources. In another attempt, transition metal dichalcogenide MoS_2 was embedded on silk nanofibres to yield functionalized silk nanofibrils (MoS_2/SNFs). Prepared multilayer assemblies of MoS_2/SNF filters were applied as a large surface active, flexible, high surface area and photocatalytic filter for EP treatment which showed 99.2 and 99.8% rejection for Au^{3+} and Pd^{2+} with adsorption capacities of 759.1 mg g^{-1} and 425.1 mg g^{-1}, respectively. Thus, 2D MoS_2/SNF filters can easily be used for one-step water purification and precious metal recovery in a value-added process that has also led to the recovery of both active filter and value-added products. Further, the hybrid porous filters of electrospun-silk-nanofibre (ESF) in composition with MOFs were fabricated via a biomineralization process as a simplistic and reliable approach for the elimination of almost 100% of dyes and metal ions from water and wastewater sources effectively.

A multilayer ESF@MOF design structure resulted in a large porous structure for high water permeation and active sites on the fibre surface, retaining EPs from both ground water and wastewater. The multi-layered filters with active nano-sized silk self-assembly and in-situ hydroxyapatite (HAP) biomineralization were prepared with the help of a computational simulation and experimental fabrication. These nanocomposite SNF/HAP filters have shown great potential for water and wastewater contaminated with EPs like HMs, dyes, proteins and nano-colloids. The adsorption potential of fibre filters towards metal ions have been measured as 63.0 mg/g for Ni^{2+}. Also, aerogels prepared using natural amyloid fibrils, are known to be benign and cost-effective alternatives for similar contaminants removal from wastewater. Further, protein-based nanofibril aerogels showed high adsorption efficiencies of 98% for ibuprofen, 92% for bisphenol A and 78% for bentazone with maximum adsorption capacities of 69.6, 50.6 and 54.2 mg/g, respectively. Further, protein-based hybrid composite filters were prepared with a combination of inexpensive and environmentally friendly materials such as activated carbon modified lactoglobulin amyloid fibrils. These fibril filters were further used for the efficient extraction of toxic metal ions from contaminated waters. Remarkably, amyloid-based fibril filters were recovered and reused several times after the recovery of EPs like metal ions, making them sustainable alternatives for water treatment applications. Also, a ZrO_2 functionalized amyloid fibril filter was successfully prepared and effectively applied for the selective elimination of fluoride (over 99.5%) from contaminated water sources. Thus, protein-based highly selective and high permeation sustainable filters have shown greater potential for EP removal and offer ease of manufacture and scalable technology for addressing the global issue of wastewater management [13, 14].

7.4 GRAPHENE AND OTHER CARBANEOUS NANOMATERIALS FOR REMOVAL OF EPS

Historically, carbon-based treatment is the earliest reported use of prevalent nanomaterials for water purification and wastewater treatment applications. Many techniques have been followed to fabricate both carbon and carbon-based hybrid materials into device forms, such as vacuum filtration, layer-by-layer deposition, drop casting and spin coating. In each case, carbaneous materials can be made into powder cake, foam, powder deposited filters or bead form with task-specific characteristics. These unique properties and potential efficiencies of both individual and hybrid materials have shown great potential for the selective separation and complete retention of EPs from water and various wastewater sources. This section discusses carbaneous and hybrid materials used for EP removal from various contaminated water sources.

Rising environmental concerns have forced researchers to explore cost-effective treatment techniques with an easy design and environmentally friendly materials. Among many material stocks, carbon-based water filters have proven economically viable treatment media which provide a sustainable futuristic option. Thus, attempts have been made to develop more sustainable and scalable materials which have led to the use of carbon-based bulk, functional and nanomaterials, consisting of various low-dimension structures of carbon such as graphene, graphene oxide, carbon nanotubes, activated carbon and hydrothermal carbon. Carbon-based materials offer exceptional large surface areas with oleophilic and hydrophobic natures which have been shown to exhibit different morphological patterns and structural characteristics. On the other hand, carbaneous filters possess unprecedented advantages that come from their excellent physical, chemical and physico-chemical characteristics such as porosity, chemical stability, mechanical robustness, electrical and heat conductivity, large area of contact, greater ability to direct, flexibility, advanced optical characteristics and high mechanical strength. These carbon-based nanomaterials which have vastly tunable structures and properties have been identified as key materials to solve some bottle neck situations like water scarcity and environmental pollution, which is prevalent in the modern era and is frequently utilized in the development of new hybrid membranes for water remediation. However, many of the commonly employed carbon-based materials are modelled as point-of-use functional nanomaterial filters, adsorbents and devices for water and wastewater remediation.

Different terrestrial and seaweed biomasses have been explored as primary sources of carbon for water treatment applications. Consequently, carbon materials derived from biomass such as waste, toxic weeds, seaweeds, rice husk, coconut shell, human hair, auricularia, silk, chicken eggshell membranes, lotus pollen and cornstalk have received great attention because of their easy availability, inexpensive attributes, tunable properties and in-built chemical functionalities. Thus, efforts have been made to adopt a more sustainable and scalable material stock for carbon-based filter preparation, namely graphene, graphene oxide, carbon nanotubes, activated carbon, hydrothermal carbon and so on. These carbaneous and grapheneous nanomaterials possess exceptional physico-chemical properties like large surface areas with tuneable functionality to achieve oleophilic and hydrophobic interaction that play a vital role in removing EPs from wastewater sources.

7.4.1 Graphene and Graphene Oxide-Based Nanocomposites

Membranes made up of 2D nanomaterials such as graphene and GO have been regarded as next-generation membranes due to their flexibility, self-supporting properties which make them stable filter films, selectivity toward target EPs and the capacity to resist the trans-membrane pressure exerted. Of all the well-known 2D materials, graphene and its derivatives have emerged as a prominent alternative to polymeric membranes in both individual and nanocomposite formats. Another major advantage with graphene/GO-based nanocomposite filters comes from the fact that these 2D materials offer exceptional molecular separation characteristics. The film-forming property of these 2D nanomaterials comes from their good dispersibility in an aqueous medium, easy functionalization, and biocompatibility which encourages researchers to utilize them in EP treatment applications. In this direction, graphene dispersions were vacuum filtered to achieve self-assembly via close packing and functionalization to immobilized 3D nanomaterials and resulting free standing filters were used for organic and colloidal EP separation. Further, various graphene-based hybrid filters with specific functions have been fabricated through self-assembly. The low energy input vacuum filtration process has been adopted to fabricate GO-based stable water filters, employing vacuum filtration. In similar approaches, layer-by-layer deposition, drop casting or a spin coating method have also been adopted to prepare economically viable 2-D structure-based water filters. Also, GO-inorganic nanomaterial (GO/TiO$_2$)-based composite filters have also been prepared with a specific surface area of 489.2 m^2/g, which is considerably significant in the adsorption of EPs and which is higher than GO sheets (<10 m^2/g) and TiO$_2$ (118.9 m^2/g). These composite filters have been used for the simultaneous photo-gradation and removal of several EPs like dyes, pesticides and APIs from wastewater sources as shown in Figure 7.4.

Free-standing GO/chitin nanocrystal nanocomposite (GO/CHNCs) filters with good lipophilicity have been fabricated using a vacuum-assisted layer-by-layer self-assembly technique useful for the adsorption of organic moieties and the separation of an oil-in-water emulsion in wastewater. Prepared GO/CHNCs have been successfully tested for their high-water permeability at about 135.6 L m^{-2} h^{-1}, while rejection was recorded at over 99% for organic EPs such as methylene blue and Congo red, while a ~97% rejection was recorded for oil from emulsion. Similarly, efforts have been made to design graphene oxide/silica (GO/SiO$_2$)-based nanoporous membranes for EP treatment taking advantage of the distinct self-assembly and synergic properties of GO and silica. The GO/SiO$_2$ nanocomposite filter with good hydrophilicity having a contact angle of ~80.7° and thermal stability showed exceptional water permeability and adsorption capacity. The composite with a total pore volume of 2.84 cm^3 g^{-1}, specific surface area of 2897 m^2 g^{-1} and mean pore diameter of 3.97 nm was capable of rejecting organic dye, rhodamine B >99% rejection with a as high as 229.15 L m^{-2} h^{-1} bar^{-1} water flux.

In another study, ginger extract (GE)/graphene oxide was combined to create a thin-layered porous membrane with a ~250 nm thickness and varying penetrability which has been utilized for EP removal applications. GE/GO nanoporous films prepared via layer-by-layer coating showed high removal efficiencies for dyes like methylene blue and Rhodamine B with an ultra-high permeate water flux of

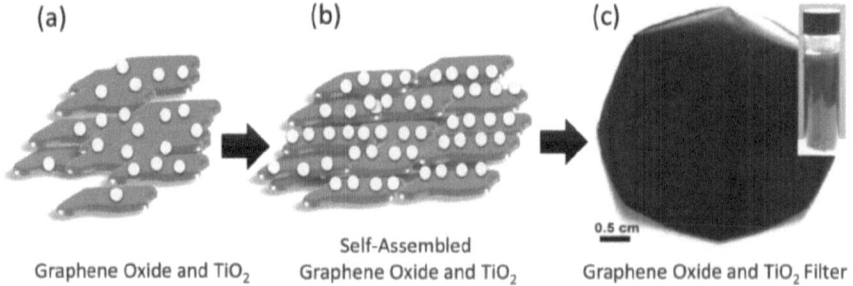

(a)   **(b)**   **(c)**

0.5 cm

Graphene Oxide and TiO$_2$ Self-Assembled
                                   Graphene Oxide and TiO$_2$ Graphene Oxide and TiO$_2$ Filter

FIGURE 7.4   (a) Schematic of a GO–TiO$_2$ composite random assembly, (b) a highly ordered self-assembled GO–TiO$_2$ film assembly and (c) a photograph of self-assembled GO–TiO$_2$ in thin-film filter form.

~420 Lm^{-2} h^{-1} bar^{-1} with >99% and 98%, for Rhodamine B and methylene blue, respectively. Interestingly, a majority of these GO-based hybrid 2D structures were found to be stable suspensions, which may make them unsuitable for a variety of applications. These dispersion states which facilitate the creation of a free-standing graphene structure with adequate functionalities have solved the issues related to traditional nanomaterials in powder form. In this direction, a simple facial technique was followed to prepare reusable polydopamine-modified graphene hydrogel (PDA-GH) adsorbents. A prepared PDA-GH had abundant functional groups with high specific surface areas which was helpful for high adsorption mechanism towards a wide range of EPs including synthetic dyes, heavy metals and aromatic pollutants. Thus, PDA-GH proved to be an efficient alternative having shown itself to be easily recyclable and a low-cost solution.

Further, magnetic nanoparticle-induced GO hydrogel was prepared as a pH responsive adsorbent. For this purpose, a 3D network of Fe^{+2} ions crosslinked with graphene oxide magnetic hydrogels (GOMHs) was prepared. A highly stable and large interpenetrating network of GOMHs showed a remarkable adsorption capacity toward pollutants due to their being a high pore connectivity porous network, thereby increasing the overall performance of the system. In this configuration, GOMHs showed a high-water flux of >111 L m^{-2} h^{-1} bar^{-1} and a retention capacity of >99% for organic EPs like methylene blue from wastewater. These results have shown the potential of these GOMHs for practical wastewater treatment applications in the field. Similarly, another set of researchers prepared phosphorylethanolamine-functionalized graphene foam (PNGF) as a 3D active filtering material for the capturing and abatement of EPs like heavy metal ions from contaminated water sources. Superhydrophilic PNGF with useful surface functional groups has successfully helped in capturing HMs like Cd(II) and Pb(II) swiftly and effectively. These PNGF filters have been noted to be a better alternative compared to others, due to their simplicity, easy recyclability, low cost, stability over a range of working pHs, and mechanical stability when applied in harsh conditions.

On the other hand, to make a graphene-based micro/nanofilter, environmentally friendly precursors have been explored. In one such attempt, bio-based graphene foams (bGFs) made from renewable materials and their use as novel adsorbents were proposed with an exceptional porous architecture and a surface area of 805 m^2 g^{-1}, which is ideal for eco-friendly water treatment applications. These ultra-high surface area 3D foam filters showed as high as 245 mg/g Cr(VI)/g on bGFs when immersed in contaminated water. Further, these foams were also used for oil–water separation with compelling performance in separating (>99%) organic solvents like toluene and petroleum oils from the contaminated aqueous mixtures [15, 16].

7.4.2 CARBON NANOTUBE-BASED NANOCOMPOSITES

Carbon nanotubes (CNTs) are nanoscale cylindrical grapheneous materials, also known as allotropes of carbon with a nanostructure that can have a very high length-to-diameter ratio. CNTs offer remarkable surface and mechanical characteristic such as increased flexibility and surface area and porosity that have the potential to be used in a variety of applications. To prepare effective CNT-based water treatment and purification filters, the modification of CNT surface chemistry is a prerequisite. The uniform dispersion of pristine CNTs is one of the main requirements to convert their dispersion into a stable and free-standing filter with a suitable polymer substrate or binder. This approach greatly increases the filtration property like water flux and pollutant rejection capabilities.

In this regard, an attempt to utilize amide functionalized multi-walled CNTs (A-MWCNTs) in sodium alginate-hydroxyapatite nanocomposite beads was made for EP removal applications, as prepared nanocomposite beads combine to create filter beds with the help of a natural polymer as a carrier medium. On the other hand, the inorganic constituents make them suitable for the removal of radioactive EPs like Co^{+2} ions from contaminated water. Further, the functionalized CNT nanocomposite filter surface area of 163.4 m^2 g^{-1} was found to be effective in retaining the radioactive EP Co^{+2} to as high as 347.8 mg g^{-1}. Thus, a functionalized CNT-based filter or adsorbent showed technical viability and low-cost adsorbent options for a wide range of metal ion removal. Further, high water permeable CNT-based filters have been prepared using novel gelatin-based magnetic adsorbents. In this approach, highly carboxylic functional (–COOH)-trapped CNTs with magnetic iron oxide nanoparticles (MNPs) embedded in filter beads were prepared and tested for the simultaneous elimination of organic EPs like methylene blue and direct red 80 dyes with a removal efficiency of 96 and 76%, respectively. These unique findings offer viable alternatives for robust and efficient adsorbents with enormous promise for wastewater treatment applications.

In another approach, efforts have been made towards increasing the adsorption capacity of CNT-based nano-filters, amino-group rich chitosan crosslinked with a suitable agent in carboxyl group functionalized multi-walled carbon nanotubes (MWCNTs-COOH) prepared for water filtration applications. These crosslinked chitosan/CNT-based nanocomposite materials were applied for the adsorption of chromate ions at an acidic pH, attaining a maximum adsorption capacity of 164 mg/g at 40 °C within 30 min. Furthermore, the chitosan/CNT nanocomposite material was also found to be effective at the removal of Cr(VI) with multiple recycling approaches, with a removal efficiency between 98 and 100% [17].

7.4.3 ACTIVATED CARBON-BASED NANOCOMPOSITES

Conventionally, activated carbon (AC) has been widely used as a competent adsorbent for the abatement of a wide array of organic and inorganic pollutants from aqueous and gaseous media. ACs are exceptionally large surface area materials with active surfaces which can vary in different precursor-based systems, lying between 500 and 4500 m^2/g. This ultra-high surface area originates in ACs from a highly organized internal microporosity and a wide variety of surface functionalities that allow it to be used for water remediation applications. Conventionally, ACs have been widely used for the removal of pharmaceuticals like antibiotics, dyes and HMs from contaminated water. In such attempts, activated carbon fibre (ACF) and granular activated carbon (GAC) have been prepared from various precursors and frequently employed as efficient adsorbents of both organic and inorganic EPs from wastewater. These high surface area ACs have been further used for the effective and economical removal of mercury and cadmium. NaOH-functionalized ACFs showed that they improved the rate of adsorption of mercury and cadmium ions in a continuous water filtration system. Further, magnetic nanoparticles (iron oxide) assisted ACs developed as a sustainable and magnetically recoverable adsorbent for series of EPs. Similarly, bentonite clay modified activated carbon adsorbents in encapsulated bead form have also been widely used in a series of EP removal. In addition to this, AC encapsulating eco-friendly sodium alginate 3D network beads have been prepared and used for the removal of various toxic cationic and anionic contaminants from water.

Carbon nanocomposite beads having a surface area of 433 m^2/g have been used for the adsorption of toxic heavy metals like As(V), Cd(II) and Pb(II) with adsorption capacities of 5.0, 41.3 and 74.2 mg/g, respectively. Further, spherical shaped novel activated carbon with a surface area of 1491.21 m^2/g derived from glucose and sodium dodecyl benzene sulfonate was used for chromate removal application with an adsorption capacity of ~230 mg/g at 25 °C. Further, researchers have recently focused on utilizing naturally abundant biomass as precursors or raw materials for the synthesis of highly activated carbon, based on economic and environmental considerations. Biomass has been selected from both agri-residue and waste of toxic origin to make the overall process sustainable. Among them are corncob, oil-palm trunk fibre, rice husk, palm shell, peanut hulls, coconut husk and toxic weed like parthenium. In this direction the eco-friendly synthesis of an adsorbent consisting of silver nanoparticles functionalized bamboo-based activated carbon (Ag-AC) and in another study, ultra-high surface area (>4000 m/g) carbon adsorbents were prepared from toxic but highly abundant parthenium to serve as a highly efficient adsorbent for a series of EPs. Interestingly, these adsorbents from naturally abundant sources have been tested both in batch form and in a continuous fixed-bed column to assess the commercial importance and potential in large scale water filtration applications. In both these approaches, activated carbon has been effective in removing formaldehyde, dye, pharmaceuticals at a high flow rate and with rejection at the same time. It was observed that the stable adsorption bead and powder can be easily converted into filter cake or a membrane with tunable flow rates. Several attempts have also been made to use exhausted high surface area carbon as an energy storage material to make it a more value-added product [18].

7.4.4 Hydro/Solvo/Iono-Thermal Carbon-Based Nanocomposites

Currently, research interests have shifted to adopt easy and cost-effective methods to prepare highly surface functionalized carbon-based adsorbents. In this approach, simple hydrothermal/solvothermal/ionothermal carbon (HTC/STC/ITC)-based carbon active materials were prepared using water, organic solvents and ionic liquids as media, respectively, to induce surface functional groups on a carbon surface. This approach also helps in making greener and sustainable water filters compared to existing conventional methods including membrane-based separation processes. HTC/STC/ITC methods adopted in combination with different precursors have been prepared with exceptional efficacy and anti-fouling properties along with a high potential to have the capacity for EP retention. However, at present, HTC materials have been characterized with a low surface area which limits their long-term sustainability. However, a tunable surface area and functionalities open up the potential tuning of the adsorption capacities with modified preparation protocols. However, improved approaches have adopted different organic solvents as media to dissolve carbon precursors to yield high surface area STCs with a distinct architecture. Using this route, carbon precursors like glucose have been converted to yield STC with the help of a eutectic system as solvent media under solvothermal conditions. This approach has been adopted to overcome the insufficient surface area and restricted functionality of HTCs. Further, a carbon surface has been finely tuned with the desired functionality using $FeCl_3$ as a co-precursor to result in Fe-doped STC. Further, these Fe-STCs have been used both batch-wise as well as in continuous filtration mode to separate a series of EPs from water samples. In many cases, functional STCs have been deposited on Whatman filter paper in the form of a filter; these filters were tested in a stainless-steel syringe. This active material (~10 mg/mL STC) was uniformly coated on filter paper support, making it a STC-nanomaterials-based membrane, as shown in Figure 7.5. This unique filter set-up with hybrid carbon

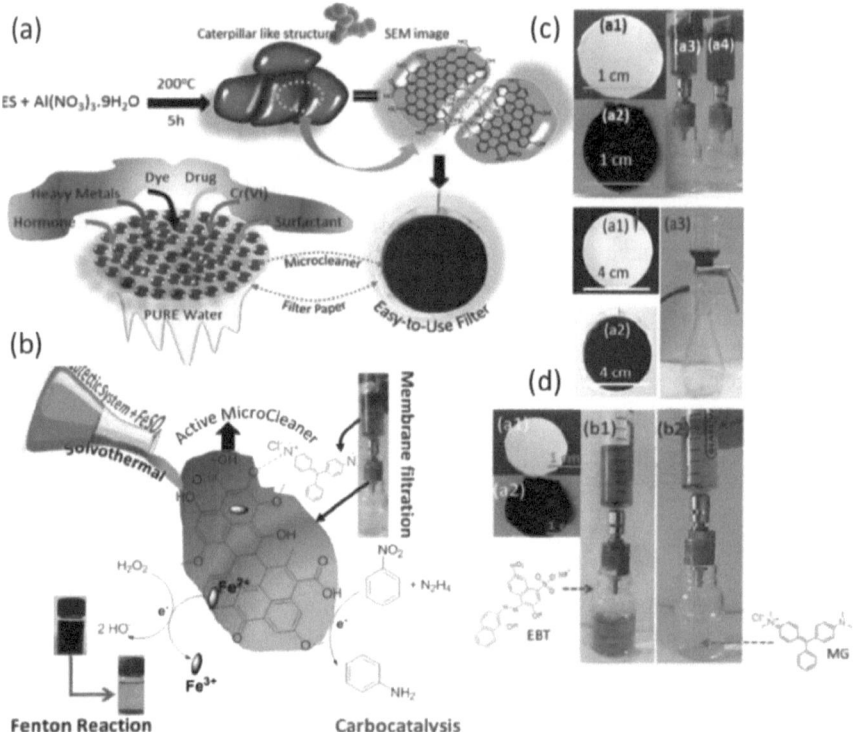

FIGURE 7.5 (a) Schematic of Fe/STC-based bionanocomposite-based microcleaner prepared using glucose and a eutectic solvent as solvothermal precursors, (b) schematics of the application of a carbon-based easy-to-use filter for water purification using a syringe and vacuum filtration membrane in water filtration and its renewal via the Fenton reaction and (c) and (d) Fe-STC-based vacuum filter used for the removal of EPs like MB and MG dyes under a continuous flow method.

materials having a surface area of about 26–175 m^2/g and abundant surface functionalities showed a high removal efficiency (>96%) for organic pollutants with a flux rate of 602 Lm^{-2} h^{-1}. In all cases, series of other EP-like hormones, HMs and biological contaminants have been tested with varying levels of adsorption capacities. Interestingly, these STC carbon filters were easily regenerated and reused for many cycles [19].

Further, similar filter assemblies have been prepared using the highly oxygenated/Al-functionalized solvothermal carbon (Al-STC) nanocomposite deposited on a support paper in powdered form via a simple vacuum filtration technique. These stable deposited filters have been shown to be a sustainable alternative for robust water purification in a dead-end filtration mode (Figure 7.5). The Al-STC-based filters demonstrated high retention efficiencies for organic EPs like dyes, pharmaceutical ingredients and heavy metal ions from contaminated waters. The STC filter was successful in rejecting >99.9% of methylene blue and malachite green dyes and pharmaceutical drugs, namely ciprofloxacin and paracetamol which were rejected with >99.9 and 53% efficiencies, respectively. Interestingly, STC filters also showed high rejection efficiencies for hormones (oxytocin), chromate ions and surfactants (CTAB) with rejection rates of 88.6, >99.9 and 94.9% from contaminated sources, respectively. Therefore, STCs have also been characterized as easy to use and easy to regenerate using a simple ethanol wash, making them cost-effective and sustainable EP treatment solutions.

7.5 NANOCLEANERS

Currently, a wide range of nano and bulk materials are in use as part of chemical and physical water treatment techniques, such as adsorption, coagulation, oxidation, photo degradation and membrane separation. Among these, adsorption has been proven as one of the most adaptable techniques owing to its advantages of high efficiency, convenient operation and comparably low process cost. Adsorption techniques offer the capture of pollutants with ultra-high capabilities. On the other hand, nanomaterials have shown great promise in capturing dangerous pollutants from contaminated sources, ensuring safe drinking water. Nanomaterial-based cleaners or adsorbents can be easily designed to provide innovative centralized or decentralized water treatment systems. However, even though the ultra-fine or small particle size of powdery adsorbents provide ultra-high surfaces which result in high adsorption capacities, the nano size also poses unavoidable limitations in practical application and causes intrinsic problems such as agglomeration, leaching of nanomaterials and a complex post-separation process or regeneration step for separating and recovering the adsorbents from water. Nevertheless, nanomaterials have been effectively utilized for water treatment applications with the help of a carrier support to make them sustainable. In this direction, functionalized carbonaceous materials have been prepared following hydrothermal or solvothermal or ion-thermal carbonization procedures [20].

Recently, hydro/solvo/iono-thermal carbonization systems have been widely used for the production of numerous and a wide variety of functional nanoparticles and hybrids using the facial method and template approaches. Prepared nanomaterials and nanocomposites were subsequently used as an effective media of importance in science and technology to capture especially series of EPs from water and wastewater sources. To this end, several main approaches were adopted: (i) direct facial synthesis of oxide or hydroxide-based nanomaterial, (ii) metal/composite encapsulation within carbonaceous materials and (iii) support assisted surface functionalization using nanocomposites.

A model nano-adsorbent or nano-cleaner would have high porosity and high surface area with specific adsorption sites. The porous structure of nanomaterials contributes to the improvement of their surface area and adsorption capacity when used for water treatment. Metal–organic frameworks (MOFs) are one such class of porous nanomaterial significantly developed and used for water treatment over the last few years. MOFs are attractive owing to their enduring porosity, large surface area, their offering ion-exchangeable sites and controllable pore sizes or shapes, both in microporous and mesoporous regimes. Such extraordinary properties make MOFs an important nanocleaner for the treatment of different EPs in various contaminated sources. Recently, MOFs, ZIFs, LDHs and composites of these basic nanostructures have also been explored as a new platform for the adsorptive removal of EPs from an aqueous medium. Apart from this, researchers are focusing on the development of water-stable polymer-based nanomaterials to explore their applicability in water treatment.

On the other hand, noble metal nanoparticle-functionalized hydrothermal carbon with various morphologies has been obtained from glucose and biomass as a template for co-precursors. Furthermore, these functional nanomaterials prepared with the help of metal-oxide/hydroxide, MOF, ZIFs termed as Nanocleaners and have been utilized for EPs removal in various wastewater samples. Interestingly, nanomaterials have been encapsulated in the carrier or template to provide stable non-leaching properties when used in water treatment applications. Interestingly, the carbonaceous nanocleaners have the ability to encapsulate nanoparticles in their cores with the retaining of the surface functional groups. Subsequently, a considerable number of nanocleaners have been reported as either having carbon or polymers as active materials.

7.6   NANOMATERIAL-BASED FILTERS FOR CAPTURING EPS

Among various composites carbon and polymer-based systems in combination with inorganic precursors have been widely used as nanocomposite adsorbents for EP removal from contaminated water sources. Among these, iron, alumina, zirconia, manganese, nanocarbons and MXenes have been brought together to create nanocomposite filters for the effective treatment of EPs in wastewater. Among many, Fe-based adsorbents are frequently studied due to their magnetic property and relatively insoluble nature which allows them to be used under various conditions; the magnetic property of the iron oxide system makes it easy to recycle and regenerate the adsorbent. Moreover, multi-component systems of iron-aluminium (Fe/Al)-based composites have been proven to be promising hybrid nanomaterials and are widely applied composites in water purification owing to their easy preparation, low cost and their super adsorption capacity to selective EPs like microorganics, APIs,

EDCs, PCCPs, heavy metal ions and biological substances like synthetic hormones. The present study integrates Fe/Al complexation in combination with surface functionalized polymer and/or carbon residues as an effective and stable nanocomposite cleaner.

On the other hand, MOFs with improved selectivity, hydrophilicity, surface characteristics and fouling resistance surface functionality have been considered because of their versatility and potential interaction with EPs. As discussed in earlier sections, MOFs have been widely used as an active component in various membrane-based filtration processes. However, now the unique activities and characteristics of MOFs have been directly utilized as effective filters in association with carriers or support materials. On the other hand, bare MOF particles are difficult to convert directly into a stable water filter due to their poor dispersibility and binding property into thin-film or multi-layer filters. Thus, recent research focusses on preparing nanomaterial-based filters in the form of powder-packed cakes or beds to mimic polymer membranes. With this approach, generally, a vacuum filtration is used to deposit nanomaterials into a filter form with a fabric or paper support. In another approach, nanomaterials are mixed with polymer or carbon binders to create different thickness of filters with various levels of porosity and surface activity. These two approaches have been adopted when considering self-cleaning and the regeneration of the nanomaterial-based filter on a sustainable basis.

7.7 NANO-ADSORBENTS FOR REMOVAL OF EPS

Nanomaterials have been increasingly used due to their unique physical and chemical properties which offer a wide range of tunability. Nano-adsorbents can be produced using the high chemical activity and adsorption capacity of a series of inorganic precursors on the surface of the nanomaterials. In such situations, the surface atoms can be tuned to be unsaturated and thus can be subjected to combine with pollutants in the water or wastewater systems via various interactions to serve as efficient adsorbents. Therefore, nanomaterials can be prepared as a strong adsorbent with the help of suitable precursors, media and templating agents. Remarkably, materials reduced to the nanoscale or nanosized can suddenly show a variety of unique properties, compared to their bulk versions, due to two main reasons:

1. The surface properties are explained by: (i) in nanoscale dimensions material will have more surface atoms compared to inner atoms, (ii) the surface will carry high free energy due to the increased surface area and more accessible surface atoms results in the increase of surface energy associated with the particles at the nanoscale, and (iii) the increased surface area provides an increase in the rate of a chemical reaction which helps in fastening the many water treatment processes.

2. The volume effects originate from: (i) a lower wavelength, higher frequency and higher energy, (ii) the reduced material size allows a blue shift of atoms for optical absorption spectra which may help in interaction with various EPs, (iii) super-paramagnetism which originates when the particle size is less than the magnetic domain in a material, and (iv) a free electron model

which predicts the average energy spacing that increases as the number of atoms is reduced and this propagates improved catalytic properties of nano-adsorbents.

It is well documented that the size-dependent properties in nearly all inorganic nanoparticles changes the material properties significantly above the critical particle size of about 30 nm. Nanomaterials possess key physical characteristics and are classified into condensed materials and nano-dispersions. For water treatment purposes, a variety of precursors have been adopted. Among many categories, nanomaterials are mainly divided according to their characteristic surface properties and functionalization into inorganic, organic and composite assemblies for adsorption applications. Nano-adsorbents include metallic nanoparticles like gold and silver nanoparticles (NPs), metallic oxide NPs such as MnO_x, zirconia, aluminium trioxide or titanium dioxide, nanostructured mixed oxides like Fe/TiO_2, Al/Fe mixed oxide nanoparticles and iron-based magnetic NPs.

On the other hand, carbonaceous nanomaterials and polymeric nanomaterials are another important category of nano-adsorbents that include CNTs, carbon nanoparticles and carbon nanosheets, and grapheneous nanomaterials. Similarly, silicon nanomaterials include silicon nanotubes, silicon nanoparticles, silicon nanosheets, bio-polymer nanoparticles and polymer nanocomposites. Further, wet spin or dry electrospun nanofibres, nanoclay, polymer-based nanogels, xerogels and aerogels make another advanced class of nano-adsorbents. All these end products with a size of approximately 1–100 nm have shown significant influence on EPs through their unique surface chemistry and electronic properties.

Nevertheless, the physical, material and chemical properties of NPs are directly related to their intrinsic compositions, from which their preparation methods follow. Thus, the design, preparation, characterization and application conditions play a critical role in addressing EP abatement. Moreover, with the increase in awareness to develop sustainable nanocomposites, biomaterials and biopolymers have attracted wide research attention to develop high performance composite-nano-adsorbents as promising alternatives for the future. These developments are revolutionizing the nano-based treatment techniques in addressing EP risk in wastewater and the environment. These natural origins have also shown great potential in reducing impending toxicity and help in enabling easy recyclability and surface activity regeneration.

7.8 FENTON AND FENTON-TYPE CATALYSTS FOR DEGRADATION OF EPS

Recently, the scientific community has established the suitability of advanced oxidation processes (AOPs) for degrading EPs which are found in great quantities in wastewater samples and categorized as potentially dangerous for the environment. The Fenton catalytic reactor is an advanced water treatment technique increasingly recognized as effective alternative to remove trace organic EPs from water and wastewater treatment. The Fenton process is generally categorized as an AOP, which utilizes ferrous ions and hydrogen peroxide for the generation of the second most powerful oxidant, hydroxyl ($^•$OH) radicals in an aqueous solution. These radicals further degrade organic EPs from contaminated sources. This process aims to address the limitations of the classical water treatment techniques by reducing toxicity in the water samples. An advanced Fenton process uses a specially designed reactor in which the exhausted ferrous catalyst turns into a heterogeneous catalyst to aid in the oxidation process thereby degrading the EPs in the samples without getting converted into sludge [20, 21, 22].

Initially, the Fenton reaction was designed to be used as an iron-based catalyst in the form of Fe(II) and H_2O_2; however, recent advances have discovered many redox-active metals such as Mn, Cu and Ni also display Fenton-like reactions. Conventionally, a Fenton process can be represented via a general mechanism as follows:

$$Fe(II) + H_2O_2 \rightarrow Fe(III) + {}^•OH + OH^- \quad k = 40 - 80\,M^{-1}\,s^{-1} \tag{7.1}$$

$$Fe(III) + H_2O_2 \rightarrow Fe(II) + HO_2^• + H^+ \quad k = 0{\cdot}001 - 0{\cdot}01\,M^{-1}\,s^{-1} \tag{7.2}$$

$$\text{Organic EP / matter} + {}^•OH \rightarrow \text{degradation products} \tag{7.3}$$

Interestingly, Fenton reactions are light sensitive, so the process can utilize a low-cost catalyst as an economical alternative to degrade toxic organic EPs. In this process, [Fe(OH)]$^{2+}$ is the crucial photoactive Fe(III) complex under sunlight and mainly occurs at the low pH of 2.8. Conventionally, in the presence of light irradiation, [Fe(OH)]$^{2+}$ species gets reduced to Fe(II), which in turn generates an $^•$OH radical followed by the degradation of the contaminant.

In an advanced Fenton reaction, hydroxyl radicals produced in situ by the reaction between ferrous ions and hydrogen peroxide via the classical Fenton process (Eq. 7.4) interact with organic pollutants by redox reactions (Eq. 7.5), dehydrogenation reactions (Eq. 7.6) and an electrophilic addition to π systems, hydroxylation (Eq. 7.7). These reactions produce organic radicals and further oxidation reactions of these radicals with hydroxyl radicals result in complete mineralization of organic EPs.

$$Fe^{2+} + H_2O_2 \rightarrow Fe^{3+} + OH^- + OH^• \tag{7.4}$$

$$OH^• + RX \rightarrow RX^{+•} + OH^- \tag{7.5}$$

$$RH + OH^• \rightarrow R^• + H_2O \tag{7.6}$$

$$RHX + OH^• \rightarrow RHX(OH) \tag{7.7}$$

Interestingly, Ferrous ions utilized for the degradation process during the Fenton reaction are regenerated by the additional reaction of ferric ions with hydrogen peroxide, as in Eqs (7.8) and (7.9).

$$Fe^{3+} + H_2O_2 \rightarrow Fe^{2+} + H^+ + HO_2^{\bullet} \tag{7.8}$$

$$Fe^{3+} + HO_2^{\bullet} \rightarrow Fe^{2+} + O_2 + H^{\bullet} \tag{7.9}$$

More recently, studies have confirmed the added advantages of using external energy on the AOP process in the form of UV or sunlight, electricity and ultrasound which greatly influence the Fenton process to boost its pollutant degradation efficiency.

Generally, the typical chemical feature that connects the AOPs is the creation of the hydroxyl radicals ($^{\bullet}$OH). In advanced processes, the definition of AOPs also comprises the techniques that include oxidants such as $SO_4^{\bullet-}$ and $Cl^{\bullet}$. However, it is observed that a higher oxidative capacity of the catalytic system is highly dependent on its components. In addition to individual metal-based Fenton catalysts, nanocomposite catalysts are gaining increasing importance in AOP EP degradation. In this direction, graphene oxide nanosheets have been combined with a series of metal precursors like Fe, Cu and Ni to obtain improved Fenton activity. Among these, catalytic membranes prepared using graphene oxide/Fe(III)-based MOFs were utilized to treat a series of organic EPs like dyes and bisphenol A. Interestingly, GP/Fe owing to its synergistic separation activity showed an improved proficiency with a flux of 26.3–30.6 L m^{-2} h^{-1} bar^{-1} with an EP retention efficiency of >97% for MB and 87.27% for bisphenol A. More significantly, a composite Fenton catalyst showed remarkable regeneration and reusability up to 12 cycles after dye and bisphenol A degradative separation.

In another approach, hydrophilic catalytic membranes were fabricated using poly(acrylic acid) (PAA) functionalized UiO-66-NH$_2$ (UiO-66-NH$_2$@PAA) MOFs through a vacuum filtration method. These catalytic membranes showed >99.9% separation efficiency for oil–water separation with a pure water flux of 2330 L m^{-2} h^{-1} bar^{-1}. Significantly, after each separation study, composite MOFs showed a potential reusability, maintaining a >80% flux retention. Another strategy, a BiVO$_4$/ferrihydrite (BiVO4/Fh) system, has been utilized for the photo-Fenton degradation of a series of organic dyes like acid red-18, MB and MG to produce decolorized water at high efficiency contaminated with a near-neutral pH. However, composite materials of oxidized MWCNTs and ferrihydrite (CNTs/Fh) have been evaluated for the successful degradation of toxic bisphenol A from industrial wastewater samples. Here, a 3% CNT/Fh system showed multi-fold degradation efficiency compared to simple Fh (Fe) in degrading the EP. Further, an environmentally benign Fenton catalyst of manganese ferrites (Mn–Fe$_2$O$_4$) has been prepared following a sol-gel method to degrade pharmaceutical EPs like ciprofloxacin and carbamazepine. The successful degradation of these pharmaceutical EPs via the photo-Fenton action of visible-light-active Mn–Fe$_2$O$_4$, also has shown remarkable regeneration properties after the degradation process. More significantly, the magnetic property of a Fenton catalyst plays a crucial role in their separation and reusability. On the other hand, to improve the ferromagnetism in α-Fe_2O_3 (hematite) sol-gel-assisted manganese ferrite (Mn–Fe_2O_4) was synthesized for the treatment of EPs in various contaminated sources [23].

In another strategy, a heterogeneous Fenton catalyst was prepared using a hybrid system of iron minerals incorporating numerous supporting materials like zeolites, MOFs, clays, graphene oxide (GO), silica and LDH. This strategy is followed not only to increase catalytic activity but also to induce long durability against highly reactive radicals. For instance, layered structures of photo-Fenton Al-Fe smectite pillared clay have been prepared for the treatment of winery wastewater having high recalcitrant polyphenolic compounds. These photo-Fenton catalysts were prepared by intercalating poly-hydroxy aluminium ($Al3(OH)_4^{+5}$) with ($Fe_3(OH)_4^{+5}$) species between the layers of natural clay. Further, a photo-Fenton performance under the influence of UV-C light radiation showed >75% degradation of total organic carbon (TOC) from winery wastewater. Similarly, a $BiVO_4$ semiconducting active nanomaterial was loaded with interlayers of the hydroxy-iron montmorillonite (Fe/Mt) to use the resulting $BiVO_4$/Fe/Mt photo-Fenton catalyst. In this study, an 8% $BiVO_4$/Fe/Mt nanocomposite demonstrated a >85% TOC, originated from the presence of acid red-18 under visible light irradiation. The remarkable ˙OH generation capacity of this composite catalyst was attributed to the synergistic effect between $BiVO_4$ and Fe/Mt. Similarly, Cu-doped $LaFeO_3$ and $BiFeO_3$ were developed as a visible-light active photo-Fenton catalyst for removing organic EPs like dyes and 2-chlorophenol under visible light. Further, Cu-doping, along with Fe(II) and Cu(I), also acted as a dynamic species in a Fenton-like manner in splitting the H_2O_2 into ˙OH. Also, an Ag-doped $LaCaFeO_3$-δ ($AgLaCaFeO_3$-δ) perovskite was converted to a synergetic peroxymonosulfate (PMS) activating agent for the effective removal of biological EPs like bacterial pathogens.

7.9 RECOVERY AND REUSE OF NANOMATERIAL-BASED ADSORBENTS

The advances in nanotechnology findings have provided several alternative technologies and improved material performance for many industrially important applications. In particular, a lack of efficient water treatment techniques pose a serious threat to water bodies and put stress on fresh water bodies. Hence, there is an urgent need for an efficient high-performance, sustainable and greener point-of-use water treatment system. Among various advanced technologies, methods and/or materials used, an advanced oxidation process has proven a relatively easier and cost-effective method to treat contaminated water. Significantly, catalysts used in AOPs exhibit robust degradation and agglomeration attributed to their catalytic activity and separation potential. In particular, nano-sized functional materials have exhibited robust catalytic activities owing to their high surface area and good dispersibility in water media [24].

As discussed in the previous sections, several adsorbents of inorganic, polymer or composite nanomaterials, Fenton and photo-Fenton catalysts based on transition metal oxides like Fe, Ni, Co and Cu have been studied extensively due to their inexpensive, environmentally benign and magnetically recyclable nature. On the other hand, composites of $CoFe_2O_4$ nanoparticles with an average particle size of 25 nm and saturation magnetization (77.3 emu/g) have been successfully used for the abatement of EPs and have been recycled and reused several times. Significantly, the saturated magnetic property of the $CoFe_2O_4$ catalyst helped to yield excellent recyclability. Furthermore, magnetic $N-CoFe_2O_4$ nanoparticles have been used for EP removal and can be quickly separated from the solution using an external magnet. The magnetic properties helped in the recovery and reuse of an $N-CoFe_2O_4$ nanoparticle catalyst. Several polymer-based nanoparticles and composite systems have also shown chemical stability when used for the catalytic degradation of EPs under varying pH, concentration and temperature conditions. Also these polymer-based composite adsorbents have been tuned to be easily recycled by changing the pH or surface properties. Therefore, the adsorption or degradation activity of a series of polymer, inorganic or nanocomposite nano-adsorbents or nanocleaners have shown significant recyclability and reusability. In most of the cases, adsorbents have been prepared using green and sustainable materials via simple facile strategies.

The advanced methods have also used a series of new heterogeneous catalysts with the doping of Ti, Pd, Nb, Cr, Mn, Co and Ni which have also enhanced the catalytic activity of magnetite in Fenton and photo-Fenton activities for the removal of EPs. The magnetic nature of doped heterogeneous catalysts in Fenton reactions enhances the activity of magnetite by increasing the rate of formation of hydroxyl radicals. However, in all the cases the magnetic nature of catalysts helped in recovering and regenerating the catalyst, making the overall process greener and sustainable. A particular benefit of nano-adsorbents with magnetic and surface functionality permits the post-treatment separation of the nano-adsorbents and their recycling and reuse.

7.10 LIMITATIONS OF NANOMATERIALS FOR EP REMOVAL APPLICATIONS

As discussed in previous sections, various nanomaterials with a large surface area and a large number of surface functional groups have been utilized in recent years. Nonetheless, these nanomaterial-based adsorbents or nanocleaners when used for water treatment for different contaminated waters or when added to aqueous environments, often pose regeneration/recycling problems. Another major disadvantage of the Fenton or photo-Fenton process is the production of a large amount of iron-contaminated sludge. Due to the sluggish regeneration rate of ferrous ions in the Fenton process, the number of ferric ions increases with the reaction time. This elevates the concentration of ferric ions and results in the formation of insoluble complexes. Also, the concentration of hydroxyl ions increases with Fenton reactions which may further increase the separation of secondary pollutants. Consequently, the complexities increase with the pH of the solution which results in further production of iron sludge. Therefore, supplementary downstream treatment of the generated sludge is essential in the conventional Fenton process. The removal of EPs in the conventional Fenton process is a combination of degradation and separation processes. Even though the rate of sludge production can be decreased by increasing the regeneration rate of ferrous ions, the Fenton reaction can be stabilized by adjusting the pH to 3.

Although nanomaterial-based water treatment techniques, methods and protocols have been successfully prepared, tested and evaluated for the abatement of EPs from water and wastewater samples, nanomaterial-based adsorbents pose a serious risk at various stages of their use. Even though several studies have proven that the nano-sized adsorbent has potential as a 'point-of-use' material in future water purification systems, these systems still suffer from several gaps which need more attention. Largely, futuristic water treatment systems require chemically and thermally stable, efficient, inexpensive and reusable eco-friendly materials to address the decontamination of real wastewaters. Nevertheless, a series of studies have shown that multifunctional nanomaterials in water treatment and environmental remediation have the potential to deliver customized solutions. While nanotechnology will certainly deliver efficient water treatment systems or solutions, before that several apprehended issues must be properly solved before nanotechnology can be considered harmless and accessible. Some of the concerns researchers, innovators and technologists should consider during the development of nanomaterial-based water treatment systems are listed as follows [25, 26].

7.10.1 ATOM ECONOMY AND SUSTAINABILITY

Currently, nanomaterial-based water treatment techniques are being evaluated as relatively economical, compared to conventional treatment techniques like chemical methods and physical methods such as membrane technology approaches. However, the use of greener and sustainable approaches by adopting renewable material sources can make nanomaterial-based water treatment solutions much more economical and cost-effective. The cost of raw materials or precursors and processes adopted generally defines and contributes to the final product cost. These parameters also influence the end product's physical and chemical nature which directly influences the productivity of treated water or the quality of recovered water. Therefore, the near future will see the increasing use of bio-based precursors and a simple one-step preparation route requiring less energy. Further, several studies have already proven the advantages of nanocomposites derived from biomaterials for EP treatment due to their inherent functionalities, low cost and biodegradable aspects. In particular abundant, renewable polymers with tunable chemical properties act as host materials for active nanocomposites and provide enhanced active sorption sites, easy processability and some intriguing improvements in the performance of composites achieved by nanoparticle structure association.

7.10.2 STABILITY AND LEACHING OF NANOMATERIALS

One of the major concerns associated with the nanomaterial-based water treatment system is its stability and shelf life. Once the nanomaterial has been used in water for EP abatement purposes, most of the nano-sized active species disperse into the water system; however, their poor mechanical and chemical stabilities may pose a serious threat to product water. This may lead to the formation of undesired by-products or secondary pollutants in the water media. Another major concern with the nanomaterial water system is the nano-toxicity associated with the traces of residual adsorbents or treatment media. Further, polymer-based nanomaterial suffers from surface fouling when used in the long run which drastically decreases the shelf life and efficiency. According to a series of reports, upon disposal or usage several of these persistent trace nanomaterials collect in the environment and will eventually settle out, exposing the species in the sediment phase. Nevertheless, efforts are in place to stabilize nanomaterial systems and standardize the evaluation of nanomaterial toxicity to predict or prepare for impending risk. Nanomaterials in filter form also pose unfavourable conditions like fine particle leaking or leaching which permeates from the system due to the nano-sized filter bed. The release of nanoparticles may occur through desorption from composite materials or dissolution of metals likely to be persistent in the medium and which are not biodegradable. In both cases, nanomaterials end up as a secondary contamination in purified water which again poses serious environmental and occupational health risks upon consumption or reuse of purified water. In particular, nanomaterial adsorbents have been increasingly characterized for their hazardous nature rather than their constituents compared to previously encountered chemicals, which necessitates a specialized risk assessment.

7.10.3 REGENERATION AND DURABILITY

Another major challenge associated with the pilot scale implementation of nanoscale filters is the regeneration and reuse of active surfaces. Conventionally, many materials have been used as one-time, irreversible cleaning agents through a permanent physical or chemical interaction with conventional pollutants. However, the growing demand for sustainability and greener approaches is forcing researchers to find new ways to develop multiple uses of materials. Even though recent advances have shown promising material stocks in use as easily regenerative and self-cleaning nanomaterial stocks. In this direction, several biopolymer and protein-based composites have been explored for the efficient treatment of EPs in various contaminated water sources. Further, to induce high surface functionality, recyclability and self-cleaning properties, surface modification approaches have been adopted. These approaches have resulted in inducing excellent filtration properties to nanomaterials at varying pH and end use conditions.

REFERENCES

1. María Pedro-Monzonís, Abel Solera, Javier Ferrer, Teodoro Estrela, and Javier Paredes-Arquiola, A review of water scarcity and drought indexes in water resources planning and management, *Water Journal of Hydrology*, 2015, 527, 482–493.
2. Cecilia Tortajada, Contributions of recycled wastewater to clean water and sanitation Sustainable Development Goals, *NPJ Clean Water*, 2020, 3, 22.
3. Runnan Zhang, Yanan Liu, Mingrui He, Yanlei Su, Xueting Zhao, Menachem Elimelech, and Zhongyi Jiang, Antifouling membranes for sustainable water purification: strategies and mechanisms, *Chemical Society Reviews*, 2016, 45, 5888–5924.

4. Chella Santhosh, Venugopal Velmurugan, George Jacob, Soon Kwan Jeong, Andrews Nirmala Grace, and Amit Bhatnagar, Role of nanomaterials in water treatment applications: a review, *Chemical Engineering Journal*, 306, 2016, 1116–1137.

5. Shabnam Taghipour, Seiyed Mossa Hosseini, and Behzad Ataie-Ashtiani, Engineering nanomaterials for water and wastewater treatment: review of classifications, propeerties and applications, *New Journal of Chemistry*, 2019, 43, 7902–7927.

6. Neeta Pandey, S. K. Shukla, and N. B. Singh, Water purification by polymer nanocomposites: an overview, *Nanocomposites*, 2017, 3(2), 47–66.

7. Alexis Wells Carpenter, Charles François de Lannoy, and Mark R. Wiesner, Cellulose nanomaterials in water treatment technologies, *Environmental Science & Technology*, 2015, 49(9), 5277–5287.

8. Alexis Wells Carpenter, Charles-François de Lannoy, and Mark R. Wiesner, Cellulose nanomaterials in water treatment technologies, *Environmental Science & Technology*, 2015, 49(9), 5277–5287.

9. Nishil Mohammed, Nathan Grishkewicha, and Kam Chiu Tam, Cellulose nanomaterials: promising sustainable nanomaterials for application in water/wastewater treatment processes, *Environmental Science: Nano*, 2018, 5, 623–658.

10. Peng Liu, Charles Milletto, Susanna Monti, Chuantao Zhu, and Aji P. Mathew, Design of ultrathin hybrid membranes with improved retention efficiency of molecular dyes, *RSC Advances*, 2019, 9, 28657–28669.

11. Anshu Kumar, Parimal Paul, and Sanna Kotrappanavar Nataraj, Bionanomaterial scaffolds for effective removal of fluoride, chromium, and dye, *ACS Sustainable Chemistry & Engineering*, 2017, 5, 895–903.

12. Manohara Halanur Mruthunjayappa, Vibha T. Sharma, Kalpana Dharmalingam, Nataraj Sanna Kotrappanavar, and Dibyendu Mondal, Engineering a biopolymer-based ultrafast permeable aerogel membrane decorated with task-specific Fe–Al nanocomposites for robust water purification, *ACS Applied Bio Materials*, 2020, 3(8), 5233–5243.

13. Wenyuan Xie, Fang He, Bingfang Wang, Tai-Shung Chung, Kandiah Jeyaseelan, Arunmozhiarasi Armugam, and Yen Wah Tong, An aquaporin-based vesicle-embedded polymeric membrane for low energy water filtration, *Journal of Materials Chemistry A*, 2013, **1**, 7592–7600.

14. Yu-Ming Tu, et al. Rapid fabrication of precise high-throughput filters from membrane protein nanosheets, *Nature Materials*, 2020, 19, 347–354.

15. *A New Generation Material Graphene: Applications in Water Technology*, Mu Naushad (ed.), Springer International Publishing, Print ISBN: 978-3-319-75483-3, Electronic ISBN: 978-3-319-75484-0, 2019, Springer Nature Switzerland AG.

16. Ahmed M. E. Khalil, Fayyaz A. Memon, Tanveer A. Tabish, Deborah Salmon, Shaowei Zhang, and David Butler, Nanostructured porous graphene for efficient removal of emerging contaminants (pharmaceuticals) from water, *Chemical Engineering Journal*, 398, 2020, 125440.

17. Ihsanullah, Carbon nanotube membranes for water purification: developments, challenges, and prospects for the future, *Separation and Purification Technology*, 2019, 209(31), 307–337.

18. R. Mailler, Removal of a wide range of emerging pollutants from wastewater treatment plant discharges by micro-grain activated carbon in fluidized bed as tertiary treatment at large pilot scale, *Science of The Total Environment*, 2016, 542(Part A), 983–996.

19. Manohara Halanur, Supratim Chakraborty, Kanakaraj Aruchamy, Debasis Ghosh, Nripat Singh, Kamalesh Prasad, D. Kalpana, S. K. Nataraj, and Dibyendu Mondal, Engineering Fe-doped highly oxygenated solvothermal carbon from glucose-based eutectic system as active microcleaner and efficient carbocatalyst, *Journal of Materials Chemistry A*, 2019, 7, 4988–4997.

20. Mostafa Khajeh, Sophie Laurent, and Kamran Dastafkan, Nanoadsorbents: classification, preparation, and applications (with emphasis on aqueous media), *Chemical Reviews,* 2013, 113, 7728–7768.
21. P. V. Nidheesh, Heterogeneous Fenton catalysts for the abatement of organic pollutants from aqueous solution: a review, *RSC Advances*, 2015, 5, 40552.
22. Nishanth Thomas, Dionysios D. Dionysiou, and Suresh C. Pillai, Heterogeneous Fenton catalysts: a review of recent advances, *Journal of Hazardous Materials*, 2021, 404, 124082.
23. Peter A. Ajibade and Ebenezer C. Nnadozie, Synthesis and structural studies of manganese ferrite and zinc ferrite nanocomposites and their use as photoadsorbents for indigo carmine and methylene blue dyes, *ACS Omega*, 2020, 5(50), 32386–32394.
24. Asim Ali Yaqoob, Tabassum Parveen, Khalid Umar, Mohamad Nasir, and Mohamad Ibrahim, Role of nanomaterials in the treatment of wastewater: a review, *Water*, 2020, 12, 495.
25. Sayali S. Patil, Utkarsha U. Shedbalkar, Adam Truskewycz, Balu A. Chopade, and Andrew S. Ball, Nanoparticles for environmental clean-up: a review of potential risks and emerging solutions, *Environmental Technology & Innovation*, 2016, 5, 10–21.
26. Ankit Nagar and Thalappil Pradeep, Clean water through nanotechnology: needs, gaps, and fulfillment, *ACS Nano*, 2020, 14(6), 6420–6435.

8 Micro-Plastic Pollutants: Identification, Detection and Removal Techniques

8.1 INTRODUCTION

As of 2020, researchers have estimated that there are ~6 trillion macro-plastic fragments and microplastic in our oceans; these distributions have been identified as consisting of up to 46,000 pieces in every square mile of ocean, weighing up to 269,000 tonnes. Also, it is estimated that every day nearly 8 million pieces of microplastic enters our oceans. Even though several studies have warned of the potential risk of the over-exploitation of plastics and their unregulated discharge into the environment for the last two decades, regulatory authorities have only begun to assess the potential harms of microplastics very recently. Recent studies have found microplastics everywhere in soil samples, surface water, in deep oceans, in Arctic snow and Antarctic ice, reverse osmosis (RO) treated water, shellfish, table salt, conventionally treated drinking water and beverages. More worryingly, these microplastics in the environment could take several decades to degrade fully.

The initial investigations were limited to focusing on detecting microplastics of a large size regime, like microbeads found in personal-care products and plastic residues left over from moulding and manufacturing processes, disintegrated parts from discarded bottles and the residual debris from large scale plastic dumps. It is estimated that with time gradually all these plastic particles will wash into rivers and oceans. As early as 2015, a team of oceanographers estimated there were between 15 trillion and 51 trillion microplastic units floating in surface waters worldwide. During same period, environmental agencies and regulators stepped up their efforts to assess the environmental risk and find ways to quantify the health hazard upon exposure. As a result of these efforts, hundreds of studies have identified the presence of microplastics in animal feedstocks, aquatic organisms and almost in every water body. Also, studies have identified the presence of various types of microplastics in different size, shape and chemical compositions. Further, initial reports have also confirmed that the micro- or nanoplastics with sizes less than 1 micrometre enter living cells easily, potentially disrupting cellular activity. On the other hand, this problem is anticipated to grow exponentially, as the annual production of plastic products has crossed 400 million tonnes and a major portion of it is one-time use and throwaway plastic which is also projected to double by 2050. Even if the worldwide production of plastics stopped with immediate effect, it is estimated that more than 5 billion

DOI: 10.1201/9781003214786-8

tonnes will still remain in the environment available for disintegration into micro-plastics for decades to come [1–4].

Even though exact research is still assessing the precise health risks associated with the unintentional or accidental consumption of microplastics in humans, the sheer presence of microplastics in cells or tissues is enough to irritate just by being a foreign material. Nevertheless, the long-term presence and added accumulation of microplastics, microfibres or dust can inflame lung tissue and lead to cancer. Yet, the larger microplastics are also expected to exert negative effects on humans, aquatic life and environments at large through their chemical toxicity. These trends may also continue to pose a potential threat as commodity plastics, such as one-time use plastic bags, plasticizers, stabilizers and pigments in plastics and packaging materials, are inadvertently entering daily life and discharging into the environment. Many of these substances have been characterized as hazardous and interfering with endocrine systems. Another assumption is that microplastics in the environment might act as favourable substrates or carriers to attract other EPs and then transport them into animals or aquatic life through food chains. Therefore, currently microplastics are well recognized as the most hazardous EPs which frequently evade all conventional and advanced filtration systems, holding potential threats for marine organisms and human beings. In the marine environment, fish and other aquatic creatures confuse microplastics with food, and their chemical inertness and colorless properties deceive humans and enter the human food chain.

8.2   DEFINITION AND THREAT OF MICROPLASTICS

At present various water bodies are characterized by the presence of ubiquitous microplastic pollutants. However, technically, there was no prescribed definition of a nanoplastic until 2018, when researchers proposed an upper size limit of 1 µm which is tiny enough to remain dispersed through a solvent (water) media where organisms can more easily consume them, instead of sinking or floating as larger microplastics. Lately, microplastics also have been detected in several marine animals. So far, several studies have focused on microplastic risk assessment in marine organisms which show growth inhibition and reproductive disorders in several marine life forms. Marine biologists have also noted that due to microplastic toxicity, marine animals are laying smaller eggs, which are less likely to hatch. Further, the presence of microplastics in the marine environment have altered the food habits of marine animals. This situation has influenced reproduction complications in aquatic life. However, ecotoxicologists have yet to assess the specifications of the microplastics that exist in aquatic environments. These assessments consider the type of polymer, the nature of the additive, and the physical dimensions and concentrations in the environment [5, 6].

On the other hand, researchers have shifted their focus towards environmentally realistic circumstances using standard polymer systems in simulation with different physical dimensions, such as fibres or traces of plastics, rather than original polymeric spheres. Some observations have considered microplastic assessments using chemical compositions that mimic biofilms, which appear to make animals more likely to eat microplastics. Observations also conclude that microplastic fibres are more likely to enter marine life compared with spheres, as fibres take longer to pass through zooplankton upon discharge into the environment. Further, zooplanktons exposed to microplastic fibres produce half the typical number of larvae; the subsequent adults were smaller. Remarkably, microplastic fibres are not ingested into marine life; however, they interfere with swimming which over a period induces deformations in the organisms' bodies. Further, adult marine animals such as Pacific mole crabs(*Emerita analoga*) were short lived upon exposure to microplastics. Lately, computer simulations have been used to predict interactions between fish and microplastics, the nature of these interactions, how often fish encounter microplastics, the microplastic consumption capacity of a fish, and the probability of a fish eating enough microplastic pieces to affect its growth. Accordingly, researchers noticed that, at present microplastic pollution levels, fish run a risk of 1.5% in locations characterized by microplastics, but at higher microplastic concentration hotspots the fish are likely to experience higher risks. This is due to microplastics concentrations being marked at different locations in different weather and seasons throughout the year.

When it comes to microplastic toxicity in humans, no authentic reports exist at present, yet circumstantial examinations show the potential effects of microplastic specks in people. These studies considered the cells or human tissues exposed to microplastics, and animal studies conducted using mice or rats as experimental species. One such study concludes that mice fed on large quantities of microplastics exhibited inflammation in their small intestines. Also, controlled experiments conducted on mice exposed to microplastics in two studies showed a lowered sperm count and fewer, smaller pups, compared with an unexposed group of animals. Some of the *in vitro* studies using microplastics on human cells or tissues also suggested the existence of plastic toxicity. However, the effect of microplastic concentrations on humans is not yet clearly similar to that seen in the case of marine animals. Nevertheless, most of the controlled studies used polystyrene spheres as the standard microplastic model, which does not represent the diversity of microplastics that people ingest, such as commonly used polypropylene (PP), polyesters (PEs), cellulosic fibre, polyethylene (PE) or polyethylene terephthalate (PET). However, these studies are still at a preliminary stage and require more *in vitro* studies than animal ones to assess the potential risk of various microplastics in humans. But researchers have yet to design protocols to extrapolate the effects of solid microplastics as EPs on tissues to probable health complications in entire living beings [7, 8].

Interestingly, the latest studies have considered the risk associated with the microplastic-associated parameters in animals in which these tiny inert particles could remain in the human body, hypothetically accumulating in some tissues. Control studies conducted on mice have found that microplastics (~5 μm) could stay in the intestines or reach the liver. One study focused on gathering data on how quickly mice excrete microplastics and the assumption that only a fraction of particles 1–10 μm in size would be absorbed into the body through the gut. Based on this, researchers have extrapolated the possibility that an individual adult might accumulate several thousand microplastic units in their body over their lifetime. Remarkably, researchers have devised sophisticated experiments to explore the possibilities of the presence of microplastics in human tissues. For this purpose, researchers extracted six samples of placenta tissues using suitable chemical reagents and surprisingly found 12 pieces of microplastic in 4 of those placentas. However, researchers have also given the benefit of doubt to human error or sample contamination during the extraction or analysis process.

Nevertheless, the most worrying thing about microplastic particles is that these EPs are small enough to penetrate and persist for a long time in tissues, or even in cells which warrant more attention in environmental sampling. To confirm this assumption, researchers used animal studies, in which pregnant mice inhaled extremely tiny particles, and later detected the microplastic particles in almost every organ in their foetuses. However, it was also noted that the microplastic particles typically need to be less than a few hundred nanometres to enter cells.

Until very recently, researchers did not define plastics as nanoplastics; however, lately tiny invisible particles (<1 μm) have been detected under optical microscopes and spectrometers and been distinguished from microplastics. Even though there is no such distinction between micro- and nanoparticles, nanoplastics can be distinguished by their interaction pattern with light, which can measure the length, width and chemical composition to a few micrometres. In addition to this, below 1 μm, plastic pieces or particles become difficult to distinguish from non-plastic particles like marine sediment or biological cells or suspended dust. Nevertheless, nanoplastic presence in oceanic environmental samples has been confirmed during seawater desalination and through other controlled sampling studies conducted in different geographic locations. These analytical procedures have been conducted with indigenous and task-specific approaches in which colloidal solid extractions from seawater samples were filtered and particles larger than 1 μm were rejected, the filtrate collected through mesh of less than 1 μm and the concentrates prepared for further spectroscopic analysis. These concentrated sediments were further used for mass spectrometer analysis for the qualitative analysis of molecules and molecular weight, thereby determining the polymer origin. However, throughout, it is difficult to distinguish nanoplastic by their sizes or shapes [9, 10].

8.3 MICROPLASTICS IN MARINE AND GROUND WATER

Plastic and microplastic pollution or contamination in the marine environment is defined by fragments of plastic debris originating from the unregulated discharge of plastic products. Microplastics are now categorized as an EP due to their abundance in natural and aquatic environments. Due to this abundance and growing health risk concerns, the United Nations Environment Programme (UNEP) has included microplastics in the list of top 10 environmental concerns. Essentially, plastic EPs are broadly categorized by their origin as primary microplastics which are micro in size and released into the marine environment from essential plastic products. Currently, two types of plastics and microplastics have been identified as EPs in the world's ocean: (i) primary and (ii) secondary microplastics. Primary microplastics are directly released into the environment in the form of small particulates. They can be a voluntary addition to products such as personal and cosmetic products like scrubbing agents, toiletries, cosmetics, toothpaste, facial products, hair gels, shower gels, cosmetic lotions and composite surfactants. Microplastics may also originate from the scratch of large plastic objects or products during manufacturing, use or maintenance, such as the erosion or corrosion of tyres when driving or the abrasion of synthetic textiles during washing and discarding. Secondary microplastics occur from the disintegration of larger into smaller plastic fragments and pieces upon exposure to a harsh marine environment and weathering. Further, secondary microplastics are disintegrated pieces of plastic resulting from the weathering of bigger plastic debris, like water bottles, one-time use carrier bags, straws, cups, plates, car fenders and household plastic products due to exposure to sunlight, temperature and humidity over time. On the other hand, once the plastic debris reaches seawater, ocean waves can also disintegrate larger plastic pieces by repeatedly washing and grinding them into the sand and by heat and wind friction over a period of time. Now, every ocean has been polluted with large amounts of microplastics; most of them as independent debris and some as supported pollutants which makes them distinguishable based on their composition, shape and size. More worryingly, ocean currents and circulation patterns provide a harmonious environment to distribute microplastics across the different oceanic environments. This uninterrupted transportation of microplastics across oceans makes them difficult to track and remove at specific points. This situation also creates a difficult scenario for which to design and develop sustainable microplastic collection and removal plants at fixed points or sites. However, recent attempts have been made to track hot spots of debris to identify their movement and density. This method adopts a satellite-assisted imaging technique to create models which help to predict the entry and disintegration points in oceanic environments [11].

At present several million tonnes of microplastics enter the world's oceans each year; however less than 1% is found floating at the surface in visible form. This suggests that a large amount of microplastics in the ocean are likely to be buried in the sediments or suspended in the highly dense environment. Nonetheless, plastics in the marine environment are found in all sizes and shapes mainly due to marine litter. More significantly, the high density of plastic debris reported until now is found in the central areas of the North Pacific and North Atlantic Oceans. Furthermore, oceanic currents and circulation models suggest that all five subtropical ocean gyres act as convergent regions making them the most likely accumulation regions. On the other hand, surface ocean currents are spatial, temporal and weather dependent variables and the concentrations of plastic debris are continuously fluctuating. However, there is a clear lack of studies in many oceanic basin regions where elaborate statistical information on plastic debris remains unknown.

Furthermore, a United Nations Environment Assembly (UNEP) study in 2017 estimates that annually there is ~12 million metric tonnes of plastic debris released into the oceans. This problem is expected to persist in the decades to come as the modern life style has adopted products where virtually everything is made of plastic due to its lightweight, strength, low cost and durability. However, more than 60% of the plastic debris originates from one-time use plastics via inappropriate disposal, dispersion and accumulation in the marine environment. Moreover, the COVID-19 pandemic has exponentially promoted the one-time use and disposal of face masks which are mainly made up of non-woven plastic fibres, adding to the massive plastic and microplastic waste in the environment. Accordingly, nearly 90 million plastic medical face masks have been used every month and disposed into the environment creating a new challenge for marine plastic litter. The most significant sources of microplastic contamination in oceanic environments occurs through coastal dumping sites, coastal cities, ports, coastal landfills and shipping activities [12].

Further, recent scientific evidence shows that microplastic particles are infiltrating the marine ecosystem from human consumption or from lower trophic levels. Once the microplastics slip from human activities near coastal cities or dump yards of landfills, these EPs either float on oceanic waves or sink when they encounter enclosed bio-film, and settle onto the sediments. These microplastic pollutants include obstinate, bioaccumulative, hydrophobic and hazardous substances such as polychlorinated biphenyls (PCBs), cellulosic fibres, polyester, polypropylene and PET debris and dioxins. In addition to self-toxicity, microplastic particles also act as carrier substrates to other conventional and emerging pollutants.

Hence, due to continuous plastic manufacturing and the usage and discarding of other polymer-based products, we may anticipate that the number of microplastics in ecosystem vary each day passing. These EP categories mainly consider primary microplastics, putting the overall quantity of both micro- and macroplastics in the ocean at greater than the presently estimated 8 million metric tonnes. Nevertheless, large plastic waste is readily visible in every neighbourhood and aquatic sources which one day are ingested, then injure, entangle or suffocate marine life. While microplastics in water may not be easily visible to the human eye, potentially undesirable impacts are less obvious and their increased release into the oceans will have far reaching consequences. Nonetheless, the accumulation of microplastic toxicants in the food chain and/or sorption will directly affect human health.

There is a general belief that most of the plastic and microplastic contamination in aquatic bodies and at large in the world's oceans occurs from mismanaged plastic waste. Therefore, most recent reports on plastic and microplastic pollution focuses almost entirely on the quantification of secondary, invisible and untraceable sources and on waste reduction and management. However, coordination and consorted efforts on the worldwide releases of primary microplastics (polymer or plastic products) into the world's oceans is mostly lacking.

Groundwater remains in a stationary aquifer for tens to hundreds of years, or even longer. On the other hand, landfills and plastic dump yard erosions increasingly contaminate ground water. A combination of these longer reservoir times with plastics which may temporarily resist degradation means that those chemical effects could effectively build up in ground water. Also, these increasing concentrations of plastics may facilitate bioaccumulation of microorganisms, thereby increasing the possibility of toxic bioaccumulation. Together, biological EPs and microplastics could result in long-term threats to ground water contaminations which are currently not well understood for their health and environmental risks. However, there is still a lack of information on the origin of microplastics in groundwater sources. This is also due to large external influencing factors such as low-rainfall conditions, flooding events and ground water extraction rates which constantly affect microplastic concentrations and density in groundwater.

Recently, data suggest that global annual plastics production has crossed a record of 350 million tonnes with only a small percentage (6–26%) being recycled after use.

Additionally, microplastics in ground water have a propensity to act as a vector for other contaminants like pathogens, organic pollutants and heavy metals through their strong dispersion and diffusion mechanisms. Thus, currently several research groups are considering the gap concerning the physical, chemical and biological mechanisms that govern and control the existence, accumulation and transportation of microplastics in groundwater. Therefore, future studies need to focus on understanding the transport behaviour of various kinds of plastics and microplastics such as types, shapes, sizes and surface morphology and analyse these microplastics as potential supports or substrates or vectors for microorganisms in soil, ground water and the marine environment.

8.4   MICROPLASTICS IN WASTEWATER

Even though research attention on the fate of plastic and microplastics in wastewater has increased only during recent years, today approximately 80% of the 8 billion metric tonnes of plastic products produced and discarded is in landfills, dump yards or the environment. The release pattern of microplastics and their pathways entering aquatic environments remain understudied. Most wastewater treatment techniques initially consider detailed investigations of samples for conventional pollutants, biogenic organic matter (BOM), colloidal pollutants, biological EPs and lately microplastics, where the challenges associated with the separation of microplastics from wastewater is designed. In spite of the several limitations associated with the detection, identification and analysis of microplastics in the presence of BOM-rich and biological matrices, there is a growing trend towards examining the fate of EPs, in particular, microplastics in wastewaters. Even though there is a lack of a proper microplastic removal technique in the practices, some simple protocols have been used to detect and quantify microplastic abundance in different wastewater samples; methods have also been developed to remove microplastics during the final stages of settlement.

Since the beginning of the plastic age, some 70 years ago, the modern life style has adopted almost every product made of polymer/plastic precursors. Even though fears have erupted across the world regarding the need to minimize plastic production and reduce the dependency on plastic-based materials and an emphasis on the production of reusable polymer/plastics, instead the world has seen an increasing growth in plastic production, which upon use consequently spreads into the environment to such a point that we have now termed it as a toxic and emerging pollutant. Nowadays, microplastic particles have been ubiquitously detected in various wastewater samples, in particular in domestic and municipal industrial wastewaters, and have been characterized by their high concentrations in a broad range of shapes, sizes and densities. At present it is estimated that over 60 million plastic and microplastics could be being released into natural waters every day from domestic and municipal wastewater treatment plants and this number may increase exponentially in coming years as the production and discarding of plastic products is ever increasing [13].

The main source of plastics and microplastics are petroleum-based synthetic polymers. These polymers start their journey to become plastic products in the form of pellets (85%), which goes into the production of thousands of household and personal care products, synthetic textiles (12%), synthetic rubber in tyres (2%) and the remaining 1% have been categorized as speciality polymers. Textile and rubber industries generate plastics and microplastic EPs as accidental or residual waste; however, the domestic use of large numbers of polymer-based personal care products releases microplastics into wastewater sources which can be categorized as intentional and inevitable waste. A major share of microplastics is known to be produced as an unintentional pollutant through abrasion, weathering or inadvertent spills during the production, transport, usage, maintenance or recycling of products containing plastic. The major portion of the intentional release of microplastics in the form of personal care products largely enters wastewater from every house and municipal channel [14, 15].

Therefore, the uncontrolled and untenable use and disposal of plastic is causing stubborn and widespread environmental pollution. On the other hand, microplastics in wastewater also act as a host for dangerous EPs by adsorbing harmful agents, such as APIs, HMs, pesticides and pathogenic organisms. Even though information on microplastics in wastewater and their origin, fate and transformation is yet to be understood in detail, current knowledge gives a fairly good trend and forecast of the impending threat from microplastics in various wastewaters. As discussed earlier, personal care products and cosmetics may comprise up to 5% of primary microplastics with an average size of 250 μm. These microplastics in surfactants/soaps have replaced natural exfoliants such as powdered walnut husks in exfoliant washes as they irritate and damage the skin less. In toothpaste, polymeric micro-ingredients remove plaque and stains through the process of their abrasive action. In a single use, an exfoliant wash can release 4500–94,500 pieces of microplastic while toothpaste releases around 4000 pieces. However, the majority of these released microplastics have been predicted to account only of polyethylene-based products in both the hygiene as well commodity goods category.

On the other hand, nearly 35% of detected microplastics in the oceans have been traced to fibres from synthetic textiles that are released during the washing of clothes. It is estimated that from one thousand to one million microfibres are released by a single garment. Similar studies conclude that in a washing load of 5–6 kg, polyester fabrics release over 6 million microfibres while acrylic fabric releases 700,000 microfibres into the drain which eventually reach the marine environment. In Finland, an estimated 154,000 to 411,000 kg of cotton and polyester microfibres annually are released through washing machines. These numbers may well vary with textile properties, like polymers, thread density, knitting style, washing conditions (e.g. temperature, speed, friction, velocity, washing time, type of detergent and softener), and garment strength and weathering. Other household sources of microplastics in wastewater systems may include glitter, contact lens cleaners, small buttons and jewellery.

On the other hand, non-domestic sources of plastics and microplastics in wastewater may include: (i) polymer paints and polymer induced pigment microplastics that are eroded from air blasting process or during cleaning routines; (ii) polymer pellets lost during manufacturing or transport; (iii) extensively used packaging material 'Styrofoam' is used and discarded during packaging or shipping or unpacking; (iv) microfibres from synthetic textile industries; and (v) fine plastic dust from drilling and cutting plastics. Therefore, microplastics may enter the environment through drains or sewage systems from all these sources, intentionally or accidentally. Therefore, it is highly expected that high concentrations of plastic particles and microplastics enter wastewater treatment plants every day. Interestingly, contrary to what was earlier believed, wastewater treatment plants likely retain most of the microplastics. It is believed that most microplastics (>75%) are removed during the primary treatment of wastewater, whereas secondary treatment is responsible for only a small decrease in concentration (on average ~15%) and tertiary treatment has almost no effect at all.

In the primary treatment stage, microplastics are generally retained during the removal of wastewater-suspended solids through skimming processes and sedimentation due to their own settling or entrapment. However, at this stage, fine microplastics have been detected as escaping, while fibres are readily retained. Though treated effluents have been characterized by trace concentrations of microplastics per litre, the high volumes of effluents released every day leads to considerable accumulation in the aquatic environment. Therefore, generally, high concentrations of nano- and microplastic contamination have been recorded downstream of wastewater treatment plants. This situation may lead to disastrous conditions in countries or regions lacking proper wastewater treatment protocols and regulations.

On the other hand, several studies conclude that microplastic accumulation occurs largely in the solid fractions or sludge of wastewater through grit and grease entrapment. This leads to the separation of up to 45% of microplastics while the sedimentation step retains nearly 34%. Thus, overall, nearly 80% of microplastics is separated from effluent up until the sludge fractionation process, which is at present generally considered suitable as a fertilizer for field application; however, this approach has been lately regarded as a potential cause of terrestrial contamination. However, researchers are devising new mechanisms and protocols to re-evaluate the efficiency of existing methods and technologies in removing microplastics smaller than 5 µm by developing more sensitive detection methods. Efforts are also being made to assess the discrepancies in concentration measurement and protocols followed to distinguish the size and nature of microplastics released through treatment plants over a period of time [16–18].

8.5 MICROPLASTICS IN THE ECOSYSTEM

A major portion of the microplastics polluting the ecosystem is derived from different polymer origins and compositions such as polypropylene (PP), polyethylene (PE), poly(vinyl chloride) (PVC), nylon (PA), polystyrene (PS), cellulose acetate (CA), polysterol polystyrene, thermoplastic polyester, polyethylene terephthalate (PET) and foamed polystyrene. More significantly, microplastics originated from these polymers are insoluble in water, are non-biocompatible, non-degradable, and possess diverse physicochemical properties that determine the bioavailability and accumulation in organisms when entering a conducive ecosystem. This concept of microplastics categorized as a persistent or emerging pollutant is widely characterized in both terrestrial and aquatic ecosystems. It is now evident that polymers or plastics are frequently used indiscriminately in millions of products and discarded carelessly due to the lack of regulation. Most of the microplastics in the ecosystem are from inhabited household discards in which personal care products, like paste, lotion, gel, facial cleansers, facial scrubs and bath foam that constitute the major portions of nano/micro-plastics. On the other hand, landfills, construction residue, factory polymer dust, agricultural package material in farmland, ship breaking, paints, coatings, medical patches and adhesives have also contributed immensely to the increased concentration of plastic-based EPs in the environment. Further, these microplastics under the influence of weathering transfer to soils, ponds, lakes, rivers and oceans. Unusually, storm water runoff has also been observed to contain high quantities of plastics and microplastic contamination. For instance, several municipal wastewater treatment plants in various cities across the world have extracted plastic beads and microplastic debris which entered the plants through storm drains. Although it is rare that runoff may carry plastic and microplastic residues to treatment plants, lately these observations have concluded that storm water is also contributing to the overall plastic load. This abundant contamination of oceans and water bodies in general by plastic debris or microplastics is becoming a major concern. The scale, magnitude, nature and uniqueness of ocean and fresh water reservoir contamination is now referred to as plasticene and the world's oceans as a plastic soup [19].

However, present concentrations of nanoplastics and microplastics in the environment and various ecosystems are estimated to be too low to affect human health. Nevertheless, with time the level and concentration of micro- and nanoplastic will rise due to the rapid increase in the amount of plastic in the ecosystem. A study estimated that the amount of plastic disposed into the ecosystem in sealed landfills was around 188 million tonnes in 2016 and this is estimated to double to 380 million tonnes by 2040. At least 10 million tonnes of this total discarded plastic will be in microplastic and nanoplastic form. Likewise, microplastics have contaminated a wide range of ecosystems around the globe, adversely affecting almost every organism. Lately, microplastics have been detected in various soil environments. Soil is not conceived as a large reservoir of microplastics as most of the microplastic waste is landfilled or discarded from wastewater treatment plants as sediment solids. Another major source of microplastics in soil is sewage sludge applied as fertilizer for farming and these penetrate deep into environment as a consequence of air and precipitation. Therefore, microplastics in soil might pose a serious threat to soil biodiversity, food productivity and ecosystem functioning, and contaminate the food chain. Earthworms are measured as a key bio-indicator of the health of earth fertility; they also act as ecosystem engineers that regulate soil health. The presence of microplastics may also pose a potential threat to earthworms and be poisonous. Further, an increase in the concentrations of microplastics in soil fauna will make them vulnerable to further stress. More worryingly, studies conducted on European agricultural land reveal that plastic particles ranging from 700 to 4000 per kilogram of soil were recorded, which was nearly 7% microplastic fragments in dry soil weight [20].

Due to their micro- and nanometre size, microparticles can easily enter the human body in various pathways. In humans, the consumption of food or drinking water that is contaminated with microplastic is the main route of human exposure. There is a high risk of consuming microplastics when intaking food which originates from marine products. On the other hand, aquatic organisms such as plankton and filter feeders at the lower trophic levels can reluctantly or accidentally ingest microplastics due to their small size. Upon human consumption or entering the food chain, these substances can accumulate in the food chain and were detected in various concentration at higher trophic level organisms such as crustaceans and mollusks. Theis type of bioaccumulation of microplastics in marine life may potentially contain harmful additives which can result in poisoning seafood products, eventually reaching the human end-user.

A recent WHO study conducted on surface and purified or bottled drinking water samples confirmed that 83% of tap water samples and 93% of bottled water samples in eight geographical regions around the world was contaminated with microplastics. These studies have subsequently raised severe concern and attention to microplastics in the aquatic environment, particularly in purified drinking water. On the other hand, daily used food packaging materials, such as polypropylene used in yogurt containers, PET, polystyrene single-use cups and polyethylene used in plastic bags, are generally safe and inert. Nevertheless, there is increasing concern about the disintegration or breakdown of these complex macromolecules, the leaching of additives in particular toxic plasticizers and colorents could cause subsequent toxic effects on humans and the ecosystem at large. For years, baby care products and food containers were prepared using plasticizer additives, like BPA and phthalates, known as endocrine disrupting chemicals which upon accumulation in an ecosystem can potentially cause adverse effects in reproduction and growth in living beings, marine life and in humans, including carcinogenesis. Therefore, microplastics with a large surface area to volume ratio and hydrophobic nature allow or create their own suitable microenvironment which further promotes a favourable environment for the evolution or replication of microorganisms. This leads to the uncontrolled growth of bacterial colonies, leading to biofilm formation on water bodies. This further restricts sunlight and oxygen access to aquatic life causing drastic water quality degradation. In particular, biofilms mainly enable microplastics to become a suitable substrate for the other EPs to accumulate and further act as vehicles for disease by facilitating dispersal and transport throughout the environment via soil, water and in some cases air. Nonetheless, these microplastics inadvertently transform into a promising substrate by providing a means for pathogenic accumulation and transmission throughout the ecosystem. However, the ecotoxicological implication of some emerging micropollutants, in particular microplastics, remains largely unknown; yet the report which horrifyingly captured tonnes of floating plastic and entangled marine life indicates that the entire ecosystem is under enormous threat from the ever increasing use and disposal of plastic and microplastic products [21].

8.6 MICROPLASTIC SOURCE DETECTION AND REMOVAL STRATEGIES

Currently, it is estimated that several million microplastics are discharged from wastewater treatment plants per year in the size range 1 nm to 500 μm. So far, different studies have been conducted to sample, collect, identify, extract, detect and treat microplastics present in wastewater and treatment plants. These studies have also been conducted at different wastewater discharge sites and treatment plants with various optimized protocols and tools. In one such attempt, the detection, identification and analysis of microplastics in discharged wastewater samples were tested by a sieving method followed by analysis. Further, these microplastics were purified by an enzymatic-oxidative purification process in combination with focal plane array (FPA)-based transmission micro-Fourier transform infra-red (FTIR) spectroscopy. In almost all cases, a visual selection method was used for the microscopic identification and separation of microplastics from various plastic debris before further analysis. Microscopic visual selection is expected to be a biased technique due to the similar shapes and dimensions of microplastics which project the same colour and which is comparable to the background or the adjacent medium. Therefore, microscopic visual selection may be inaccurate for distinguishing different polymers and types. Though it may be possible to ignore the analytical bias by presuming every solid particulate is a potential microplastic, this may not be viable or realistic in the analysis of complex samples like co-pollutant adsorbed particulates or biologically contaminated samples in wastewater.

Nevertheless, several studies have explored the use of a spectroscopic imaging method to detect microplastics in various environmental and ecosystem samples without the need for visual selection and a counting step. For instance, an FPA-based micro-FTIR imaging tool has been used to detect microplastic EPs in surface water, seawater and wastewater collected and filtered through a membrane technique. In time, this approach and technique may not be feasible due to the lack of sensitivity and handling errors. Yet, a major advantage is that an FPA-based micro-FTIR imaging technique can be effectively used for the approximate sizing of microplastics through chemical images [22, 23].

In some cases, IR imaging can be analysed in both transmission or reflectance mode and protocols can be optimized to successfully identify microplastics as small as ~20 µm. While the transmission mode of imaging provides relatively well-resolved spectroscopic analyses of microplastics with highly distinguishable particulates, this mode may not be suitably for opaque and relatively thick plastic fragments. Further, FTIR imaging in transmission mode has also been adopted to detect microplastics >500 µm in effluent treatment plants following a visual sorting and separation steps, even though reflectance micro-FTIR imaging can be used as a stand-alone technique to screen and monitor microplastic concentrations in various samples. Therefore, this technique may well serve as a preliminary one to assess the concentration and spiked particles or partial volumes of wastewater or to segregate microplastics depending on their size and distribution. Interestingly, the micro-FTIR spectroscopy tool reflectance mode can be utilized effectively to investigate the presence of microplastics in all three (primary, secondary and tertiary) stages of an effluent treatment plant. On the other hand, the use of Fenton's reagent in the pre-treatment step can be effectively used for the segregation and filtration from wastewater. This also helps in the rapid separation of microplastics from other co-pollutants and BOM-rich samples.

Currently, wastewater treatment plants have been identified as a major reservoir for microplastics in the environment due to the discharge of domestic and municipal wastewater. Even though the traditional effluent handling method in treatment plants can remove 98% of microplastics, the final treated effluent with a residual 2% still poses a huge risk to the environment due to the high volumes of daily discharge. In addition to this, improved treatment protocols and advances in the final-stage tertiary treatment step can significantly reduce the concentration of microplastics being released into the environment. Even with several limitations to and concerns with the treatment techniques, disc filtration, rapid sand filtration, dissolved air floatation filtration and membrane bioreactor filtration have been proven effective in retaining a major portion of microplastics. Significantly, the vast majority of wastewater treatment facilities across world comprise primary or secondary treatment, whereas a significant percentage of microplastics at trace level can be retained in the tertiary treatment step prior to release into the ecosystem. Therefore, adopting a tertiary treatment step to already existing conventional wastewater treatment plants is a major challenge in controlling microplastic pollution.

Coincidentally, looking at the impending threat, several countries like Canada, the USA, Ireland, and the United Kingdom have regulated or considered banning primary source microbeads from entering manufacturing units or single-use plastic. Even though the policy of a polymer or plastic ban is a most efficient way forward, it is impossible to sustain modern life without these synthetic materials. However, alternative methods like educating manufacturers, consumers and labelling products containing plastic constituents would serve as a valuable substitute for minimizing plastic waste in the ecosystem. On the other hand, a major microplastic contributing segment like textile production and their stability in use and during washing could be improved through various techniques like (i) refining knitting, for example, tightly woven fabrics will have more fibres per area and consequently may release higher numbers of microfibres during washing; (ii) reducing the synthetic fibre load in textiles by combining natural yarn with synthetic fibre, which may reduce microfibre loss by 80%; (iii) applying a protecting layer to textile fibres to protect them from disintegration and breaking; and (iv) polishing and removing loose microfibres from freshly manufactured textiles before they reach the customer.

As discussed in previous sections, microfibre/microplastic release from synthetic textiles can be reduced using suitable fabric softeners, improving microfibre filtration systems in washing machines or by fitting microfilters in the washing machine outlet or drain. These measures may also decrease clogging possibilities in pipes of wastewater plants and in transportation and pumping routes. However, several studies point out the drawbacks of wastewater treatment plants and suggest the adaptation or development of a new supplementary step to retain microplastics. In this regard, traditionally used sand filters have proven inefficient, which instead block the safe passage of water which effectively clogs the filter system. On the other hand, membrane-based processes like membrane bioreactor microfiltration and ultrafiltration are proven techniques but are expensive. Recently, adopted wool and polyfoam based filters in low-flux tubes having a back flushing maintenance facility seem to be an efficient low-cost solution for removing microplastics from the water and wastewater samples.

Another simple strategy can be followed to reduce plastic contamination by degrading polymer species using suitable catalysts which are capable of transforming and removing microplastics from wastewater or sludge. This catalytic or photocatalytic degradation process may also help in improving the retention efficiency in wastewater treatment plants. For example, the biological reduction of microplastic concentration in sludge is effectively treated by anaerobic digestion. This is confirmed by the analysis of degraded plastic debris in the treated water. Further, catalytic photodegradation can also help in reducing the microplastic load in various water samples with the help of sunlight. This strategy can be employed by considering natural and/or weathering processes with a catalyst assisted degradation step applied to mismanaged plastic wastes such as discarded plastic bags or unintentionally discarded fishing nets. Assuming the origins of secondary microplastics are difficult to trace and monitor due to their degradation into minor unknown species, it is difficult to assess implicitly how many of the macroplastics have now transformed into microplastics or smaller organic fragments. Once in water resources like lakes, rivers or oceans, microplastics can either float or sink or transform into biofilm layers. Microplastics lighter than fresh or seawater such as polypropylene will float and disperse widely across the water body, eventually accumulating in gyres resulting from waves or oceanic currents. Other high-density polymer originated microplastics like acrylics are denser than fresh or seawater which most probably amass on the ocean bed, which means that a substantial quantity of microplastics may eventually accumulate or be submerged in the deep sea and ultimately be transported into food chains with time.

However, the best strategy to reduce microplastic pollution in the environment is to control and regulate it at the source in which a drastic reduction of loose microfibres needs to be measured. In another way, regulators need to ban the microbeads which go into: plastic article production, improving textiles which is a major source of microfibres, reducing the damage of plastics and reducing public investment, penalizing polluters and regulating manufacturers to distribute and recollect used or discarded plastic products. Further, future research ought to focus on developing methods or technologies or measures in favour of source reduction and regulation of microplastics.

8.7 PHYSICAL DETECTION AND REMOVAL OF MICROPLASTICS

The collection of samples is the first step in identifying, quantifying and detecting microplastics. Water, sediment, biota and wastewater treatment plants are the main sources of microplastic collection. There have been different methods and protocols in practice for the sampling and detection of microplastics. Microplastics in various ecosystems can be identified as floating on water due to their physical and chemical properties such as low density, lighter weight and mechanical robustness. Trawl technology is the key method of water sampling, and the most commonly used for the collection of EPs like microplastics in water. For example, the 333 μm mesh manta trawl has been widely used for the collection of plastic debris. Upon collection samples are assessed to characterize the level of microplastics. However, there are several drawbacks to sampling EPs using trawl techniques such as: (i) the level and concentration of microplastics in the environment may be underrated due to the restriction in the net's mesh size; (b) if the mesh with a lower size (micrometer, μm) is used these methods may cause sample contamination by grabbing other co-pollutants which may interfere in instrument and procedure design. Sometimes the sampling device itself may be prepared from polymers such as monofilament polyamide mesh, knitted synthetic fibres and reinforced fabrics, which may cause contamination in the collected sample. Recent developments not only resolve the difficulty of sampling small-sized microplastics from various water samples but also reduces the associated contamination risk. In one such attempt, a per litre surface grab sampling method was adopted with a 335 μm neuston net tow. These procedures and results conclude that the neuston tow method is effective for sorting and capturing microplastics in the absence of a microscope. However, these results confirm that the grab sampling technique captures a larger density and more diverse microplastic samples from sources and minimizes contamination through proper laboratory segregation and field procedures. In addition to this, the pumping method is also used to collect and segregate microplastic samples from surface water [24, 25].

On the other hand, sediment samples are collected mostly from the surface of the beach or the bottom of the sea or lake or river bed. Microplastic sampling from the beach is a simple approach and particulate concentrations have been noticed, varying in different samples sourced from different parts or depths of the beach. A sediment sample collection protocol on the beach or river side is mainly divided into (i) covering the entire beach or river bank which gives the overall estimation of distribution; (ii) collection of microplastic samples at different zones of a beach; and (iii) random and systematic collection of samples in several different coastal areas. Nevertheless, most research studies have frequently considered a collection of floating debris as a model to depict or guestimate the real sample in oceans deposited at high tide lines.

The increasing production, relatively short lifespan in retaining physical characteristics and indiscriminate or accidental disposal practices have promoted the accumulation of microplastics in the environment. Even though, microplastics undergo weathering through temperature, friction, photocatalytic exposure to the sun, mechanical disintegration and biological degradation; however, due to their long degradation time, microplastics are lost in the environment. In such situations, larger plastic debris disintegrate or leach out their additives and transform into smaller fragments. In aquatic systems, microplastics can attain or display a variety of shapes, sizes and colours.

Thus, researchers have microplastic samples collected from different wastewater sources which have been derived from commodity polymers like polystyrene (PS), polyester/polyethylene terephthalate (PE/PET), polyethylene (PE), polyamide (PA), cellulose (C) and polypropylene (PP). These polymer micropollutants have been frequently collected, identified and analysed using FTIR microscope and micro-Raman spectroscopy. All these are analysed for preliminary characteristics and characterized either as fibre or particle formats as documented in Table 8.1. Most of these characteristics have been observed under the microscope and the resulting images have been carefully analysed for their shape, size, dimensions and appearance. However, among all the collected samples, the proportion of different polymers found to be diverse and the physical nature of the segregated samples also showed not much difference in appearance. However, a general trend showed polyester-based microfibres accounted for >90% and the remaining polymers accounted for equivalent to ~80% of the total amount of microplastics collected from various sources. From limited analytical samples and studies, it can be concluded that polyamide microparticles

TABLE 8.1

Type of Microplastic, Description of Polymers Detected in Wastewater Samples and Identified by Various Microscopic Techniques

Type of Microplastic	Size, Shape and Dimension
Polyethylene (PE)	Particles
	Shape: Uneven-flakes and fragments, spheres Hardness: Medium to soft
	Appearance: Dull or a bit shiny
Polyester (PES), mostly polyethylene terephthalate (PET)	Fibres
	Cross-section: Round, oval, flat
	End: Cut, frayed, thickened
	Appearance: Shiny or dull particles
	Shape: Flat, angular fragment
	Hardness: Medium
Polypropylene (PP)	Particles
	Shape: Uneven fragments
	Hardness: Medium
	Appearance: Dull
Polyamide, nylon (PA)	Fibres
	Cross-section: Round, oval, flat
	End: Cut
	Appearance: Shiny
Cellulose	Ribbon-like fibres
	Cross-section: Round, oval, flat
	End: Cut
	Appearance: Dull
Polystyrene (PS)	Particles
	Cross-section: Round, oval, flat
	End: Cut
	Appearance: Dull

account for ~3%. Most of the plastic microfibres are typically described as thick with 3D bending, which differs from cellulose-based microfibres which are generally characterized as ribbon-like in appearance.

In another feature, highly abundant polyester microfibres have been generally characterized as a flat morphology with a soft cotton-like appearance. On the other hand, the second most abundant polyethylene microfibres have been accounted for as ~64% of microplastics. The rest of the ~36% collected have been recognized to consist of polypropylene, polyethylene and polyester. Even though the appearance characteristics of various microfibres have not been divergent, most of them have been noticed as either dull or shiny, and unevenly shaped polymer fragments were frequently characterized in various wastewater effluents. Interestingly, controlled studies have shown the similarity between microplastics in wastewater and fragments originated from cleansing scrubs which indicates major microplastic contributions originate from household or domestic wastewater release [26, 27].

Interestingly, several studies also found microfibres of polyester fragments mostly collected from contaminated lake samples and final effluent treatment plants, but less frequently from influent wastewater or sludge samples. The lack of PE microfibre fragments in a particular collected and analysed sample, in a particular study, can be attributed to the diurnal variation in wastewater flow, produced on a particular day, together with the hydraulic retention time in the treatment plant. These data and observations advocate that at least part of these personal care and cosmetic product originated fragments can easily pass through treatment processes. In some cases, polyester microfibres were completely absent in a secondary treatment step of an effluent treatment plant. This indicates that a primary treatment step has been effective in retaining PE-microplastics effectively.

8.8 BIO-BASED MICROPLASTIC REMOVAL STRATEGIES

In recent times, PCCP manufacturers have shifted focused to using cosmetic ingredients from natural extracts with properties like antioxidants, biocompatibility, biodegradability and film-forming characteristics. Increasing economic growth and improved life styles have adopted huge amounts of personal hygiene and cosmetic products in the form of fragrances, moisturizers, lotions, emulsifiers, preservatives, thickeners, colours and ingredients. These PPCPs after use and application on a daily basis are discarded into the environment without treatment. Even though many synthetic polymers like polypropylene (PP), polyethylene (PE), polyamide (PA) and polyurethane (PU), and copolymers like ethylene vinyleacetate-copolymers (EVA) and acrylonitrile-co-metacrylate or other acrylates, have been widely used as main ingredients in PPCP production. However, recent studies have repeatedly characterized these synthetic polymer-based microplastics in various wastewater samples.

Manufacturers use synthetic polymers to prepare several personal care products and cosmetics for various reasons such as some of these synthetic polymers help in the distribution and dispersion of the main ingredients and help to develop and induce a film-forming property on skin, hair, face or nails. For example, PE possesses excellent film-forming and viscosity-modulating properties. However, after use these polymers end up as microparticulate debris in domestic wastewater along with impurities from the skin or teeth, making them complex EPs. Due to the impending disasters associated with the use of synthetic polymers in PCCPs, manufacturers have shifted their focus to utilizing greener and natural sources as main ingredients. For this purpose, polyhydroxyalkanoate (PHA), polybutylene succinate (PBS), polylactic acid (PLA), cellulose and its derivatives, maize starch, chitosan from chitin or casein from animal protein have been widely considered as potential candidates for the sustainable production of PPCPs. Current studies suggest that cellulose and its derivatives, agarose, carrageenan, chitin/chitosan and alginate are the most widely used biopolymers, whereas PHAs have real future potential to be used as a main ingredient in PCCPs. Recently studies have confirmed that natural thermoplastic PHAs can degrade quickly in almost any environment including the marine ecosystem. It is estimated that PBS, PLA and PHA would be potential candidates for replacing synthetic polymer sources in PCCPs and other commodity plastics. However, the greatest task is to ensure that these biopolymers undergo complete breakdown in natural conditions enabling the design and development of new nature friendly plastic goods in mass production. Even so, the oxo-biodegradability of plastics is considered a natural process to partially eliminate the microplastic menace, though degraded debris may create secondary pollutants making the process much more complicated [28, 29].

On the other hand, cellulose and its derivatives have been reinvented and revisited widely to be adopted in various applications where biocompatibility, biodegradability and sustainability are highly desired factors. The cellulose in hygroscopic biopolymers is abundant in biomass which binds and retains water efficiently. This makes them a natural and abundant source of raw material for various application developments like PCCP and replacing traditional materials. Also, nanocellulose has been widely adopted and applied in many PCCPs and this is also termed a main source of nano- and microplastics in water.

Primary microparticles from cosmetics or PCCPs make up only a small portion of the plastic EPs in the oceans. However the increased use of PCCPs and the ever-increasing littering of packaging material, both synthetic and biopolymer, used for PCCPs, are still causes of concern. Therefore, the focus is also now on addressing all kinds of polymer-based plastic EPs. PCCP manufacturers and commodity plastic suppliers are now rapidly replacing synthetic polymers with natural biopolymers which eliminate the possibilities of long-lasting plastic microparticle presence in the environment. Nonetheless, the use of biopolymers will potentially change the existing production chain due to their ease of adoption which may largely be unaltered, and the release of microparticles from these renewable sources would be a close match for the microplastics of synthetic origin previously in use. Therefore, it is important to note that, whether or not natural or biodegradable polymers are the choice of future material sources, the over-exploitation and discharge of residues will pose an important question.

8.9   ADVANCES IN DETECTION AND TREATMENT OF MICROPLASTICS

It is by now understood that the collection, identification and quantification of microplastics is a complicated and tedious process due to their small dimensions. Therefore, current methods in practice for the collection, identification, extraction and quantification of microplastics from various sources often require a high labour force, are time-consuming, and demand advanced tools and techniques. A rational examination of microplastic contamination in various ecosystems depends on the proper sampling, identification protocols and analytical strategies. At present, due to the lack of harmonized, generalized and standardized protocols, it is difficult to quantify microplastics accurately. Even though several generic methodologies have been discussed in previous sections for the quantification and characterization of microplastics in fresh water, terrestrial water bodies, oil samples and in the marine ecosystem, more sensitive and precise methods are needed. Nevertheless, several research groups have dedicated their time to developing specific practical guidelines and workflow protocols for the precise detection of microplastics in various sample sources. This section therefore will briefly list the advanced methods that are in wide practice.

Most of the work on advanced microplastics considers the basic hypothesis of macromolecular chemistry to identify them from other co-pollutants. In this direction, the repetitive fingerprint-like molecular conformation of plastic polymers allows for the strong assignment of a particular collected microplastic sample to a certain polymer origin. Therefore, following this hypothesis, some of the advanced techniques have been frequently used for both qualitative and quantitative analysis of microplastics, namely CHNS, GC-MS, Raman and IR spectroscopy.

8.9.1 DENSITY SEPARATION WITH SUBSEQUENT C:H

This method allows the use of specific densities of microplastic particles to identify the polymer source in visually segregated EPs. This process with EP samples placed in distilled water and depending on the density of the polymer, either ethanol or concentrated solutions of calcium or strontium chloride, is added to aqueous bottles until the microplastic is neutrally afloat. Further, the density of the microparticle afloat is indirectly measured by weighing a certain volume of the solution. This procedure enables the precise estimation of the density of the microplastic. Further, these samples will be subjected to C:H:N analysis by utilizing the characteristic elemental composition of the polymer. Different groups of polymers possess a distinctive elemental composition, which helps in the identification of the plastic type and origin of a microparticle in C:H:N analysis. Further, confirmation of a collected sample is carried out by comparing it with the densities and C:H:N ratios of standard pristine polymer samples. This method characterizes an estimate to the identification of microplastic particles in various samples by narrowing the search for the potential polymer type but may also require additional and rigorous chemical analysis to substantiate the claim. The main drawback of this method is that the entire process is still relatively time consuming.

8.9.2 PYROLYSIS-GC/MS

Pyrolysis-gas chromatography (GC) in combination with mass spectrometry (MS) can be directly utilized for the assessment of the chemical composition of potential microplastics. This procedure proceeds with the analysis of thermally degraded products. For this, the pyrolysis of the microplastic sample is pyrolysed using a suitable furnace and results in characteristic pyrograms used for the identification of the polymer type. This method is currently used after extraction and visual sorting of microplastics collected from wastewater sediment samples. Once the composition is estimated, the polymer origin of microparticles is then analysed by comparing their characteristic combustion products with reference pyrograms of known standard pristine polymer samples. On other hand, if a thermal desorption step come first, the final pyrolysis of the microplastic additives can be investigated concurrently during pyrolysis-GC/MS analysis. Although the pyrolysis-GC/MS method permits a relatively good assignment and estimation of potential microplastics to a polymer type it suffers from the drawback that the particle process needs to be handled manually by placing a sample in a pyrolysis tube. Nevertheless, pyrolysis-GC/MS approaches are known to be promising and the best method available currently for both qualitative and quantitative analysis of microplastics from almost every type of sample in the environment and ecosystem.

8.9.3 RAMAN SPECTROSCOPY

Raman spectroscopy is a straightforward technique that has been successfully used to identify microplastic particulate samples collected from different environments and ecosystems. Currently, the Raman spectroscopic method is regarded as a highly reliable technique. In this method, first the microplastic-contaminated sample is irradiated with a monochromatic laser source. The type of laser used for irradiation depends on the system, generally in wavelengths between 500 and 800 nm. During irradiation, the laser light interacts via vibrational, rotational or other low-frequency exchanges with the molecules and atoms of the microplastic sample, which results in differences in the frequency of the backscattered light. This backscattered light is analysed in comparison to the irradiating laser frequency to predict the sample type. In this process Raman shift is detected, which leads to substance-specific Raman spectra generation. Since each polymer possess typical characteristic Raman spectra, this technique can be used as a robust method to identify plastic polymers within minutes. Further confirmation of the sample can be carried out by comparison with standard reference spectra. Raman spectroscopy is a non-destructive surface technique. Using this advantage, large, visually segregated microplastic particulate samples can be easily analysed using this technique. Further, Raman spectroscopy can also be coupled with microscopy (Micro-Raman spectroscopy) which enables the identification of a broad size range of microplastics down to as low as 1 μm. Recent studies have also explored the use of Raman microscopy in combination with Raman spectral imaging which enables the generation of spatial chemical images. Therefore, Micro-Raman imaging hypothetically allows the spectral analysis of whole membrane filters at a 3-D resolution below 1 μm. This approach also facilitates the detection of the finest microplastic particulates collected from environmental samples; however, this hybrid technique has yet to be tested rationally in field samples [30, 31].

8.9.4 IR SPECTROSCOPY

Fourier-transform infrared (FTIR) spectroscopy, similarly to Raman spectroscopy, offers the possibility of the precise identification of microplastic polymer particulates based on their characteristic IR spectra. FTIR spectroscopy provides reliable information on polymer samples taking advantage of the fact that IR radiation excites molecular vibrations when interacting with a microplastic sample. These excitable vibrations in IR spectra depend on the composition and molecular structure of a microplastic sample and are typically wave-length specific. In these measurements the energy of the IR radiation used for exciting a specific vibration will depend on the wavelength, which upon radiation absorbs a certain incident radiation and enables the quantification of characteristic IR spectra. Interestingly, microplastic constituents (polymers) possess fingerprint IR spectra with discrete band patterns, making IR spectroscopy an ideal technique for the identification and analysis of microplastics. Further, a versatile FTIR spectroscopy technique can also provide information on the physico-chemical weathering of collected microplastic particulates by detecting the intensity of oxidation.

The FTIR and Raman spectroscopic methods are complementary techniques; however, both techniques require a standard comparison using reference spectra for the accurate identification of polymer types in collected samples. On the other hand, large microparticle samples can be easily investigated using the surface technique of FTIR, attenuated total reflectance (ATR-IR) spectroscopy with high accuracy in less than one minute. For the characterization of the finest microparticles, an advanced version of FTIR microscopy can be utilized in both reflectance and transmittance mode. However, the reflectance mode possesses the disadvantage that limit the measurements of irregularly shaped micro-particulates, whereas the transmittance mode requires IR transparent filters owing to the total absorption patterns being limited by a certain thickness of the microplastic sample. Currently, the focal plane array (FPA)-assisted IR imaging technique has been designed as an extremely promising FTIR extension which allows the detailed, sensitive and high throughput analysis of the total microplastics on a mesh filter. This advanced technique allows the instantaneous recording of several thousand spectra with chemical images within an exposed sample area with a single measurement. Therefore, with the help of FPA fields, whole sample filters can be examined via FTIR imaging, tracing each microplastic parent polymer. Nevertheless, sample preparation plays a key role in the accurate analysis of the sample in FPA imaging; prior to proceeding with samples via IR spectroscopy, a sample must be dehydrated prior to measurement as water strongly absorbs IR radiation and samples should be disinfected for the precise identification of the polymer type of microplastic particulate when using IR spectroscopy. Nonetheless, each analytical technique poses its own advantages and disadvantages during the handling and analysing of a microplastic sample. It is also important to consider suitable sample collection and segregation protocols carefully before choosing the analytical technique. Adopting comparative studies is also useful in harmonizing the results for accuracy which normally are found to be comparable worldwide and to enumerate the problem on a global level.

REFERENCES

1. Zeynep Akdogan, and Basak Guve, Microplastics in the environment: a critical review of current understanding and identification of future research needs, *Environmental Pollution*, 2019, 254(Part A), 113011.
2. Matthew Cole, Pennie Lindeque, Claudia Halsband, and Tamara S. Galloway, Microplastics as contaminants in the marine environment: a review, *Marine Pollution Bulletin*, 2011, 62(12), 2588–2597.
3. Brigitte Toussaint et al., Review of micro- and nanoplastic contamination in the food chain, *Food Additives & Contaminants: Part A*, 2019, 36(5), 639–673.
4. Mohsen Padervand, Eric Lichtfouse, Didier Robert, and Chuanyi Wang, Removal of microplastics from the environment. A review, *Environmental Chemistry Letters*, 2020, 18, 807–828.
5. Stephen Nyabire Akanyange, Does microplastic really represent a threat? A review of the atmospheric contamination sources and potential impacts, *Science of The Total Environment*, 2021, 777, 146020.

6. J. Kramm, and C. Völker, Understanding the risks of microplastics: a social-ecological risk perspective, In *Freshwater Microplastics. The Handbook of Environmental Chemistry*, M. Wagner, S. Lambert (eds.), vol 58, 2018, Cham: Springer.

7. N. Digka, C. Tsangaris, H. Kaberi, A. Adamopoulou, and C. Zeri, Microplastic abundance and polymer types in a mediterranean environment. In *Proceedings of the International Conference on Microplastic Pollution in the Mediterranean Sea. Springer Water*, M. Cocca, E. Di Pace, M. Errico, G. Gentile, A. Montarsolo, R. Mossotti (eds.), 2018, Springer.

8. Elke Fries, Jens H. Dekiff, Jana Willmeyer, Marie-Theres Nuelle, Martin Ebert, and Dominique Remy, Identification of polymer types and additives in marine microplastic particles using pyrolysis-GC/MS and scanning electron microscopy, *Environmental Science: Processes & Impacts*, 2013, 15, 1949–1956.

9. Ee-Ling Ng, Esperanza Huerta Lwanga, Simon M. Eldridge, Priscilla Johnston, Hang-Wei Hu, Violette Geissen, and Deli Chen, An overview of microplastic and nanoplastic pollution in agroecosystems, *Science of The Total Environment*, 2018, 627, 1377–1388.

10. D. M. Mitrano, P. Wick, and B. Nowack, Placing nanoplastics in the context of global plastic pollution. *Nature Nanotechnology*, 2021, 16, 491–500.

11. K. Pabortsava and R. S. Lampitt, High concentrations of plastic hidden beneath the surface of the Atlantic Ocean. *Nature Communications,* 2020, **11,** 4073.

12. L. Lebreton, J. van der Zwet, J. W. Damsteeg, et al. River plastic emissions to the world's oceans. *Nature Communications,* 2017, **8,** 15611.

13. Paul U. Iyare, Sabeha K. Ouki, and Tom Bond, Microplastics removal in wastewater treatment plants: a critical review, *Environmental Science: Water Research & Technology*, 2020, **6**, 2664–2675.

14. Carlos Edo, Miguel González-Pleiter, Francisco Leganés, Francisca Fernández-Piñas, and Roberto Rosa, Fate of microplastics in wastewater treatment plants and their environmental dispersion with effluent and sludge, *Environmental Pollution*, 2020, 259, 113837.

15. C. Bretas Alvim, J. A. Mendoza-Roca, and A. Bes-Piá, Wastewater treatment plant as microplastics release source – quantification and identification techniques, *Journal of Environmental Management*, 2020, 255, 109739.

16. Rachid Dris, Johnny Gasperi, Mohamed Saad, Cécile Mirande, and Bruno Tassin, Synthetic fibers in atmospheric fallout: a source of microplastics in the environment?, *Marine Pollution Bulletin*, 2016, 104(1–2), 290–293.

17. Fionn Murphy, Ciaran Ewins, Frederic Carbonnier, and Brian Quinn, Wastewater treatment works (WwTW) as a source of microplastics in the aquatic environment, *Environmental Science & Technology*, 2016, 50(11), 5800–5808.

18. T. Galloway, M. Cole, and C. Lewis, Interactions of microplastic debris throughout the marine ecosystem. *Nature Ecology & Evolution,* 2017, **1,** 0116.

19. Tineke A. Troost, Térence Desclaux, Heather A. Leslie, Myra D. van Der Meulen, and A. Dick Vethaak, Do microplastics affect marine ecosystem productivity?, *Marine Pollution Bulletin*, 2018, 135, 17–29.

20. L. Lebreton, and A. Andrady, Future scenarios of global plastic waste generation and disposal. *Palgrave Communications*, 2019, 5(1), 1–11.

21. Disha Katyal, Elaine Kong, and Jacit Villanueva, Microplastics in the environment: impact on human health and future mitigation strategies, *Environmental Health Review*, 2020, 63(1), 27–31.

22. K. Schuhen, and M. T. Sturm, Microplastic pollution and reduction strategies. In *Handbook of Microplastics in the Environment*, T. Rocha-Santos, M. Costa, C. Mouneyrac (eds.), 2021, Cham: Springer.

23. Valeria Hidalgo-Ruz, Lars Gutow, Richard C. Thompson, and Martin Thiel, Microplastics in the marine environment: a review of the methods used for identification and quantification, *Environmental Science & Technology,* 2012, 46(6), 3060–3075.

24. Biplob Kumar, Pramanik, Sagor Kumar, Pramanik, and Sirajum Monira, Understanding the fragmentation of microplastics into nano-plastics and removal of nano/microplastics from wastewater using membrane, air flotation and nano-ferrofluid processes, *Chemosphere*, 2021, 282, 131053.

25. Emily E. Burns, and B. A. Alistair Boxall, Microplastics in the aquatic environment: evidence for or against adverse impacts and major knowledge gaps, *Environmental Toxicology and Chemistry,* 2018, 37(11), 2776–2796.

26. Chelsea M. Rochman et al. Rethinking microplastics as a diverse contaminant suite, *Environmental Toxicology and Chemistry (ET&C)*, 2019, 38(4), 703–711.

27. Maiju Lehtiniemi, Samuel Hartikainen, Pinja Näkki, Jonna Engström-Öst, Arto Koistinen, and Outi Setälä, Size matters more than shape: ingestion of primary and secondary microplastics by small predators, *Food Webs*, 2018, 17, e00097.

28. Solange Magalhães, Luís Alves, Bruno Medronho, Anabela Romano, and Maria da Graça Rasteiro, Microplastics in ecosystems: from current trends to bio-based removal strategies, *Molecules*, 2020, 25(17), 3954.

29. Mónica Calero, Verónica Godoy, Lucía Quesada, and María Ángeles Martín-Lara, Green strategies for microplastics reduction, *Current Opinion in Green and Sustainable Chemistry*, 2021, 28, 100442.

30. Yuchuan Meng, Frank J. Kelly, and Stephanie L. Wright, Advances and challenges of microplastic pollution in freshwater ecosystems: a UK perspective, *Environmental Pollution*, 2020, 256, 113445.

31. M. Liu, S. Lu, Y. Chen, C. Cao, M. Bigalke, and D. He, Analytical methods for microplastics in environments: current advances and challenges. In *Microplastics in Terrestrial Environments. The Handbook of Environmental Chemistry*, D. He, Y. Luo (eds.), vol 95, 2020, Cham: Springer.

9 Nanocleaners, Nano-Pollutants and Nanotoxicity

9.1 INTRODUCTION

Emerging pollutants (EPs) are projected to be a cause of major health and economic and ecologically related issues in coming days. In spite of the improvement and advancement in water treatment technologies and innovation, large populations still lack access to clean and safe drinking water. Hence, extensive effort needs to be invested at developing faster and more innovative water treatment materials and technologies to address the impending EP issue that is estimated to affect human beings, various ecosystems and the environment at large. EPs in various aquatic ecosystems are increasingly damaging life and biodiversity. Micro- and nanomotors have been developed to mimic micro-organisms and their movements in the aquatic system, inspired by nature where micro-organisms swim at a low Reynolds number by using their cellular appendages. These micro-organisms generally swim enemas thereby creating turbulent currents or flows that are often strong enough to promote mixing in water bodies. Mimicking this pattern researchers have created various types of artificial micro/nanocleaner or nanomotor, which can propel at a low Reynolds number in the contaminated water source, where an active motor disinfects the water body. Therefore, micro/nanocleaners have been extensively innovated owing to their great potential to execute a broad range of complex tasks in addressing EPs in different aquatic environments.

On the other hand, the increased use of PCCPs, rapid industrialization and industrial discharges along with domestic wastewater are affecting water quality and therefore, in such a situation, water purification becomes challenging. Furthermore, surface water is being polluted with heavy metals, organics, synthetic dyes, surfactants and pharmaceutical drugs, which is alarming from an environmental point of view. In seawater on the other hand has been characterized by microplastics and nanoplastics. Industrial wastewater has been known for its complex and high concentrations of both conventional and emerging pollutants. Among several water treatment techniques, a series of physical, chemical, physico-chemical and biological methods have been used with limited success. All these techniques have been proven partially successful, and individual techniques have been applied to perform specific tasks. Therefore, innovators have adopted advanced water and wastewater treatment processes like adsorption, advanced oxidation and electrochemical techniques [1, 2].

DOI: 10.1201/9781003214786-9

However, conventional methods have been proven successful in the partial removal of EPs; however, there are several drawbacks, such as the high cost compared to the low removal rate and the production of secondary pollutants and adsorbents which are generally non-reusable. On the other hand, innovation in nanotechnology procedures have led to the creation of highly effective catalysts and engineered nanomaterials which can be directly used for EP removal. Among these, self-propelled micro/nanocleaners and micro- or nanomotors have been developed as a new class of Fenton catalyst ($Fe^{2+} + H_2O_2$) to degrade EPs. However, nanomaterial-based cleaners are currently categorized as expensive, difficult to scale-up and requiring surfactants to effect the work and regeneration. Therefore, the design and preparation of Fenton-type catalysts as efficient, low-cost and scalable active micro-/nanocleaners are still desirable for their adsorptive degradation or photodegradation of EPs. In this direction, the resulting novel nanomaterial preparations, like hydro/solvo/ionothermal methods, co-precipitation and sol-gel techniques, have been widely adopted to make nanocleaners affordable, greener and sustainable.

In other chemical methods, the polyol method of nanomaterial preparation procedure uses non-aqueous liquid as both solvent and reducing agent. In this method, the non-aqueous solvents provide an advantage in minimizing the surface oxidation of precursors during conversion and reduce agglomeration during product formation. This technique also provides greater control of size, texture and shape in the nanomaterial. On the other hand, the micro-emulsion technique is used to prepare polymeric nanocomposite functional nanocleaners. In the thermal decomposition method which is also known as thermolysis, heat is applied to break chemical bonds in an initial precursor to undergoing decomposition through endothermic reaction to yield novel nanomaterials. Further, electrochemical synthesis is the synthesis of materials in both bulk and nanosized compounds in an electrochemical cell [3, 4].

Several physical methods like plasma, chemical vapour deposition, microwave irradiation and sonochemistry-assisted nanomaterial synthesis have been recently widely adopted. In the plasma method, nanosized material is synthesized with the help of radio frequency (RF) heating coils. Even though chemical vapour deposition (CVD) is a physical method, the process involves a chemical reaction. Initially, the CVD method was mostly used for the preparation of semiconductor nanomaterials in the form of thin films of different materials. Lately, CVD has been used to prepare both carbaneous and grapheneous nanomaterials as well as their nanocomposites which are widely adopted for water treatment applications. The CVD method involves one or more volatile precursors and nanomaterials in the form of films and layers that are deposited on a suitable substrate depending upon the end use. Microwave irradiation is another simple physical method widely used for the preparation of nanomaterials of a different nature, namely organic, inorganic and inorganic/organic hybrid nanomaterials. This method offers a hustle-free, direct and simple preparation route in obtaining a high surface area and highly functional nanomaterials and their composites.

The pulsed laser method is typically used in the synthesis of silver nanoparticles. These have been widely used for anti-bacterial application for which the pulsed laser technique helps in robust synthesis (at a high rate of 3 gm/min production). This method utilizes a silver nitrate solution with a reducing agent as a precursor, poured into a blender-like device in which metal nanoparticle transformation occurs instantaneously. Similarly, a sonochemical method is widely used for the preparation of metal nanoparticles like MnO^{4-}, Au^{3+}, Au^{+} and Pd^{2+} and helps in preparing metal-nanoparticle-based nanocomposites. Interestingly, the sonochemical reduction of the metal precursor in the absence or presence of organic additives can be utilized to functionalize the end product with a well controlled size and shape to suite the particular water treatment category, such as EPs, which were investigated in relation to the synthesis of metal nanoparticles. Another physical method, gamma radiation, is the highly reliable nanomaterial synthetic technique which is preferred for metallic nanoparticle synthesis with well controlled shape and size distribution. This method can also produce monodispersed metallic nanoparticles from greener, cheap and sustainable precursors in environmentally friendly solvents like water or ethanol at close to room temperature operations.

Nonetheless, nanomaterials and their compositions have gained significant interest in the water treatment segment due to their unique chemical and physical properties. Various methods and preparation routes discussed above produce different nanomaterials with versatile functionality and tunable surface properties. These methods have been adopted to develop task-specific nanocleaners suitable for targeted EPs. Among many solid-state material preparation techniques, hydrothermal, solvothermal and ionothermal methods have been highly effective in producing relatively moderate surface areas but with a highly surface-active functionality in produced nanomaterials. Also, other advanced methods, like co-precipitation and sol-gel techniques, are now widely explored as robust and economical material synthetic routes [5, 6].

9.2   DEPENDENCY ON NANOMATERIALS

Water purification and remediation increasingly depend on the use of nanotechnology-based nanoscopic materials such as metal oxides/hydroxides, oxyhydroxides, polymer nanomaterials and their composites, carbon nanotubes and grapheneous materials in most cases, high surface area, functionality and nanoporous morphology were exploited to perform pollutant retention, degradation or separation duty when added to contaminated water. The main advantages of using nanomaterials compare to conventional systems are that: nanomaterials directly can be added to contaminated water to mitigate pollutants or require less pressure to pass contaminated water through the nanomaterial filters; the high surface area of nanomaterials provides larger accessible active sites to make the overall process more efficient; and these advanced materials offer easy recovery and regeneration with simple washing.

The present class of nanomaterials follow moderately simple, green and sustainable material resources and preparation processes. Therefore, current and near future water remediation techniques will extensively make use of the existing stock of nanomaterials with great efficiency, affordability and sustainability. For example, functional carbon and grapheneous nanomaterials remove almost all kinds of water contaminants including organic contaminants, turbidity, oil from oil/water mixtures, bacteria, viruses and polymer contaminants. Although microporosity in these nanomaterials hinders a faster flow rate due to the slow capillary phenomenon, the surface area and functionality play a vital role in retaining large amounts of EPs from contaminated sources. On the other hand, a tunable surface charge on nanomaterials can be easily altered to counter the contaminant. For instance, an alumina-based nanocomposite filter carrying a positive surface can interact with negatively charged pollutants like inorganic colloids, viruses, bacteria and organics and remove them at a faster rate than traditional water filters.

Therefore, nanomaterials and their compositions have been applied as nano/microcleaners by mixing them in contaminated water and wastewater from different sources. Compared to other conventional water treatment techniques, in particular polymer adsorbents, micro- or nanocleaners impart efficient mixing and enhance mass transfer through active diffusion in contaminated water sources. This facilitates the improved mass transfer thereby influencing interaction between pollutant and nanomaterial via chemical reactions. Nanomaterials also offer relatively easy to achieve uniform dispersion through micro-mixing with appropriate blending or simple mechanical stirring. However, effective micro-mixing of nanomaterials in favourable formulations at the molecular scale is extremely challenging. Thus, attaining a uniform mixing of nanomaterials plays an important role in the rapid interaction of molecular level contaminants. Recently, advancements also highlight the importance of nanomaterial dependency in water treatment in the form of nanocleaners, nano-photo-catalysts, nanomembranes, imprinted polymer nanoparticles and nano-sorbents of high sorption capacity. Therefore, an overview of current nanomaterial stock at effectively removing EPs provides potential and substantial water treatment advantages which can be directly adaptable, though some imperfections still need further attention for addressing impending issues like nanotoxicity [7–9].

On the other hand, nanocleaners act as artificially active material which enhances the interaction between pollutant and cleaner, thereby reducing the treatment times. This activity overcomes the diffusion-limited reactions and improves the exchanges between their active surfaces and targets EPs. Interestingly, nanomaterials having a favourable size and dimensions propel the additional mixing along with the 3D swimming or transmissions in the bulk of the contaminated water. Thus, researchers have greatly explored these attractive properties for precise sensing and the identification of target EPs. From this perspective, nanomaterials are a better alternative for obtaining instantaneous clarified water at affordable cost. Even though there are still unaddressed issues like scalability, recovery and reuse are under consideration, though the world is looking at sustainable nanomaterial sources to make use of them for practical environmental remediation like EP removal from water.

9.3 NANOMATERIAL-BASED WASTE

Nanomaterial-based water treatment technology is still considered an immature technique because it is still in its initial phases, but nanotechnology driven material stock opens a new door of research for technologists and innovators to adopt. Compared to traditional wastewater treatment approaches, nanomaterials offer a unique task-specific solution, though they have lot of restrictions in their real-world applications. In spite of all this, current approaches are inadequate to fulfil the present large demand for a large scale supply. To compete with traditional and current engineering technologies, nanomaterial-based water treatment methods need considerable transformation. These novel nanomaterials in conjunction with bulk materials have been characterized as a potential replacement for existing technologies; however, as stand-alone materials, nanomaterial-based systems lack the technological advantages to treat wastewater.

Today an increased amount of nanomaterial-based waste in the environment has reached an alarming level. Most nanomaterial waste originates from a variety of products such as surfactants, sunscreen, cosmetics, antibacterial textiles, polymeric nanomaterials, lithium-ion batteries, glass coating and various colloidal discharges. General consumer products are prepared using a variety of nanomaterial constituents and, after utilization of these products, several of these nano-constituents leach out or are discarded at various level into the environment, posing potential danger to eco-systems. Further, altering the chemical and physical characteristics of nanomaterials drives the overall constituents to transform into unknown chemical compositions which pose possible risks to environmental health and safety. In such situations, research is still yet to prove the potential impact on health and the environment. Currently, nanomaterial-based waste is handled following conventional segregation and management techniques. More disturbingly, there is very little to no information and knowledge available in the literature on the risks and impacts on nanomaterial waste in the environment. Another impending issue associated with nanomaterial waste is the lack of information on regeneration, recycling, reuse, incineration post-usage or landfilling and storage [10, 11].

Furthermore, most of the advanced and high-performance nano-photocatalysts show some advantages, like being less toxic, chemically stable, having low cost or easily manageable. These high-performance materials, during use, also present various advantages such as physical stability, low cost and sustainability. However, many nanocatalysts like TiO_2, Zirconia, metal sulfide materials, copper-based materials, MXene and their composites possess good photostability, but many exhibit relatively low chemical stability due to photo-corrosion. In this situation, upon light irradiation, these nanocatalysts oxidize or reduce, depending on the materials and their oxidation state changes, which leads to a decrease in the efficiency of catalysts and further decomposition, creating new nanomaterial waste debris in the media. Therefore, to avoid toxicity and the generation of secondary hazardous pollutants in the form of transformed nanomaterial waste, there is an urgent need to source sustainable and green materials and to devise simple synthetic routes to achieve stable nanocatalysts for long-time performance.

Nonetheless, several attempts have been made to synthesize new photocatalysts which can work in visible ranges for sustainable water treatment results. In this direction, researchers have used inorganic photoactive nanomaterials doped with carbaneous and grapheneous materials and their derivatives to reduce the toxicity and breakdown effect. On the other hand, to overcome the recovery related issues, magnetic nano-photocatalysts have been developed to directly add to wastewater treatment and magnetically recover the photocatalytic activity through external magnetic fields. This was promoted in developing a series of easily recoverable and reusable nano-catalysts.

The use of traditional bulk materials in water treatment experience a resistance to reach pollutants due to the inadequate reach of active sites through diffusion which necessitates the stimulation or stirring of the contaminated water or wastewater in a remediation process. But the use of nano/micromotors or cleaners could possibly overcome the diffusion boundary through energetic mixing and their self-propulsion competences. These energetically driven nanocleaners rigorously stimulate the interaction process in decontamination which enhances the uptake or degradation process with the help of a highly accessible surface area and surface activities. However, nanocleaners experience a negative impact when used in a wastewater system characterized with biological contaminants. In such a situation, nanocleaners experienced reduced activity when trapped inside or among biological pollutants, and this may also initiate coagulation and subsequently end up as sediment or sludge. Although a few photo-nano-catalysts which are bio-active and magnetically dynamic have been industrialized, there are still some challenges associated with the sediments containing residues or debris of nanocatalysts and which may cause an increasing risk for water treatment in the future.

Therefore, existing nanomaterial-based technologies are lacking in fulfilling several requirements to be efficient treatment media at present, including long-term stability and difficulties in scale-up. Major issues like the fate of nano-debris, the form of interaction, the type of complexation and nanomaterial transformation need to be considered prior to their application at large scale treatment plants. Also, the lifecycle of multi-functional nanocleaners is limited to the residual materials in their initial physical form before and after being added to contaminated water. Another disadvantage is the poisoning of product water as a result of the residual treatment material itself. In many cases, the higher viscosity of water and wastewater samples restricts the complete recovery of nanocleaners or adsorbents when applying an external stimulant like magnetic or electrochemical potential. Thus, nanomaterial-based treatment techniques need a revolutionary outlook to counter existing limitations in addressing the countless EP interaction and treatment possibilities. Nonetheless, nanomaterial impurities and waste can be tackled depending on the stage of water treatment and the technique used in adopting nanocleaners. Based on their response to external stimuli, techniques can be used for the removal of nano-waste from treatment tanks, chemical effluents, sediments, charged particles, bacteria contamination and other pathogen mixes.

9.4   IDENTIFICATION OF NANO-POLLUTANTS

In an ever-changing world, many industrialized and advanced countries have considered nanotechnology-based innovations and expertise as a strategic priority for economic, sustainable, social and scientific development. The research in the field of nanoscience and nanotechnology has enabled the development of several advanced and smart materials at the nanoscale with exceptional properties that are useful for a series of applications. On the one hand, nanoscience helps to systematically manipulate the properties of materials at the nanoscale with a regularity and precision that were formerly unknown. On the other hand, nanotechnology helps in engineering the nanomaterials to be adapted to real life applications. In both cases, nanomaterials play a vital role compared to the conventional materials and make them a suitable alternative from both the industrial and scientific point of views. However, nanomaterial-based products have opened up new opportunities in the field of water treatment, healthcare, environment, food processing industry, pharmaceuticals, cosmetics, agriculture, electronics and miniaturized devices due to their unique properties. But, due to their over-exploitation and non-trackability post-usage or leaching into industrial effluents, they create unknown risks and uncertainties about the subsequent toxicity in humans, biota and the environment at large. Today, the situation is deteriorating due to the ever-increasing production of nanomaterials which inevitably leads to their use and eventual discharge into the environment.

Therefore, it is important to evaluate the risk posed by nano-pollutants in different media. Nano-pollutants are generally categorized based on their size, dimensions, morphology, chemical composition, charge on the surface and potential agglomeration patterns. Further they can be subdivided based on their origin, such as organic, inorganic and composites. The organic nanopollutants include the synthetic and natural polymeric nanomaterials; natural macromolecules like lipids/micelles, carbaneous and grapheneous nanomaterials and inorganic nanopollutants that make up quantum dots; and magnetic and superparamagnetic nanomaterials. Nanopollutants have lately been classified based on structural dimensions, namely 1D (thin-films, nanobelts), 2D (CNTs, dendrimers, nanowires) and 3D systems (colloids, oxide/hydroxide nanoparticles, fullerenes and nanocomposites). Nowadays, nanomaterials have been indirectly adopted in several PCCPs and food products, dispersed aerosols, suspensions, emulsifying agents, colloids, peptizing agents, dispersants and additives. These nanomaterials leach out or are left out in the treatment techniques or in wastewater or treated water in negligible quantities. In such situations, a highly sensitive and reliable technique needs to be devised to trace and eliminate nano-pollutants from the contaminated sources [12, 13].

Today, nano-pollutants are considered as a new specific category of contaminants with serious and characteristic properties which impact humans and ecosystems at large upon accumulation over the long run. Based on their environmental implications, these pollutants have been classified as natural nanomaterials, artificial-intentional nanomaterials and artificial-unintentional nanomaterials. It is evident that nature has been the major manufacturer of several naturally produced nanomaterials: volcanic eruptions are major sources of nano-clays, combustion reactions may transform natural minerals into oxides/hydroxides and erosion causes the physical transformation of resources into nanosized architectures. These particles exist in the environment and are called ultrafine particles in the field of air pollution. The other major group of nanomaterials are originated anthropogenic (manmade) activities, which are also known as synthetic nanomaterials. On the other hand, artificial-unintentional nanomaterials with ultrafine nano-dusts originate from air pollution caused by welding fumes, industrial residues, fossil fuel emissions, incinerators, landfill sites and discharges from wastewater treatment plants, spillage and leakage during transportation and manufacturing or handling of nanomaterials.

9.5 ASSESSMENT OF NANOMATERIALS AS POTENTIAL EPS

The nanomaterial market across more than 750 product segments worldwide is projected to grow by over $70 billion, driven by an annual compounded growth rate of 13.8%. A study on selected segments of nanomaterials suggests the overall improved growth is poised to reach over $98.9 billion by the year 2025, making the release of nanomaterials during production and use a significant major concern. However, increasing environmental awareness and regulations will eventually limit the amount of nanomaterial waste that enter the environment through discharges, dumping or accidental release. Nevertheless, nanomaterial release through regular human activities and the use and discarding of nano-embedded products are posing a serious threat to the environment [14].

Today the world has industrialized to a large extent. Except for a few regions, the manufacturing of commodities is taking place in largely decentralized industrial units. As a result, the world at large and several ecosystems in particular are witnessing the consequences of human-made chemical pollution in the environment. In many ecosystems, deterioration due to EPs may be too elusive to notice. For instance, there is leaching of bisphenol-A into the environment from discarded water bottles and nanomaterials, extensive discarding of surfactants through household activities, and the proliferation of steroids, hormones and active pharmaceutical ingredients in various water sources. In any case, it is easy to see the high levels of contamination in almost all water bodies. Therefore, what is irrefutable is that the release of high volumes of industrial and domestic wastewater is impacting biota and the environment.

Interestingly increased case studies and awareness in society, communities and researchers about water pollution is forcing the taking of measures to reduce this impending threat. In this direction, the research community and society at large are interested in reaping the probable benefits of nanotechnology in developing nanomaterial-based product development that impacts day-to-day human life. This trend increasingly utilizes the synthetic material stock to fulfil this demand. For instance, the use of colloidal particles, nanoparticle-based synthetic emulsions, polymeric residuals in PCCPs and so on is replacing traditional chemical contaminants like pesticides, micro-organics and pharmaceuticals in tested water bodies. Even though the demand for greener and sustainable material sources has gained momentum, the possibility that synthetic nanomaterials will continue to find their ways into the environment raises health concerns at different levels. This is mainly due to the fact that nanomaterials or nano-pollutants may impact on the well-being, stability and health of local ecologies in ways that are difficult to predict and comprehend. Nonetheless, due to the increased awareness, researchers, manufacturers and consumers nowadays are drawing a distinction between natural and synthetic nanomaterials. Also, awareness is leading to finding the origin, sourcing, extraction, processing and transformation into nanomaterials which may actually be an important part of future product development to predict their biological and geochemical impact.

It is also important to note that just because some nanomaterials are extracted or processed from natural origins does not guarantee their exemption from being possible environmental impacts. It does not follow that the source of the raw material reduces the severity of nano-pollutant impact on the environment. Inevitably, nano-materials have been developed that cause medium to large damage due to their size and physical properties, which makes them easily associate with the constituents of the ecosystem.

Therefore, almost every type of nanomaterials may cause a detrimental impact on the environment as they facilitate chemical reactions upon interaction, causing harm to cells, plankton, bacteria and small creatures. It is well documented that metal, metal oxide, metal hydroxide and metal-composite-based nanomaterials show excellent catalytic and photocatalytic activities in a specific and targeted application. However, if the same catalytic nanomaterials are discarded into the environment, they may continue to carry the catalytic activity and initiate chemical reactions that produce toxic chemicals, like free radicals or reactive oxygen species. In many cases, these free radicals initiate uncontrolled chemical reactions on any of the vulnerable constituents in the ecosystem. For example, a titanium dioxide (TiO_2) nanocatalyst is an excellent photo-active nanomaterial, but upon exposure to UV or visible light it activates a robust chemical reaction though hydroxide ($\bullet OH$) or superoxide ($\bullet O^{2-}$) radicals in natural waters. More significantly, when these reactive oxygen species are excessively discarded or discharged, effluent that reaches water bodies initiates unwanted chemical reactions in aquatic organisms, including bacteria, virus, plankton and small fish. Similarly, ZnO, Fe/H_2O_2, graphitic carbon nitride, several quantum dots, and composites of titania like TiO_2-SiO_2, TiO_2-ZnO_2, Au-TiO_2 and $K_4Nb_6O_{17}$/CdS have recorded similar high photocatalytic activities in various chemical reactions [15].

Another main source of nanomaterial pollution comes from nanomaterial-induced consumer products. There are several thousands of such consumer products already on the market and day-to-day use includes textiles, special self-cleaning nanomaterial-coated clothing, personal care products, surfactants, emulsion paints, next-generation batteries and packaging materials which now comprise nanomaterials as one of their main ingredients. It is now estimated that the next decade will experience a sudden surge in the incorporation of nanomaterials in personal products, sophisticated gadgets, domestic products, health care products and next-generation energy conversion and storage devices like solar cells or batteries. Nanomaterial-induced advanced products have proven added advantages to their performance. However, at the end of their use and in circumstances like breakdown or failure these products also deliver several key nano-ingredients into the environment. On the other hand, domestically used products like self-cleaning textiles, microbial fabrics or UV-blocking clothing with photo-sensitive nanomaterials end up in landfills or wash away into drains when clothes are laundered. Therefore, nanomaterials that end up in domestic wastewater through drain water or in landfills, dumps or other discards will end up in different ecosystems through various routes. Nanomaterials that seep with washed water down the drain will ultimately enter the wastewater treatment plant. Hence, nanomaterial polluted water will undergo three main physical treatment steps before chemical disinfection. Unfortunately, none of these conventional treatment processes are precisely planned to eliminate nanomaterial residues from waste effluents. Consequently, even after treatment, nanomaterials will remain in the purified or treated water that is generally released to aquifers or used for non-drinking purposes.

On the other hand, partial nanomaterial is retained in the primary and secondary treatment steps in the form of sludge in which most of the time nano-pollutants are entrapped in the microbe-bearing sludge leftovers from the purification process. Generally, this bio-sludge leftover from wastewater treatment plants is used as fertilizer for farm land, which can possibly allow synthetically entrapped nanomaterials to enter soils or water bodies through runoffs. In a worst case, sludge waste is landfilled following one of the solid waste management procedures. In such a situation, a major portion of the sludge degrades into biocompatible fragments and persistent nanomaterials which leach into the soil and enter water reservoirs during rainfall and subsequent runoffs.

In modern and speciality textile manufacturing, nanomaterial-coated fabrics are widely explored. Speciality textile industries have adopted nanotechnology innovations to induce self-cleaning, UV-blocking, and non-wettable and wrinkle free running fabric by the incorporation of nanomaterials as one of the main constituents. These innovative textile fabrics adopt photo-active nanomaterials like titanium dioxide or zinc oxide nanoparticles embedded in manufacturing strategies to create modern fabrics. However, even after several robust and rigorous stabilization protocols, nanomaterial-induced textile products have been known to leach out as active nanomaterials into the environment during laundry. These nanomaterials are also known to disconnect or leach out from the substrate when exposed to weathering or external stimuli like friction and temperature. Eventually, these nanomaterials reach wastewater treatment plants and, further escaping the treatment step, end up in the ecosystem and, upon exposure to light, may cause severe damage to aquatic or marine life. Those leached into surface water or soil ultimately end up in ground water or drinking water.

9.6   NANOTOXICITY

By now, researchers and environmental regulators are convinced that nanomaterials do leach or escape into the environment through various pathways. Regrettably, we are yet to understand the detailed environmental implications caused by the presence of nano-pollutants, even though studying the health implications of nanomaterials is just one of many challenges regarding impending environmental concerns. For example, it is yet to be understood how active nanomaterials like inorganic species interact with organic matter and biological species which leads to complexation and aggregation in the ecosystem.

Several studies have already established that the risks associated with nanomaterials may not be the same as the risks associated with the bulk versions of the same materials. Nanomaterials with a high surface area to volume ratio can make nano-sized materials more reactive than their bulk materials, which makes them persistent pollutants.

Thus, initial evidence suggests that humans have been exposed to nanomaterials internally, which may be due to the consumption of drugs with polymer carriers, surgical implants and packaged or processed food ingredients. These possible routes of nanomaterial exposure are being extensively studied to understand their health implications. Significantly, the route of nanomaterial exposure or entry ultimately regulates and determines which body structure/tissue or specific system the nanomaterial interacts with. This helps in determining the effects of nanomaterial exposure like toxicity, extent of adulteration or alterations in physiological function.

As early as 2014, it was reported that over 6000 organizations from 32 countries had prepared and used nearly 2000 consumer products that contained nanomaterials. Most of these products (>40%) were categorized as health care products. On the other hand, in the same period, a nanomaterial-related database inventory showed the European market had consumed over 3000 products that had nanoparticles as a main ingredient. Most of these products were categorized as personal care or cosmetic products. Initial consumer markets used silver, titanium and silicon-based nanomaterial compositions in cosmetic and health care products. At the same time, researchers started to notice the toxic effects of nanomaterials on humans and biota. These efforts have also been part of programmes planned by the regulatory authorities and environmentalists.

Several research groups over the world are involved in delving deep into understanding and assessing the field of nanotoxicity. An initial agenda mainly included health implication studies due to exposure to ultrafine-airborne nanoparticles, mining dust, widely used asbestos and artificial mineral fibres. Soon, toxicological studies clearly indicated the fact that nanoparticles may induce a greater health risk than bulk materials of the same precursor material. These observations led to the issuing of a series of guidelines to caution manufacturers and consumers and to regulate the use of nanomaterials in various products. Among all the properties, size and dimensionality in nanomaterials play a role in determining the possible toxicity. Further, recent studies have evidently concluded that the superior chemical reactivity in nanoparticles occurs due to their ability to produce a high number of reactive oxygen species (ROS) including free radicles ($^{\cdot}OH$ and $^{\cdot}O$), which make them long lasting [16–18].

More worryingly, a large number of nanomaterials like carbonaceous nanomaterials and their nanocomposites, CNTs and metal oxides/hydroxides have been analysed for their reactive oxygen species production capacity. However, reactive oxygen species and free radical production have been characterized as the main reason and the primary mechanism of nanomaterial toxicity in the environment. These free radicals induce chemical transformations in the natural substances of human and biota that may result in inflammation, oxidative stress, and damage to proteins, DNA and membranes. Some other studies have also confirmed the effect and influence of chemical composition, high surface structure activity, surface charge and the dispersibility of nanomaterials on nanotoxicity. Due to their nanoscale size and dimensions, nanomaterials possess higher mobility in certain environmental conditions than their bulk materials. Therefore, higher activity combined with high mobility helps them to readily engage with the human body. Further, a higher mobility factor allows them to easily cross membranes and gain access to cells, tissues and vital organs. Further, their smaller size leads to the faster transportation of nanomaterials within cells, leading to cell mitochondria and the cell nucleus, which may cause major structural damage and the eventual death of cells. It is also important to note that there is no direct relation between toxicity and nanosized materials; however, the toxicity may originate from the origin of the material or substance.

Therefore, one cannot assume that bulk materials are safe and that nanomaterials possess toxicity. Principally, it can be noted that not all nanoparticles are toxic;

however, the existing studies suggest that nanomaterials impose greater risks of toxicity than the bulk materials of the same substance. Regardless of their chemical conformation and compositions, all modern and advanced engineered nanomaterials have been categorized as potent influencers of inflammatory lung injury in living beings, in particular in humans. Work-place exposure to high levels of nanomaterials has been noticed to cause constant lung inflammation which further leads to respiratory complications, fibrosis and cancer. Constant exposure to nanofibres like CNTs, asbestos and electrospun polymer nanofibres in the workplace poses the risk of lung irritation and prolonged contact may induce inflammation, leading to lung cancer.

9.6.1 RESPIRATORY SYSTEM

Some studies also noted that, other than composition and origin, surface area and surface activity are better predictors of nanotoxicity in nanomaterials, though most may not be accurate in assuming this as most of the present nanomaterial exposure standards are based on weight-based procedures. Lately, nanotoxicity studies conducted on CNTs, following weight-based permissible limit exposure in the work place, mimicking the constant graphitic particle standard, found that CNTs resulted in irritation and reduced pulmonary function due to inflammation and indication of fibrogenesis. Therefore, weight-based exposure studies concluded that CNTs are more deadly than silica dust or ultra-fine carbon black at the workplace. Further, single-walled carbon nanotube exposure (SWCNT) studies following graphite particle norms indicated that workers exposed to these nanofibrous pollutants may develop lung lesions. Also, CNTs have induced serious complications leading to the death of kidney cells and reduced cellular adhesive capability.

9.6.2 PULMONARY SYSTEM

Prolonged exposure of nanomaterials through ingestion, skin adsorption and inhalation can advance and access the blood stream. Once in the blood stream, these nanomaterials can move throughout the body and contaminate organs and tissues such as the kidneys, bone marrow, brain, spleen, heart, liver and the nervous system. Upon contamination nanomaterials induce increased inflammation and oxidative stress. However, these effects depend on the concentration, accumulation and distribution of nanomaterials within the body. However, it is important to mention the fact that the key distribution locations and the target tissues or organs for nanomaterials are yet to be completely established or confirmed. However, the liver has been increasingly characterized with nanomaterial contamination, followed by the spleen. A low concentration of fullerene exposure has shown potential toxicity towards human lung cells, brain damage in fish, caused death in water fleas and has induced proven bactericidal effects. The effects of nanosized materials on pulmonary toxicity have been reported in recent studies. Accordingly, rats induced with nanomaterials like ultrafine carbon black and TiO_2 nanoparticles have recorded enhanced lung inflammation. Further, higher loads of nanoparticles in rats resulted in inflated lung responses, ultimately resulting in the development of lung tumours. These short studies conducted on animal samples reveal that the pulmonary toxicity of nanomaterials depends on the nanomaterial size, density, surface activity, degree of aggregation and method of nanomaterial production.

9.6.3 Cardiovascular System

Recent studies have indicated that cationic nanomaterials like gold and polymeric nanoparticles of polystyrene cause haemolysis and blood clotting, whereas typically anionic particles have been known for their harmlessness. However, regular exposure to a high concentration of carbon particles originated from diesel exhaust particles (DE) have been known to cause altered heart rates in hypertensive rats. These animal study results have been extrapolated and interpreted as a direct effect of DE nanoparticle exposure on the pacemaker activity of the heart. Further, continuous exposure to SWCNTs has also resulted in cardiovascular disorder symptoms.

9.6.4 The Reticulo-Endothelial System

Nanomaterial-contaminated livers experience inflammatory gastrointestinal tracts leading to cardiovascular system contamination, since all blood exiting the gastrointestinal tract drives into the hepatic portal vein that directly distributes through the liver. Even with low toxicity, nanomaterials like fine carbon black and polymeric nanoparticles like polystyrene beads stimulate the macrophages through reactive oxygen species free radicals. Further, nanomaterials stimulate calcium signalling, to induce proinflammatory cytokines like tumour necrosis factor alpha, causing oxidative stress and liver disease.

9.6.5 Central Nervous System

Some of the studies reveal that nanomaterials inhaled can enter the brain, possibly through two different mechanisms, namely transsynaptic transport which happens through the olfactory epithelium, and via the blood-brain barrier. A number of pathologies like high blood pressure and allergic encephalomyelitis lately have been related to the amplified permeability of nanomaterials through blood-brain barrier studies in several controlled experimental setups. Also, inflated symptoms have been directly correlated to nanomaterial surface charges which alter blood-brain integrity. Further, higher production rates of reactive oxygen species and oxidative stress are known to induce neuro-degenerative ailments like Alzheimer's and Parkinson's diseases.

9.6.6 Dermatological Effects

Several personal care and cosmetic product manufacturers utilize thousands of colloidal nano-formulations with precursor sizes ranging between 50 and 500 nm. Even though PCCPs contain a low percentage (<3%) of nanomaterials, the extensive use of end products leads to the accumulation of nanomaterials in humans.

Some of the unique features of colloidal nanomaterials comes from their exceptional properties like their light scattering ability which increases the optical path of UV photons entering the upper part of the skin layer. Through this pathway, skin absorb more photons which adversely affects skin health. Also, the dermatological effects of nanomaterials like photo-active creams, lotions and skin-creams are estimated to penetrate the skin. Recent *in vitro* studies have shown that MWCNTs are capable of accumulating on specific skin sites and initiating an irritation response in human epidermal keratinocytes, which are a primary route of occupational exposure.

9.6.7 SILVER TOXICITY

Recent *in vitro* studies have revealed that Ag nanoparticles are likely to induce toxicity in cells. In recent times, the use of Ag nanoparticles has increased a lot in day-to-day life. In particular, their use in cosmetics and textiles has considerably grown, which has subsequently amplified the potential exposure to human skin. Being known for its anti-bacterial nature, Ag even in trace amounts may directly harm plants, microbes and animals. Also, increased concentrations of Ag nanoparticles in the environment could distress local bacterial populations inducing extreme consequences in ecosystems. The uncontrolled release of Ag into the environment is related to the stability of Ag in products under different conditions such as fabric quality, pH, preparation routes, sweat formulation and washing. Medical products exposing keratinocytes to extracts of Ag-containing wound dressings were observed to induce the most cytotoxic damage. On the other hand, fibroblasts exposed to Ag have more sensitivity to activity than Ag nanoparticles than keratinocytes. Also, Ag nanoparticle exposure induces cell death and oxidative stress in human fibrosarcomas, skin carcinoma cells, DNA damage and apoptosis in fibroblasts and liver cells. Other studies have also demonstrated Ag toxicity in the lungs and liver in which a substantial depletion of the antioxidant glutathione condensed the mitochondrial membrane potential and increased ROS. Nevertheless, there is inadequate evidence on the *in vitro* toxicity of silver nanoparticles in lung cells.

Latterly, several studies have been conducted to measure the toxic effect of silver nanoparticles on non-mammalian models. Hence, a real-time study conducted on Ag nanoparticles and their transportation and biocompatibility in zebrafish provided additional knowledge about the delivery and effects of Ag nanoparticles *in vivo* which further gave newer insights into molecular transport mechanisms during early embryonic development. Further, single Ag nanoparticles induced into zebrafish embryos were further investigated for their effects on early embryonic development. These studies establish that Ag nanoparticles were observed inside embryos at each developmental stage. However, Ag toxicity and the resulting types of abnormalities were known to be highly dose dependent. Nevertheless, studies reveal that higher concentrations of Ag increases the mortality and hatching delay in zebrafish embryos. Successive studies found that Ag nanoparticle exposure induced DNA damage and initiated genes associated with oxidative stress and metal detoxification/metabolism in experimental models.

Nonetheless, several recent studies reveal that Ag nanoparticles induce skin disease, induce reproductive failure, cause developmental malformations and morphological deformities in a number of non-mammalian animal models. Therefore, excessive Ag nanoparticle exposure induces oxidative stress, apoptosis and DNA damage.

9.6.8 Toxicity of Carbon-Based Nanomaterial

Researchers around the world consider serious the threat posed by nanomaterials, in particular carbon-based nanomaterial, the use of which increased a lot in the last decade. However, new investigations on nano-carbon toxicity continues to support a positive outlook in which the chronic toxicity of carbon nanomaterials such as CNTs has been seen to be negligible in a sample study using mice as a model in four-month long experiments. Further, studies confirm the changes in the neutrophil count in mice induced with PEGylated-oxidized SWCNTs which were greater in counts compared to those mice treated only with PEGylated SWCNT. This suggests that varying functionalities in carbon nanomaterials, in particular CNTs, can alter toxicity in living beings. On the other hand, a recent *in vivo* cancer therapy study used CNTs as drug delivery enhancers and was able to demonstrate that tumour cells respond to toxicity differently. In this study, a noticeable decline in breast cancer tumours was recorded due to the use of SWCNT as conjugated carriers for paclitaxel, rather than controlling for paclitaxel alone in a mice model. But, detailed analysis shows that a few tumour-bearing mice treated with non-functionalized SWCNTs had a comparable rate of tumour development as that of the untreated or control model. This study concluded that cancer cells might develop a resistance mechanism against CNTs in short-term testing conditions. However, larger study data are yet to be implemented to ascertain the long-term effects of carbonaceous nanomaterial toxicity on pulmonary functions, respiratory inflammation and bronchoalveolar lavage of living beings.

9.6.9 Toxicity of Silica Nanoparticles

Silica nanoparticles have shown relatively high biocompatibility when used in different products and exposures. In other words, silica nanoparticles have been shown to be more useful in biological applications with low cytotoxic potentials in living beings when used in moderate quantities. A major issue with silica nanoparticles comes from their vulnerability to agglomerate which influences the complexation among protein macromolecules leading to aggregation accumulation *in vitro* above certain doses (>25 µg/mL). However, at higher doses silica is known to induce oxidative stress and cause cytotoxicity both *in vitro* and *in vivo* which was confirmed by increased lipid peroxidation, reactive oxygen species and reduced cellular glutathione. However, more significantly these effects were inconsistent when silica nanoparticles were administered with different particle sizes and doses. Yet again, researchers are still to establish the long term exposure of silica nanoparticles and are working on determining the implications of nano-silica with different sizes, distributions, functionalities and exposure limits on living beings.

9.6.10 TOXICITY OF COPPER NANOPARTICLES

Copper is a vital trace element widely used in several products as an important ingredient. On the other hand, Cu-nanoparticles have shown great potential in osteoporosis
treatment drug formulations, as antibacterial materials, widely used additives in livestock and poultry feed, and Cu-metal systems have been used as intra-uterine contra-
ceptive devices. Also, Cu-nanoparticles have been extensively industrially important
products like lubricants and reactive catalyst electrochemical transformational reactions with various sizes, surface areas, surface energy and activity. Additionally,
these properties also provide Cu-nanomaterials with great dispersibility and compatibility with application media. Even though Cu-nanoparticles in appropriate doses
are known to cause less harm, though it is important to distinguish between the
effects of nanoparticulates from dissolved metals and dispersed ones to assess the
toxicity effects. In spite of a growing application interest in implementing copper
nanoparticles in various products and formulations, there is a serious lack of information on their nano-toxicity impact on human health and the environment. Copper
is considered to induce potential health risks because Cu-nanoparticles are capable
of creating toxic effects in humans and animals when ingested frequently in high
concentration. Numerous studies increasingly highlight the potential risk involved
in using Cu-nanoparticles and copper-based compounds at large. Most of the time
Cu-based substances or products or nano-formulations are used in food or health care
products. Therefore, sensitive copper toxicity in humans occurs from long term oral
exposure which is known to induce harmful effects in the gastrointestinal system,
kidneys and liver. Also, toxicity due to Cu-nanoparticles or Cu-containing product
exposure is mainly categorized as hepatocellular necrosis, acute tubular necrosis in
the kidney, disruption of the epithelial lining of the gastrointestinal tract, drowsiness
and anorexia. A recent study report reveals that when mice were exposed to high
concentrations, Cu-nanoparticles induced severe impairment in the liver, kidney and
spleen. Additional studies discovered that the toxicity of nanosized Cu-nanoparticles
was highly interrelated with their size, surface area and surface energy. Nonetheless,
even though the potential risks of Cu-nanoparticles on human health have been
identified and estimated, sub-acute toxicity due to Cu-nanoparticles has not been
described or confirmed.

9.7 HEALTH RISKS ASSOCIATED WITH EXPOSURE TO NANOMATERIALS

The extensive use of nanomaterials in recent times has initiated intense debate on
nanomaterial exposure and nanotoxicity in living organisms. The previous section
has provided detailed accounts on the toxicity associated with several nanoparticles and nanomaterial-incorporated products. However, today the toxic patterns of
pharmaceutical formulations with chemotherapeutic drugs, nanomedicines, delivery
systems and nanocarriers have been brought into serious discussion. It is estimated
that the nanomaterial toxicity involving nanomaterials in PCCPs comes with direct
oral and skin exposure in humans. On the other hand, pharmaceutical drug formulations with a delivery system and nanocarriers in addition to the active ingredient

have been shown to produce several adverse bioactivities upon consumption. These include protein and peptide drugs, and chemotherapeutic drugs, like anti-cancerous agents, which upon ingesting are presumed to cause toxicity and the selective initiation of cytotoxicity in normal cells and organs. It is well established by now that the nanomaterial constituent substance, particle size, shape and surface area contribute directly and significantly to toxicity in humans. In many organic/inorganic nanocomposite systems, apart from physical dimensionalities, the structure of the nano-system also contributes immensely to nanotoxicity. Most of the detailed studies on nanotoxicity conclude with a set of observations, outcomes and interlinkages between nanomaterial-based products and nanotoxicity associated with their exposure in order to understand the future of nanotechnology-based products and developments. Also, most of the nanotoxicity has been directly correlated to its ability to form reactive oxygen stress, radical formation and the subsequent DNA damage mechanism. Based on a large set of data, several other studies also conclude that the major reason for nanotoxicity can be overcome using green and sustainable nanoscience and technology approaches [19, 20].

Further, to assess the accuracy and relevance of currently adopted methodologies for the toxicity screening of nanomaterials, several factors need to be considered, and outcomes need to be revaluated considering: (i) whether there is any specific toxicological end point or points that are of greater concern for nanomaterials, such as respiratory, cardiovascular, neurological or immunological effects as well as endocrine disruption; (ii) revisiting testing protocols, procedures and tools to replace the currently used methods in closure association with types of organisms, exposure regimes, possible variables in testing media, advanced analytical methods and test outlines to overcome the limitations of existing methodologies to standardize the toxicity; and (iii) revaluating the efficiency and accuracy of current analytical methods that lack precise quantification to generate dose–response relationships. Currently there is no standard protocol, procedure or one-point solution for nanotoxicity testing; however, it will suffice to mention that the three key essentials of a toxicity screening strategy and approach should include in the near future studies that are physicochemical characterizations of nanomaterials, *in vitro* assays both cellular and non-cellular, and *in vivo* studies to design rational and mechanism-based assessment protocols.

9.8 PREVENTION AND METHODS TO TREAT AND MANAGE NANOMATERIALS

Nanomaterials have, since their origin, preparation, usage and discarding into the environment, attracted special attention. This is due to their unique properties associated with size, morphology and their distribution. For the same reason, nanomaterials also pose several questions about their practical use compared to their bulk material over the long run. Even though when used as a treatment medium, these materials can enhance effectiveness against the removal of EPs better than conventional methods, owing to their increased reactivity and selectivity of EPs. Yet, there still remain several serious questions awaiting conclusive answers, mainly concerning the potential nano-toxicity and its fate and transportation in the environment. On the other hand, efficiency in waste management using nanomaterials mainly depends on the selection of suitable media, methods and routes for the targeted EPs under prevailing conditions. Nonetheless, nanotechnology-based solutions can be employed either for *in situ* or *ex situ* EP management in wastewaters. Through an *in situ* approach, nanotechnology can be used in the form of a membrane bed or filter through which a reactive or interactive zone is created vertically for the flow path in which the retention of EPs occurs with minimal concentrate waste in the form of rejectants. On the other hand, an *ex situ* waste management approach is followed in the form of adsorption, advanced oxidation and photocatalysis. In this multi-step approach, managing waste is a complex process, and the nanomaterial and pollutant interaction may yield a secondary pollutant. This approach is generally categorized as feasible but non-realistic for field and industrial application.

Even though several ecotoxicological studies discussed in previous sections indicated the toxic effects of nanomaterials on mammalian cell types and to some aquatic organisms, there is still poor understanding and a lack of data for determining toxicity thresholds. These are the main reasons why environmental regulatory authorities and controlling agencies are enforcing laws concerning nanomaterials, internationally. Yet, in spite of the increased awareness only some general guidelines on the management of nanomaterial are being practised. Thus, it is essential for the scientific community to generate specific data of the potential toxicological effects of nanomaterials and to impose their widespread use at the industrial and community level in general and for waste management in particular. Also, it is important to assess the real impacts of nanomaterials on ecology through a series of case studies, which also helps in creating a rational legal framework concerning environmental implications [21].

Nevertheless, it is largely projected that the application of nanomaterials in EP treatment is going to decrease waste management cost, energy stresses and increase the process's efficiency in the near future. However, currently existing high production costs and nanotoxicity vulnerabilities and lack of information on their consequences, limits their industrial production for the time being. More significantly, there are still substantial knowledge gaps concerning their fate, and transportation must be addressed and clarified before nanotechnology is widely applied in EP wastewater management. Therefore, strategies targeting the development, application and disposal of nanomaterials hold great potential for further developing nanotechnology-based tools for the future.

REFERENCES

1. Ken Donaldson, and Craig A. Poland, Nanotoxicity: challenging the myth of nano-specific toxicity, *Current Opinion in Biotechnology*, 2013, 24(4), 724–734.
2. Hans C. Fischer, and Warren C. W. Chan, Nanotoxicity: the growing need for *in vivo* study, *Current Opinion in Biotechnology*, 2007, 18(6), 565–571.
3. José García-Torres, Albert Serrà, Pietro Tierno, Xavier Alcobé, and Elisa Vallés, Magnetic propulsion of recyclable catalytic nanocleaners for pollutant degradation, *ACS Applied Materials & Interfaces*, 2017, 9(28), 23859–23868.

4. Samsul Rizal et al. Cotton wastes functionalized biomaterials from micro to nano: a cleaner approach for a sustainable environmental application, *Polymers*, 2021, *13*(7), 1006.

5. Nancy A. Monteiro-Riviere, and Alfred O. Inman, Challenges for assessing carbon nanomaterial toxicity to the skin, *Carbon*, 2006, 44(6), 1070–1078.

6. Pavan P. Adiseshaiah, Jennifer B. Hall, and Scott E. McNeil, Nanomaterial standards for efficacy and toxicity assessment, *Wiley Interdisciplinary Reviews: Nanomedicine and Nanobiotechnology*, 2010, 2(1), 99–112.

7. Martin Meyer, What do we know about innovation in nanotechnology? Some propositions about an emerging field between hype and path-dependency, *Scientometrics*, 2007, 70(3), 779–810.

8. S. K. Sahoo, S. Parveen, and J. J. Panda, The present and future of nanotechnology in human health care, *Nanomedicine: Nanotechnology, Biology and Medicine*, 2007, 3(1), 20–31.

9. Michiko Sato, and Thomas J. Webster, Nanobiotechnology: implications for the future of nanotechnology in orthopedic applications, *Expert Review of Medical Devices*, 2004, 1(1), 105–114.

10. Henning Wigger, Stephan Hackmann, Till Zimmermann, JanKöser, Jorg Thöming, and Arnimvon Gleich, Influences of use activities and waste management on environmental releases of engineered nanomaterials, *Science of The Total Environment*, 2015, 535, 160–171.

11. Tanushree Dutta et al., Recovery of nanomaterials from battery and electronic wastes: a new paradigm of environmental waste management, *Renewable and Sustainable Energy Reviews*, 2018, 82(Part 3), 3694–3704.

12. H. Tang, D. Wang, and X. Ge, Environmental nano-pollutants (ENP) and micro-interfacial processes, *Water Science and Technology,* 2004, 50(12), 103–109.

13. Roberto Scandurra, Anna Scotto d'Abusco, and Giovanni Longo, A review of the effect of a nanostructured thin film formed by titanium carbide and titanium oxides clustered around carbon in graphitic form on osseointegration, *Nanomaterials,* 2020, 10(6) 1233, doi:10.3390/nano10061233.

14. https://www.researchandmarkets.com/r/ageks0

15. Mitra Naghdi, Sabrine Metahni, Yassine Ouarda, Satinder K. Brar, Ratul Kumar Das, and Maximiliano Cledon, Instrumental approach toward understanding nano-pollutants, *Nanotechnology for Environmental Engineering*, 2017, 2(3), 1–17.

16. Peter P. Fu, Qingsu Xia, Huey-Min Hwang, Paresh C. Ray, and Hongtao Yu, Mechanisms of nanotoxicity: generation of reactive oxygen species, *Journal of Food and Drug Analysis*, 2014, 22, 64–75.

17. Amedea B. Seabra, Amauri J. Paula, Renata de Lima, Oswaldo L. Alves, and Nelson Durán, Nanotoxicity of graphene and graphene oxide, *Chemical Research in Toxicology*, 2014, 27(2), 159–168.

18. Thomas R. Pisanic et al. Nanotoxicity of iron oxide nanoparticle internalization in growing neurons, *Biomaterials*, 2007, 28(16), 2572–2581.

19. Puja Khanna, Cynthia Ong, Boon Huat Bay, and Gyeong Hun Baeg, Review nanotoxicity: an interplay of oxidative stress, inflammation and cell death, *Nanomaterials*, 2015, 5, 1163–1180.

20. K. Ray, Clearance of nanomaterials in the liver. *Nature Reviews Gastroenterology & Hepatology*, 2016, 13, 560.

21. O. Tsehmistrenko, The biological methods of selenium nanoparticles synthesis, their characteristics and properties, *Tehnologìâ virobnictva ì pererobki produktìv tvarinnictva*, 2020, 2(158), 6–20, doi:10.33245/2310-9289-2020-158-2-6-20

10 Policy Efforts to Curb the Use of Microplastics and Remove Emerging Pollutants

10.1 RECOGNITION OF THE EMERGING POLLUTANT THREAT

Emerging pollutants (EPs) are substances and compounds that have lately been widely recognized as hazardous to the environment, and subsequently to the health of human beings. Until very recently, the main emphasis on the implications of widely used chemical substances in the environment has been related to heavy metals, micro-organic pollutants, conventional active ingredients in pesticides and pharmaceutical products. However, several detailed environmental risk assessment studies on the so-called EPs have included a new class of modern products including personal care products, veterinary medicines, nanomaterials, paints, colloids, industrial additives and modern drugs like antibiotics, steroids, endocrine disruptors, hormones, polymer microbeads and microplastic and coating sprays. They have been termed 'emerging' since their increasing production, usage and discharge were earlier ignored; however today the number of rising environmental concerns are being linked to them. Moreover, many of these new classes of widely used consumer products which are now a source of EPs have not been regulated under currently followed national or international legislation, hence they pose an unknown greater risk to our livelihood. Also, several other studies established inextricable links between these EPs and the severity of wastewater. Many case studies have established that municipal, domestic and industrial wastewater is a primary pathway for their wider transportation, distribution and diffusion in the aquatic environment. Though more research and case studies are needed to established the specific toxicity patterns of EPs in the environment, it is widely recognized that these pollutants are increasingly becoming a threat [1–3].

Other than widely synthesized consumer products, some EPs, such as the natural toxins and naturally degraded secondary ones within the various ecosystems involving animals, plants and microbes, may generate EP-like pathogens with varying level of toxicity. These concerns have been reflected in the rapid increase in the numbers of scientific publications, environmental monitoring activities, implementations of stringent discharge norms, commissioning case studies, regulatory guidelines exploring the environmental and ecological impact analyses due to EPs, and

DOI: 10.1201/9781003214786-10

further dispensations and propagations of this knowledge among industrial manufacturers, policy makers, researchers and the public through the popular press across the globe.

On the other hand, water pollution both from conventional pollutants and emerging contaminants is a severe worldwide problem that immediately requires ideas for monitoring, regulating and implementing plans to deliver attainable solutions. Each day, several million tonnes of domestic, municipal, agricultural and industrial waste is discharged into global water bodies, which is estimated to be the equivalent of the weight of the entire anthropological population. United Nations' studies estimate that the amount of wastewater produced annually across the globe is about 1500 km^3, six times more water than occurs in all the rivers of the world. The release of EPs into the environment has probably been happening for a long time, but it may not have been noticed due to a lack of awareness and tools, and not observed until new detection methods were developed [4, 5].

Even after several years of suspicion and study reports, EPs are currently not included in routine national and international monitoring programmes and their fate, actions and ecotoxicological properties are often not well quantified. However, it has also not been established that the EPs released from sources or points of contamination like drainage, industrial effluent plants, municipal and domestic wastewater treatment plants or from diffuse bases through atmospheric deposition or from crop and animal farms have specific characteristic effects on ecosystems. Further, these entry points have been categorized into more than 20 classes associated with their origin, like urban or domestic discharges, stock farming which releases pharmaceuticals, agricultural activities that over-exploit pesticides, industrial cleaning and municipal and domestic disinfection activities and their by-products, wood preservation and industrial chemicals.

Several national governments and organizations are formulating regulations and framing policies associated with safe and clean drinking water, considering the EPs listed by world health organizations. This action has prompted several health agencies and organizations to seriously consider the threat associated with EP-contaminated water at different end use sites. However, water quality norms, standards and guidelines vary among federal governments, provincial states and regions, mainly due to changes in the frameworks of local social and economic conditions and cultural beliefs related to water bodies and environmental conditions. However, the global organization which has almost every country on the planet as a signatory, the World Health Organization (WHO), has laid down guidelines for drinking-water quality, which it amends from time to time with the support of case studies. Similarly, the Council of the European Union (EU) has adopted directives on water quality and safety intended for human consumption. Further, the United States and most European nations have parallel regulatory systems managed by federal environmental agencies and local governments. At the local level, many European countries, state governments in the USA and India either follow federal guidelines or set up their own regional legislations to regulate their water quality and the health of water reservoirs. However, England and Wales have established a unique multilateral (tripartite) regulation arrangement involving the Drinking Water Inspectorate (DWI), the Environment Agency and the Office of Water Services (Ofwat). In all cases, guidelines and regulations are framed with the intention to protect the public against exposure to conventional contaminants, EPs like HMs, microplastics, nanomaterial contaminants and polymeric pathogens such as coliforms, *Cryptosporidium spp.*, *Giardia lamblia* and viruses like *Legionella spp.*

Earlier, global health organizations and environmental regulatory authorities including the WHO focussed on framing guidelines to tackle waterborne disease like diarrhoea which is a major cause of concern in developing countries with an estimated 4% of total daily global disease and which is responsible for the death of millions of people each year. However, the number of global waterborne-disease cases and deaths have been increasing due to the deterioration of the aquatic ecosystem caused by EPs having new and unpredictable symptoms. However, the true occurrence of waterborne illnesses due to definitive EPs in a specific region is not reflected by the any of the currently reported statistics. However, there can be multiple reasons for the lack of statistics at present on the causes of new diseases, which may or may not be from EPs and for which it is also important to note new symptoms need yet to be studied, investigated and reported, specifically in connection with the new adversary that is EPs.

In normal conditions, public health authorities across the world operate reactively, in some cases proactively, contrary to the background of inclusive public health policy and in collaboration with all stakeholders. Now the situation is forcing us to create awareness of EPs in drinking water, wastewater and domestic discharges; the regional, national and global public health authorities should sporadically conduct case studies and produce reports outlining the state of EPs in various water samples, ecosystems and reservoirs. Further, based on these study reports, authorities should highlight public health concerns and priorities in the context of the overall health of the ecosystem and the environment, keeping public health as a priority. This suffices the need for actual exchange and conversion of information between local, regional, national and global agencies regarding the implications of EPs on public health. To achieve this target, regional and national health agencies should lead or contribute in the preparation and implementation of policies to guarantee access to some form of consistent water safety measures.

10.2 WHO GUIDELINES ON HANDLING EPS IN DIFFERENT SECTORS

New and emerging pollutants present a new global drinking water and wastewater treatment quality challenge with potentially serious threats to human health and ecosystems. As it is important to care about good quality water to sustain human well-being, on the other hand it is also important to monitor wastewater discharge, treatment and reuse measures to improve livelihoods and the overall health of the environment for sustainable future development. In this direction several international agencies and organizations including the WHO under its flagship project, the UNESCO-IHP International Initiative on Water Quality (IIWQ), have further devised programmes to initiate scientific, technical and socio-economical case studies to address water related policy issues in close coordination with UNESCO member states. These programmes also aim to strengthen policy building capacities to manage human health and ecological and environmental risks instigated by EPs in water and wastewater. Therefore, these strengthening efforts are projected to yield fruitful results in building organizational structures and policies among UN member states to achieve improved drinking water quality standards, wastewater treatments, management with safety protocols in place to reuse the treated wastewater, and food security [6, 7].

It is also imperative to acknowledge that other than currently used or in practice EPs, new contaminants will occur in the future due to a range of factors, including changes in waste disposal practices, such as generally adopted composting process and anaerobic digestion methods to prepare compost for their subsequent application in agriculture, landfill management methods, demographic change, changes in land use, and the implications of climate change. Among these, the study of climate change is expected to impact on the scattering of the chemicals of the EP category in the environment. In addition, climate change is likely to affect the type and amount of emerging chemicals used in agriculture and industrial purposes. The future risks of compounds and chemicals could thus be much more diverse than today. Therefore, it is important that we initiate the assessment of EPs by keeping their future potential transformations and by-products in mind, focusing on their current and future implications on a pacific target. Overall, several study results anticipate that climate change will result in an increase in risks of EPs or constituent substances currently in use, especially in agricultural products which ultimately end up in the human food chain. However, the magnitude of the risk from EPs will be remarkably dependent on the contaminant type, level and their physico-chemical properties. Nonetheless, climate change and the ever-increasing population and demand will drive the increased use of pesticides and biocides in farming practices which are categorized among the most dangerous EPs. Also, at the local and regional level, extreme weather events will assemble and accumulate EPs from soils, land-fills and faecal matter, possibly increasing their bioavailability in different ecosystems. Climate change will also disturb the fate, presence and transport of EPs in agricultural feeds, the food chain and aquatic ecosystems. However, the increased temperature and varying moisture content are expected to decrease their presence in the atmosphere, but due to their conducive hydrological characteristics may help in diffusing EPs into water supplies. With this, the risk of many nanomaterials and low density microplastic particulate pollutants could therefore surge significantly.

The currently practised water-related guidelines and associated documents are used to source information on conventional pollutants and their implications on water quality, health and management approaches. However, an overview of currently practised water and wastewater related guidelines shows that they only deliberate primarily on drinking water and conventional pollutant matters. Therefore, it is high time that policy makers, water and health regulators, technologists, researchers and their advisors review and revisit national, regional and global standards for the production, use and disposal limits of EPs. However, it is important to note that these guidelines are recognized as representing the position of the UN system, mostly on recent drinking water safety-related issues and conventional pollutants under 'UN-Water', the body that coordinates the 24 UN agencies and programmes concerned with water-related issues [8].

Therefore, a futuristic approach demands a new mechanism to provide specific and high-quality monitoring evidence on the presence, concentration levels and accumulation of the pertinent EPs for a risk assessment across the ecosystems. For instance, in the European Union, a watch list of an EP category is expounded in detail from national monitoring programmes, since it was made obligatory by the Water Framework Directive. Similarly, several other regulatory authorities have listed the sources of EPs or other substances to monitor the frequency of occurrence, the expected risk for human health, aquatic ecosystem, and developing efficient and reliable analytical techniques. In a recent report the WHO listed more than 3000 synthetics as well as naturally occurring substances or compound EPs of present and future concern. In this context, future studies also need to consider the several factors associated with the management of substances that have been listed in the WHO EP document and in this current scenario it remains a challenge.

Therefore, current ecological risk assessments are followed in aquatic environments and in human health to consider either direct or indirect measurement techniques, particularly in water samples by comparing contaminant concentrations with standard ones. This is currently done with an analysis of EPs in environmental compartments with concentrations marked as below that which is acceptable or pose adverse effects on organisms. However, these procedures have significantly complex protocols. The method for human health risk assessment is slightly different, as human exposure to EPs is generally estimated to some extent through different pathways prior to comparison with threshold levels; for instance, the potential of an EP is estimated to be either a non-carcinogenic or a carcinogenic risk in comparison to reference doses and slope factors, respectively. On the other hand, these methodologies of risk assessment typically depend on the robustness and eminence of databases, as toxicity and ecotoxicity end points may vary depending on the sources. Therefore, future protocols and guidelines for the precise evaluation of ecological and health risks associated with EPs, including uncertainty and variability features, must be considered as essential parameters.

On the other hand, the Environmental Protection Agency (EPA) currently practises a series of guidelines formulated in 1985 to derive an ambient water quality criterion (AWQC) for aquatic life. The guidelines address acute risks like short-term implications such as the survival and growth of aquatic life and the chronic risks associated with longer term consequences like reproduction due to the presence and consumption of traditional pollutants. The EPA's previous and widely adopted 1985 'Guidelines for Deriving Numerical National Water Quality Criteria for the Protection of Aquatic Organisms and Their Uses' defines specific objectives like appropriate, internally reliability and a viable way of springing national criteria for the guarding of aquatic ecosystems.

Nevertheless, deriving numerical national and global water quality criteria for the safety of aquatic organisms from EPs and their real-world uses is a complex process which utilizes data from many areas of aquatic toxicology. However, a systematic overview of the relevant information on EPs in various sources and water treatment plants indicates that there is sufficient acceptable data available to predict rational water quality criteria for the protection of aquatic organisms from exposures to high concentrations of pollutants. Likewise, the EPA in a recently published revision guideline recommended procedures for deriving site-specific water quality criteria for the safeguarding of aquatic life from any category of pollutant.

Thus, the worldwide identification, detection, analysis and monitoring of the management aspects of EPs in surface water, wastewater and wastewater treatment plants have received substantial consideration by the WHO and national and international regulatory organizations. All of these organizations and agencies have invested their resources, efforts and expertise in assessing, documenting and publishing exclusive reports on the presence of EPs in groundwater and wastewater. So far, global agencies including the WHO, EPA and European agencies and organizations have documented several exhaustive scientific study reports on EPs such as hydrocarbons, pharmaceuticals, organic matter, series of substances in PCCPs and heavy metals in wastewater; however, there are very fewer numbers of in-depth systematic investigations conducted on the presence of EPs like microplastic, nanomaterials, polymeric microbead, surfactants and colloidal pollutants. Also, there is a lack of information documented on the handling and management of suitable remediation technology for these EPs.

10.3   INTERNATIONAL TREATIES AND CONSENSUS ON EPS

Way back in early 1990, United Nations' member states came together to formulate the 'Water Convention' as an effort to initiate a regional convention of the United Nation's Member States on Economic Commission for Europe (UNECE). The Convention declarations were accepted in Helsinki, Finland, in 1992 and came into force in 1996. Further in 2003, the Convention's Parties decided to amend the treaty to make it possible for any United Nations' member state to consent to this agreement. Further in 2016, the agreement officially became a part of the global legal agenda for trans-boundary water collaboration and cooperation, accessible to all member states. In this convention, among many, two agreements adopted are fully constant and coherent, and the few differences between member states provide useful complementarities. There is thus an important benefit in promoting and implementing the two water resolutions as a package, and numerous countries are parties to both instruments. Amongst the signatories, the Water Convention offers a unique legal and inter-governmental platform for trans-boundary water cooperation. Among more than 110 signatory countries from all over the world, parties are obliged to take measures to prevent, control and reduce any trans-boundary influences and implications on the environment, human health, safety and socio-economic conditions. Parties are obliged to use water resources sustainably, considering ecological and socio-economical requirements and traditions. These measures mainly include a responsibility to monitor environmental impact assessments and other means of evaluation, preventing and reducing water pollution at its source, accrediting and monitoring wastewater discharges contaminated with both conventional as well as emerging pollutants, formulating and applying best environmental practices to diminish or lessen inputs of nutrients and hazardous substances from agriculture, industry and other recently recognized diffuse sources like domestic, municipal wastewater and wastewater treatment plants. Thus, signatories are also required to establish water-quality objectives, norms and criteria, draw up contingency strategies, and to plan to minimize the risk of accidental water pollution.

In order to achieve these goals, the Convention requires signatories to arrange trans-boundary agreements and set up joint regulatory agencies to cooperate on the management and protection of their trans-boundary waters. The Convention encourages the protection of the health of a river basin from contamination and of its level. Joint agencies or regulatory authorities like lake or river commissions are tasked to (i) provide or create a stage or forum for the exchange and sharing of information on prevailing and planned uses of waters, as well as on pollution sources, pollutants, level of pollution and the ecological conditions of waters; (ii) serve as a forum for regular and uninterrupted consultations; (iii) set up joint monitoring programmes; (iv) conduct joint or coordinated testing, evaluation and assessments of the conditions of their shared waters and of the efficiency of the actions taken to address trans-boundary impacts; (v) decide on emission or discharge limits for wastewater and set up joint water quality objectives including formulating pollutant detection mechanisms, analysis and management; (vi) develop rigorous action plans and strategies for the reduction of pollution loads; and (vii) create cautionary or warning systems and alarm procedures [9–13].

Therefore, access to clean water or sustaining high quality prevailing water or recovering or regenerating high quality reusable water from wastewater sources is one of the critical topics of the 21st century. Though demands for water continue to surge, the availability of fresh water resources is declining. Today, water resources are under tremendous stress due to over-exploitation and pollution. To add to these natural calamities, floods and droughts are becoming more frequent and intense, further affecting water quality in various aquatic regions. Irrefutably, water resources have spread across political borders, covering nearly half of the earth's land surface and accounting for an estimated 60% of global freshwater flow. Regardless of their disputed quality, level of pollution, type of contaminant dissolved in it or dispersed in it, fresh resources support the incomes and livelihoods of more than 3 billion people and play a crucial role in innumerable ecosystems. Therefore, several international monitors and regulatory authorities emphasize cooperation on shared water resources, which is vital to secure peace and stability, economic growth, development and well-being, the protection of natural resources and to achieve sustainable development goals.

It is clear from the above that the world is extremely cautious about paying attention to EPs within their larger sustainable goal target of protecting the environment. All effort is being made to protect nature from EP release by preventing their entry into the environment or by retracting or confining them temporally and spatially. Today, the world probably may not have the resources, techniques or information to experimentally assess the risks of every potential EP, and there is thus an urgent necessity for methods that can be used pragmatically by environmental agencies, regulatory authorities and policy makers within a particular region or state to identify the specific concerns over EPs. Further, these EPs would then be considered for monitoring, identification, detection, evaluation and fate and effect assessment studies. Over the past few years, national and international agencies organizing community events, public consultations and research related to resource mobilizing to prioritize awareness and the identification of EPs have been of most concern. Among them, a number of studies have been commissioned to identify priority EPs in the environment. The focus includes a range of different pharmaceutical ingredients used in veterinary, human medicines, pesticides and most importantly their degradants, by-products and complexes formed in combination with co-pollutants. Further, commissioned studies also consider different exposure pathways and have been aimed at different protection goals to effect the UN's 17 Sustainable Development Goals (UNSDGs) in relation to sanitization, clean drinking water, socio-economical uplift and climate action.

Accordingly, SDGs, also known as the Global Goals, have been adopted as a part of a worldwide call to act to end poverty, protect the planet, supply clean and safe drinking water, create a pleasant climate and ensure that all people enjoy peace and prosperity. Therefore, the UN formulated 17 SDGs to be achieved by 2030. Among 17 goals, the sixth was given to achieving universal and impartial access to safe and affordable drinking water.

This priority also targets several important policy objectives, namely: (i) access to adequate and impartial sanitation and hygiene for every stakeholder and to end open-defecation by considering that the rural, semi-urban world population still lacks proper infrastructure and with special attention to women and girls and those in vulnerable situations; (ii) improve water quality by reducing pollution, controlling, minimizing and eliminating the dumping, abating and releasing of hazardous chemicals, substances, materials and products by splitting the proportion of untreated wastewater and significantly increasing recycling, regenerating and safe reuse worldwide; (iii) adopting measures to significantly increase water-use competence across all sectors of industry, segments of domestic life and ensuring sustainable and minimal extractions of ground water and the supply of adequate freshwater to address water scarcity and considerably reduce the number of people affected by water scarcity; (iv) to design, innovate and implement integrated water resource management technologies at all levels including through trans-boundary cooperation as appropriate to reduce the conflict caused by over-exploitation and its potential outcomes; (v) adopt methodologies and protocols to protect, restore and rejuvenate water-related ecosystems, including aquifers, lakes, rivers, wetlands, forests, mountains and marine life; (vi) aim to achieve international cooperation, infrastructure, capacity-building and sustenance in developing countries in water and sanitation-interrelated activities and programmes such as water rejuvenation of fresh water bodies, harvesting, surface water disinfection, ground water desalination, increasing water usage efficiency, upscaling domestic wastewater treatment facilities, recycling and reuse technologies; (vii) aim to support and strengthen the participation and contribution of local communities in improving water and sanitation management; (viii) adopt innovative infrastructure or upgrade retrofit manufacturing facilities to make them sustainable, targeting increased resource-use competence with the larger intention of adopting clean and environmentally inclusive technologies and industrial processes. The UN, also seeks global participation and involvement in achieving these goals from all member state countries with their respective capabilities and resource investments.

Nonetheless, from the technology and implementation point of view, the above goals and targets seek to integrate approaches with the backing of solid data on the potential hazard of an EP to either humans, organisms, aquifers, ecosystems or the environment at large.

Interestingly, in a recent trend, several local and national governments are framing site and theme specific advisories and legal frameworks to protect local biodiversity. These actions have been initiated mainly due to the major threat from the industrial discharge of wastewater either without or with partial treatments. This welcome development is due to increased awareness by adopting a combination of relevant regulatory frameworks from global agencies and organizations to meet international standards. Therefore, a combination of regulations and management measures are essential to influence and achieve efficient, competent and well-organized water resource management.

10.4   NEED FOR STRINGENT REGULATIONS

Today, due to several commercial interests at stake, several technologies are competing to capture the wastewater treatment and management space which slightly deviates from the focus of framing stringent protocols. Also, there is a lack of legal and legislative back-up to implement the wastewater treatment related regulation effectively. On the other hand, popular demand and increasing market demand for bottled drinking water deflects considerable attention from wastewater treatment and management. Another reason that is plausible is that much converted or reclaimed water from wastewater sources do not attract value, considering the fact that investing in wastewater treatment infrastructure is a costly operation. Until the world comes together and collectively changes this approach it will be difficult to accept responsibility for a better future. Generally, large polluters like industries, municipal effluent treatment plants and domestic medium to large treatment units can be controlled and regulated. However, millions of micro, small and medium scale industries leave large amounts of unmanaged and untreated wastewater from undesignated zones and illegal discharges into aquifers, rivers, drains and so on; if managed well, wastewater treatment and recycling of higher quality reusable water can be a good business prospect. In this direction, regulatory authorities should join hands with executives and legislators to stimulate the political will toward stringent implementation of wastewater treatment norms.

However, it is clear that industrial wastewater is the most neglected of all components of water management. As discussed in the previous section, several UN, European, WHO/UNICEF and DG reports commissioned by global organizations like the UN/WHO/UNICEF have shown how important it is to focus on wastewater management to achieve sustainable development goals, though when they have been developed, they are driven by national priorities. Water is critical to many other development challenges and a more holistic water agenda, including water resources and wastewater management, is needed. Regardless of the existence of several disengaged water resource management approaches, the large-scale reliable data stimulate political commitment, determination, well-placed investment, and independent enforcement authorities can only underpin sector advocacy. Unsympathetically, monitoring and regulatory authorities must be stringent in enforcing frameworks and at the same time must be grounded on what is measurable, affordable and applicable across a wide range of volumes and severity. It is clear that to achieve SDGs, wastewater needs to be fully recognized and included within the overall water cycle assessment as one of the important steps forward to enhance sustainable develop-ment. Though structural framework can be put in place to handle point sources, even then wastewater municipal and agricultural runoff make a significant contribution.

It is estimated that nearly 20% of total fresh water is consumed by the industry sector. The global status quo concerning the regulation of industrial wastewater treat-ment varies from 'highly effective' to practically 'non-existent'. Largely 'highly effective water treatment technique' regulations have been developed and adopted over long periods, backed by stringent institutional and legal frameworks. Using these institutional and legal frameworks regulatory authorities implement consider-able influence on maintaining high quality discharges for industrial processes and effluent treatments. Further, there have been several directives put in practice with their respective industrial wastewater discharges worldwide. Global authorities have already supplied general norms to manage industrial wastewater at the point source of pollution. However, the presence of specific and stringent regulatory directives plays a vital role in maintaining the health of discharged wastewater into the environ-ment, such as the Industrial Emissions Directive (IED), the European Pollutant Release and Transfer Register (E-PRTR), and the Urban Waste Water Treatment Directive (UWWTD) [14, 15].

By contrast, much of the world, especially the developing and under-developed region, have little or no institutional or legal framework or provision; or if it is avail-able it fails to enforce effectively on industries. It is estimated that 70% of industrial wastewater in developing and under-developed countries is discharged untreated. Many countries even lack a basic legal requirement to register industrial discharges, pollutant nature, awareness of the EP type and level. Such a situation leads to worry and, despondently, to quantifying the problem as bad or severe and, in some case, catastrophic. More worryingly, in many places industries use obsolete technologies and methods to treat and discharge highly toxic substances and a new class of EPs using processes that raise great concern. Even so, currently there are several effective technologies and approaches that are available to be implemented for industrial wastewater control and treatment of both conventional and emerging pollutants. However, the lenience on the part of implementing authorities and the absence of stringent review processes give them space to skip an efficient utilization of technol-ogy. This lenience on the part of authorities is generally misused by industries to get issuance of 'permits' or 'consents' for discharges either to drains or straight to water-courses, with or without pre-treatment. However, increasingly wastewater manage-ment awareness trends at the source level are gradually changing, and a new generation of product manufacturers are adopting new measures to avoid the use of potential contaminant resources regardless of their general type or classification. Even with positive developments generally followed and innovative approaches reported in recent research and scientific studies, stringent legislative back-up is required to implement for efficient water resource management. Nevertheless, based on the type and level of contamination, due to EPs, the specific procedures can be selected.

But all these efforts require resources and a large-scale infrastructure with the support of computational simulations which will provide precise or feasible quantita-tive evaluation of their effectiveness in reducing environmental levels and risks in large-scale catchments for specific EPs. Besides, a cost-benefit analysis for handling specific EPs in particular wastewater samples or catchment areas should be carried out to analyse the economic viability of the selected measures or procedures or techniques for effective purging. Therefore, a combination of stringent regulations, legal frameworks, innovative technologies and transboundary co-operation will resolve issues associated with EPs in ever-evolving data monitoring, analysis and application of tools for successful removal and sustainable water resource management.

From the previous sections, it is clear that over the last few years there has been increasing interest in the risks of EPs in the environment. There are also examples where these new pollutants have caused catastrophic impacts on ecosystems, for instance pharmaceutical drug accumulation in water resources and their adverse effects on the bird population. It is therefore critical that the research community, technologists, conservationists and policy makers continue to work together to identify EPs of most concern in a logical and realistic manner. In order to achieve sustainability in addressing issues related to EPs, there are many questions and queries that need to be addressed, namely: (i) What are the known number of substances that pose a risk to humans and the environment? – What we know today is through limited resources of information which indicates that only a small percentage of substances in use today have been investigated. (ii) How can we analyse specific EPs in infected or environmental media? Although advances in analytical chemistry and sophisticated tools revolutionized the understanding of chemical substances in various media and forms, there is still no specific tool or protocol in place to detect and quantify a class of EP at low levels in complex media or samples. (iii) Are all global monitoring agencies considering all the main exposure pathways? Currently, regulatory authorities have been divided into local, national, regional and global levels, even though this gives a holistic approach to the eco-centric risk assessments; however, an overall global environmental risk assessment requires extended information sharing on the exposure of EPs, on leaching to groundwater and runoff to surface and trans-boundary transmission across countries and from soils. (iv) What do we know about the ecotoxicity effect of EPs? It is likely that currently followed standards and ecotoxicity tests may not precisely quantify the effects of selected EPs; however, several studies have indicated that toxicity effects originated from EPs are real in humans and aquatic life. However, in-depth studies on ecotoxicity due to EPs from the point of behavioural impact, physiology and biochemistry might help in understanding the toxicology and pharmacology effects of EPs on the overall environment. (v) How will EPs interact with each other and other co-pollutants? Today, every ecological system has been exposed to a number of pollutants. Each and every water source and living being is susceptible to a mixture of both conventional and EPs. In such a situation, conventional pollutants may act as hosts for EPs, resulting in collective effects of these mixtures that are likely to create greater risk than single compounds alone. However, there are still studies that need to be conducted to understand how EPs interact with each other and with other co-pollutants. This also requires advanced and innovative approaches to be adopted in addressing unknown implications. (vi) What does the ecotoxicity of an EP mean? Several studies have established a number of subtle side-effects in tested marine and fresh water animals following exposure to selected EPs at trace concentrations. However, further studies need to be commissioned to establish specific ecotoxicity on aquatic life and ecosystem functioning. (vii) What are the mechanisms to evaluate the fate and behaviour of EPs in various sources? As established in the case of many traditional pollutants, our understanding of fate and behaviour in the environment is more or less recognized and models have been accepted for predicting a range of important fate parameters. However, the mechanism of transformations and degradation for many EPs and their co-pollutants need to be further understood to develop improved models. (viii) What are the mechanisms available to mitigate EPs against any identified risks? Thus far, the environmental risks associated with EPs have been well established and potential risks have been identified; however, it may be necessary to introduce treatments, measures or protocols as mitigation options. Therefore, studies are needed to further focus on a better understanding of the mitigation of EPs, against those factors controlling the fate and behaviour in different EPs. (ix) What will EP-contaminated food, and agricultural and different components, in the food chain look like in the future? More than half of the world's population depend on marine or aquatic livestock. In view of the increased contamination in agricultural products, it is anticipated that aquatic ecosystems are likely to look very different in the future with their composition and contamination. This will also be further influenced due to the increased rate of disposal, landfill, and demographic and climate changes. Therefore, these changes are likely to affect the risks of contaminants due to EP adulteration in agricultural products, foods and aquatic life. It is therefore significant that we initiate elaborate and comprehensive studies to assess how the food chain might alter in its composition.

10.5 SUMMARY AND FUTURE PROSPECTS FOR EP TREATMENT TECHNOLOGIES

EPs are largely unregulated anthropogenic substances that occur in almost all constituents of the environment including air, soil, water, food and human or animal tissues in trace concentrations. EPs are perceived to be persistent in the environment for a long time and are capable of disturbing the physiology of target receptors. Therefore, EPs are regarded as a great risk to humans, living beings and the environment. The prominent classes of EPs include pesticides, active pharmaceuticals ingredients and personal care products, polymeric microbeads, surfactants, plasticizers, fire retardants, nanomaterials and micro/nanoplastics. Even though a detailed understanding of EPs' harmful effects began a decade ago, intensive research has only recently been initiated on the detection, occurrence, ecological effects, fate and treatment technologies of EPs. However, a comprehensive treatment technique to handle and remove EPs from the various sources has yet to be devised successfully. Even though there have been several attempts made to remove some EPs from simulated mixtures and contaminated fresh water there is limited information available on comprehensive EP removal techniques, particularly from domestic, municipal, industrial wastewater and conventional wastewater effluent treatment plants. A detailed overview of the literature and technological assessment reports reveals that conventional water and wastewater treatment techniques are inefficient when it comes to handling EP removal.

However, some of the conventional techniques have been found useful in screening EPs at an initial stage which prevents them from spreading across vulnerable eco-systems. Interestingly, primary and secondary treatment techniques have been highly efficient in retaining larger EPs like microplastics and some nanomaterial-based pollutants. Further, a detailed overview of water and wastewater treatment techniques like physical, chemical, physico-chemical, biological, advanced treatment techniques like membrane-based separation process, adsorption techniques, advanced oxidation processes and Fenton/photo-Fenton degradation suggest the effectiveness of individual techniques in addressing particular EPs. However, the overall performance of these techniques has shown limited success. In particular, none of these treatment techniques have shown promising prospect for EP removal from industrial wastewater and effluent treatment plants at an individual level. Interestingly, a combination of conventional and advanced water treatment processes has positively shown great promise at removing EPs from industrial and treatment plants.

Therefore, effective treatment of EPs in wastewater requires an entirely new approach and complete upgrade of the existing process designs. Further, a study report suggests that there is a strong correlation between the properties of EPs and potential treatment techniques such as the nature of an EP's parent substance, occurrence, size, distribution, fate and biodegradability. Interestingly, these properties and their extent in EPs can be approximately estimated based on physico-chemical properties like surface area, surface charge, surface energy, partitioning coefficient, solubility, dispersion capacity, solid-water distribution coefficient and photo-sensitivity. On the other hand, the limited biodegradation potential of EPs can be estimated from the nature of a pollutant and its biodegradation constant values and efficiency of treatment techniques. In wastewater and effluent treatment plants, the EP removal efficiency varies in the range of 20–40%, 30–70% and ~90% during the primary, secondary and tertiary treatment steps, respectively. More significantly, these techniques are effective in retaining EPs like polymeric microbeads, microplastics and colloidal nanomaterial pollutants. Nonetheless, recent advanced tertiary treatment technologies are showing considerable efficiencies in addressing EPs and are estimated to be most suitable alternatives for EP treatment; however, complete EP removal from complex wastewater sources is yet to be achieved. Besides, continued leading edge research, adopting novel design strategies and process developments for EP treatment will be indisputably essential for a sustainable future.

Present studies extensively examine and elaborate the environmental risk assessments associated with EPs in direct correlation with their origin and potential toxicity. EPs originate from a variety of starting material, natural, synthetic, hybrid, composite, blend, polymeric, as a consequence of natural/man-made disasters. Some EPs of natural origin like clay (volcanic ash), biopolymer (seaweed) and pathogens and their transformation products may be shaped in the environment by biochemical processes in mineral, plants, animals and microbes. On the other hand, synthetic and hybrid materials have emerged as a result of industrial innovation and growth, consumer demand and ever-changing life-styles. In all cases, environmental risk assessment and their treatment schemes already exist to a certain extent. However, to design and develop efficient EP treatment techniques the following observations need to be considered:

- EPs are essentially not new chemicals or substances, rather they are widely used and considered ultra-safe today. However, EPs are those substances that have been in the various ecosystems and environment for a long time but whose presence, significance and adverse effects are only now being studied and recognized.
- EPs are being released into the air, water, land, soil and agricultural environment via a number of routes and pathways. EPs are increasingly released directly to the environment through domestic, municipal, industrial wastewater and landfill. Once in soil EPs are transported to water bodies and ground water by leaching, runoff and drainage processes. The extent of the transportation is dependent on the tenacity of the EP and its potential interactions with another component in the environment.
- EPs' main source and release is from wastewater treatment plants. A major source of EPs in the environment comes from non-agriculture and landfill sources. Most of these EPs have their origin in consumer goods and industrial discards that are more significant than emissions from agriculture activities.
- Many EPs appear to behave differently in environmental systems. Many of them carry unique physical and chemical properties which make them complex to understand compared to their bulk version. These unique properties in EPs pose technological challenges in their detection and efficient treatment techniques. This means that computational modelling methods and strategies generally followed for predicting their fate properties and exposure to conventional pollutants are not always appropriate for EPs.
- Detection and quantification of EPs in the environment is challenging. However, over the years, researchers and innovators have developed robust methods and adopted conventional tools to analyse some classes of EPs in various samples. Even though currently followed methods suffer from various disadvantages for their precise detection and characterization of EPs, advanced tools have emerged to precisely quantify a few EPs in water samples.
- Although a great number of research reports have explored the occurrence of a wide range of EPs in various samples in various ecosystems across the globe, only a few reports have appeared with specific information on the occurrence of EPs in wastewater systems. The studies that have looked at domestic, municipal and specific industrial wastewater systems have detected a range of EPs including APIs, HMs, drugs, human pharmaceuticals, personal care products, synthetic hormones, nanomaterials, microplastics and transformations and by-products of man-made chemicals in trace concentrations.
- As of today, most EPs are known to cause acute toxicity at medium to high concentrations. However, EPs even in trace concentrations are known to cause serious health risks in aquatic life and humans. In many cases, like microplastics, nanomaterials, polymeric microbeads, HMs, pesticides and

surfactants are presently found in much higher than permissible concentrations measured in the environment, indicating that EPs are unlikely to cause acute effects on ecosystems upon direct release and accumulation.

- New EPs are posing serious challenges to removal techniques. Currently several technologies have been adopted to treat or remove EPs from wastewater sources with limited success. However, EPs are likely to persist in the environment for a long time posing serious technological challenges.
- Advanced treatment techniques. Even though almost every available water treatment technique, method, tool and protocol has been tried and tested in an attempt to find suitable treatment solutions for EP removal from wastewater sources, regulated standard procedures are absent from the practice. Among these, a combination of traditional and advanced treatment techniques has shown potential in eliminating EPs from wastewater sources. However, it is still important to work towards the precise identification, detection and quantification of EPs so that resources can be mobilized for solving larger issues associated with EP problems. Nevertheless, currently a number of prioritization tactics and strategies already exist for the prospect of scanning for different classes and types of EPs in the environment. However, these approaches require further innovation and need to be applied widely in a larger context. Also in order to achieve sustainable goals by adopting suitable EP removal techniques, stringent water quality management mechanisms with the backing of legal and legislative support is the need of the hour.

Although currently published research, case studies and expert opinion on risk assessments and evaluation indicate that trace concentrations of EPs in drinking water are unlikely to pose risks to human health, information and knowledge gaps still exist in terms of evaluating risks associated with long-term exposure to low to medium concentrations of EPs such as pesticides, HMs, pharmaceuticals, micro-organics, microplastics, nanomaterials and the combined effects of mixtures of these pollutants. All the evidence available on EPs is currently inconclusive and limited, therefore extensive future research on the occurrence, fate and treatment techniques in these areas may be beneficial to understand and devise better characterizing tools and treatment techniques to minimize long-term risks. One of the key challenges in assessing exposures to various EPs in drinking water and evaluating the potential risks to human health is the limited data. Therefore, the design of extensive EP monitoring, identification and treatment programmes is a resource-intensive exercise; however, it is necessary for preparing human resources and establishing an infrastructure for minimizing future risk. Therefore, future research should emphasize filling these knowledge gaps. This can be achieved by providing support to researchers and innovators to develop cost-effective methods and protocols for the effective management of EPs within the context of an overall risk assessment and evaluation for all water and wastewater related hazards, noting that EPs concern both drinking water and the larger environmental emergency. In this context, a global organization like the WHO will likely mobilize resources and continue to support relevant studies to frame guidelines accordingly.

REFERENCES

1. Violette Geissen, Hans Mol, Erwin Klumpp, Günter Umlauf, Marti Nadal, Martine van der Ploeg, Sjoerd E. A. T. M. van de Zee, and Coen J. Ritsema, Emerging pollutants in the environment: a challenge for water resource management, *International Soil and Water Conservation Research*, 2015, 3, 57–65.
2. Maria Gavrilescu, Kateřina Demnerová, Jens Aamand, Spiros Agathos, and Fabio Fava, Emerging pollutants in the environment: present and future challenges in biomonitoring, ecological risks and bioremediation, *New Biotechnology*, 2015, 32(1), 147–156.
3. Mohamed Chaker Ncibi, Borhane Mahjoub, Olfa Mahjoub, and Mika Sillanpää, Remediation of emerging pollutants in contaminated wastewater and aquatic environments: biomass-based technologies, *CLEAN, soil Air, Water*, 2017, 45(5), 1700101.
4. O. Erdogan and M. Kara, Analytical approach to the waste management of nanomaterials in developing countries, *Frontiers Drug Chemistry Clinical Res*, 2019, 2, 1–5.
5. Carlos Peña-Guzmán et al., Emerging pollutants in the urban water cycle in Latin America: a review of the current literature, *Journal of Environmental Management*, 2019, 237, 408–423.
6. Nadeem A. Khan et al., Recent trends in disposal and treatment technologies of emerging-pollutants – a critical review, *TrAC Trends in Analytical Chemistry*, 2020, 122, 115744.
7. S. M. Abdelbasir, K. M. McCourt, C. M. Lee, and D. C. Vanegas, Waste-derived nanoparticles: synthesis approaches, environmental applications, and sustainability considerations, *Frontiers in Chemistry*, 2020, 8(782), 1–18.
8. Publications of the World Health Organization are available on the WHO web site (www.who.int) or can be purchased from WHO Press, World Health Organization, 20 Avenue Appia, 1211 Geneva 27, Switzerland.
9. Directive 2008/105/EC of the European Parliament and of the Council of 16 December 2008 on environmental quality standards in the field of water policy. Official Journal of the European Union; L 348, 24.12.2008, 84–97.
10. Directive 2009/128/EC of the European Parliament and of the Council of 21 October 2009 establishing a framework for Community action to achieve the sustainable use of pesticides. Official Journal of the European Union; L 309, 24.11.2009, 71–86.
11. Directive 2010/75/EU of the European Parliament and of the Council of 24 November 2010 on industrial emissions (integrated pollution prevention and control). Official Journal of the European Union; L 334, 17.12.2010, 17–119.
12. UNICEF. Water, sanitation and hygiene annual report 2008, 2009.
13. M. Van der Ploeg, W. Appels, D. Cirkel, M. Oosterwoud, J. Witte, and S. van der Zee, Microtopography as a driving mechanism for ecohydrological processes in shallow groundwater systems, *Vadose Zone Journal*, 2012, 11, vzj2011.0098. https://doi.org/10.2136/vzj2011.0098.
14. Emmanuel Thanassoulis, DEA and its use in the regulation of water companies, *European Journal of Operational Research*, 2000, 127(1), 1–13.
15. Council Directive 91/271/EEC of 21 May 1991 concerning urban waste-water treatment. Official Journal; L 135, 30/05/1991, 40–52.

Index

Page numbers in **bold** indicate tables.